Symmetry/Asymmetry Studies in Modern Power Systems

Symmetry/Asymmetry Studies in Modern Power Systems

Guest Editors

Tao Zhou

Cheng Wang

Zhong Chen

Lei Chen

MDPI

Basel • Beijing • Wuhan • Barcelona • Belgrade • Novi Sad • Cluj • Manchester

Guest Editors
Tao Zhou
School of Automation
Nanjing University of Science and Technology
Nanjing
China

Cheng Wang
School of Automation
Nanjing University of Science and Technology
Nanjing
China

Zhong Chen
School of Electrical Engineering
Southeast University
Nanjing
China

Lei Chen
Department of Electrical Engineering
Tsinghua University
Beijing
China

Editorial Office
MDPI AG
Grosspeteranlage 5
4052 Basel, Switzerland

This is a reprint of the Special Issue, published open access by the journal *Symmetry* (ISSN 2073-8994), freely accessible at: https://www.mdpi.com/journal/symmetry/special_issues/774A97LOG7

For citation purposes, cite each article independently as indicated on the article page online and as indicated below:

Lastname, A.A.; Lastname, B.B. Article Title. *Journal Name* **Year**, *Volume Number*, Page Range.

ISBN 978-3-7258-6438-6 (Hbk)
ISBN 978-3-7258-6439-3 (PDF)
https://doi.org/10.3390/books978-3-7258-6439-3

Contents

About the Editors

Tao Zhou

Tao Zhou is an Assistant Professor at the School of Automation, Nanjing University of Science and Technology, China. His primary research interests focus on the stability and control of power systems, encompassing renewable energy grid-connected control and AI in power systems. He actively contributes to the academic community as the Deputy Secretary-General of the Power Grid Stability Control Technology Subcommittee of IEEE PES Power System Relaying & Control Satellite Committee, China. His distinguished contributions have been recognized with awards such as the Key Young Science and Technology Talent by the Ministry of Industry and Information Technology and a "Double Creation" Doctor in Jiangsu Province.

Cheng Wang

Cheng Wang is an Associate Professor at the School of Automation, Nanjing University of Science and Technology. He earned his Ph.D. in Electrical Engineering from Huazhong University of Science and Technology in 2018 and was a joint Ph.D. student at Lehigh University, USA. His primary research focuses on modular power electronic converters, including main circuit topology and performance control, reliable grid integration and flexible interconnection of photovoltaic power generation systems, and fault diagnosis of power electronic converters. Dr. Wang serves in editorial roles for several prestigious international journals, including as an Editorial Board Member for the top-tier SCI journal *IEEE Transactions on Power Electronics* and as a youth Editorial Board Member for *Power System Protection and Control.*

Zhong Chen

Zhong Chen is a Professor and Doctoral Supervisor whose research specializes in power system planning, operation, and control, with a particular focus on the application of artificial intelligence in smart dispatch and maintenance of power systems, as well as the intelligent interaction between electric vehicles and the energy internet. He holds a prominent leadership position as the Vice Chair of the Power Grid Stability Control Technology Subcommittee of the IEEE PES Power System Relaying & Control Satellite Committee, China. His work bridges cutting-edge computational techniques with critical power engineering challenges, driving advancements in the stability and intelligence of modern power grids.

Lei Chen

Lei Chen is a Researcher in the Department of Electrical Engineering at Tsinghua University. As an IEEE Fellow, his research is centered on the core areas of power system stability, dynamic analysis, and control. His work addresses the fundamental challenges in understanding and ensuring the secure operation of modern power grids. His expertise is pivotal in developing advanced methodologies for analyzing and mitigating dynamic phenomena, contributing significantly to the field of power system engineering.

Preface

It is our distinct pleasure to present this Reprint, "Symmetry/Asymmetry Studies in Modern Power Systems," which compiles the twelve insightful articles originally published in the corresponding Special Issue. The motivation for curating this collection stems from the critical and evolving challenges in operating and designing contemporary power grids, where the high penetration of inverter-based renewable resources and power electronics is fundamentally redefining traditional concepts of system balance and stability. This Reprint is dedicated to providing a comprehensive exploration of both theoretical advancements and practical applications for managing symmetry and asymmetry, covering topics from advanced control strategies and renewable integration to system protection and equipment operation. Our primary aim is to offer researchers, engineers, and advanced students in the field of electrical power systems a consolidated and insightful resource that not only highlights the current state of the art but also stimulates future innovations. We extend our sincere gratitude to all the authors for their valuable contributions and to the reviewers for their rigorous efforts, which have been instrumental in shaping a cohesive and high-quality volume. We also gratefully acknowledge the support of the Natural Science Foundation of Jiangsu Province under Grant BK20241481, which contributed to the completion of this work. We are confident that the research contained herein will significantly contribute to the development of more resilient, efficient, and sustainable power systems.

Tao Zhou, Cheng Wang, Zhong Chen, and Lei Chen
Guest Editors

symmetry

MDPI

Editorial

Special Issue: Symmetry/Asymmetry Studies in Modern Power Systems

Tao Zhou and Cheng Wang *

School of Automation, Nanjing University of Science and Technology, Nanjing 210094, China;
zhoutaonust@njust.edu.cn
* Correspondence: chw714@njust.edu.cn

Abstract

This Special Issue, "Symmetry/Asymmetry Studies in Modern Power Systems," presents a curated collection of research addressing the critical and evolving role of symmetry in the context of energy transition. The contributions, selected through a rigorous review process, collectively advance the understanding and management of power system balance, stability, and resilience amidst the increasing integration of renewables and power electronics. The published papers offer innovative solutions across several interconnected areas, including advanced control for active power symmetry, optimized renewable integration and inertia support, intelligent equipment operation, system-wide dynamic analysis, scheduling under uncertainty, and enhanced protection and power quality. By synthesizing advanced computational techniques with core power engineering challenges, this issue provides both theoretical insights and practical methodologies. It underscores a paradigm shift towards actively orchestrating system stability within inherently asymmetric conditions, laying a foundation for the design of more resilient, efficient, and sustainable future grids. Finally, key future research directions are outlined to further integrate adaptive control, physics-informed machine learning, and standardized metrics for holistic system design.

Keywords: power system symmetry; renewable energy integration; power system stability; power symmetry and balance control; power system optimization and scheduling

1. Introduction

Concepts of symmetry and asymmetry have long been recognized as fundamental to the analysis and operation of power systems [1]. However, with the increasing integration of renewable energy sources, widespread adoption of power electronic devices, and the emergence of complex grid architectures, the roles of symmetry and asymmetry have become more pronounced and multifaceted [2]. This Special Issue, "Symmetry/Asymmetry Studies in Modern Power Systems," addresses this evolving landscape by presenting cutting-edge research that explores the intricate relationship between symmetry principles and modern power system performance. The contributions in this issue provide valuable insights into how symmetry and asymmetry influence fault diagnosis, system planning, operational efficiency, and stability in contemporary power systems. By examining these phenomena across various dimensions—from fundamental theory to practical applications in renewable integration and grid-forming control—this Special Issue offers a comprehensive perspective that is essential for advancing the design and operation of resilient, efficient, and sustainable power systems in the era of energy transition. The collected research not only deepens our theoretical understanding but also provides actionable methodologies for addressing the complex challenges posed by modern power system dynamics.

2. Thematic Overview of Contributions

Following a rigorous peer-review process, this Special Issue "Symmetry/Asymmetry Studies in Modern Power Systems" received a total of about 20 high-quality submissions from researchers across the globe. After thorough evaluation by experts in relevant fields, 12 papers were accepted for publication, reflecting the diversity of approaches and perspectives in the field. These selected contributions present significant advancements in understanding the critical role of symmetry and asymmetry in modern power system operations, analysis, and design. The accepted papers included different symmetry/asymmetry studies on techniques for modern power systems:

- Power symmetry and balance control (contributions 1, 2, 8, and 9);
- Symmetry and asymmetry in renewable energy integration (contributions 6 and 12);
- Power equipment operation optimization (contributions 3 and 5);
- Power system dynamic analysis and stability (contribution 4);
- Power system optimization and scheduling (contribution 7);
- Power system protection and power quality (contributions 10 and 11).

The integration of renewable energy sources and power electronic devices has significantly reduced system inertia, resulting in unstable frequency and threatening the active power symmetry and balance essential for grid reliability [3]. Maintaining power symmetry—ensuring balanced generation and load dynamics—is critical for preventing frequency deviations, oscillations, and cascading failures, particularly in modern systems with the high penetration of variable resources [4]. This direction addresses these challenges by developing advanced control strategies that enhance dynamic stability, optimize resource coordination, and support the transition to resilient, low-carbon power networks [5]. Contribution 1 proposes a knowledge-aggregation-based deep reinforcement learning method for load frequency control in isolated microgrids, combining improved whale optimization with LSTM and self-attention mechanisms to optimize power symmetry and balance under uncertainties. Contribution 2 introduces a coordinated control strategy for multi-type flexible resources and under-frequency load shedding, leveraging voltage variation effects to improve active power balance and frequency support capabilities. Contribution 8 develops an adaptive voltage reference-based multi-objective optimal control approach for multi-terminal DC systems, employing normal boundary intersection to ensure power flow symmetry with large-scale offshore wind integration. Contribution 9 presents a stability control method using grid-forming converters and virtual synchronous generator technology to achieve active symmetry in distribution grid elastic balance regions, enhancing inertia and damping under disturbances.

Maintaining the delicate balance between variable renewable generation and load demand is paramount for modern power systems, as inherent asymmetries can severely compromise frequency stability and grid reliability [6]. This research direction focuses on developing advanced control and evaluation frameworks to enhance active power symmetry, mitigate the impacts of source–load imbalances, and harness the innate support capabilities of inverter-based resources [7]. Contribution 6 systematically evaluates the inertia support capability of wind turbine generators operating symmetrically, quantifying their potential to provide crucial grid stabilization services comparable to conventional synchronous machines. Contribution 12 introduces a distributed active support method for photovoltaic systems, utilizing a state–disturbance observer and a dynamic surface consensus algorithm to ensure robust frequency stability amid source–load asymmetry.

The optimization of power equipment operation is critical for enhancing the reliability, efficiency, and longevity of electrical systems, particularly as the integration of renewable energy sources and fluctuating loads introduces new challenges to grid stability [8]. Transformers, as key components in power networks, require advanced control and prediction methods to maintain power symmetry and prevent failures under dynamic conditions [9]. This direction focuses on developing intelligent strategies for transformer capacity regulation and temperature monitoring, which are essential for reducing operational costs, ensuring safety, and supporting the transition to sustainable energy systems [10]. Contribution 3 proposes an on-load capacity-regulating control method for power transformers that combines load forecasting with hesitant fuzzy control to dynamically adjust transformer output based on real-time demand, optimizing resource utilization and minimizing losses. Contribution 5 introduces a predictive model for transformer top-oil temperature using an LSTM neural network enhanced with a self-attention mechanism and optimized by an improved whale optimization algorithm, improving accuracy in temperature forecasting to prevent overheating and extend equipment life.

The research direction of power system dynamic analysis and stability is pivotal for ensuring the reliable operation of modern power systems, particularly as the integration of renewable energy sources introduces greater variability and reduces system inertia, leading to increased frequency deviations and instability risks [11]. This field focuses on modeling, analyzing, and controlling dynamic behaviors to maintain active power symmetry and balance between generation and demand, which is essential for preventing blackouts and enhancing grid resilience under disturbances such as load fluctuations or faults [12]. Contribution 4 addresses this by proposing a small-signal modeling approach for power-load frequency response that incorporates voltage variation effects, enabling a more accurate analysis of how loads can contribute to frequency support and improve system stability under dynamic conditions.

The research direction of power system optimization and scheduling is critically important in the era of energy transition, addressing the challenges posed by the large-scale integration of intermittent renewable energy sources and the increasing complexity of multi-energy flows in modern grids [13]. Its significance lies in enhancing the economic efficiency, reliability, and sustainability of power systems by achieving optimal resource allocation and maintaining active power balance between generation and demand under various uncertainties [14]. Contribution 7 specifically advances this field by developing a bi-level optimization scheduling strategy for Park-Level Integrated Energy Systems (PIESs) that effectively tackles the uncertainties associated with price-based demand response to improve scheduling decisions.

The research direction of power system protection and power quality is fundamental to ensure the security, reliability, and stability of modern electrical grids, especially with the increasing integration of power-electronics-interfaced resources like renewables and electric vehicles that introduce new types of faults and harmonic distortions [15]. Its significance lies in developing advanced methods to quickly isolate faults, prevent equipment damage, and maintain voltage and current waveforms within strict standards, thereby guaranteeing the safe operation of critical infrastructure and the delivery of clean power to end-users [16]. This field is pivotal for mitigating the risks of cascading failures and enhancing the overall resilience of the power system against disturbances [17]. Contribution 10 proposes a quantitative state evaluation method for relay protection equipment using an improved Conformer model optimized by a two-stage Artificial Physics Optimization

(APO) algorithm to accurately assess equipment health and predict failures. Contribution 11 introduces an integrated approach combining an I-ADALINE neural network with selective filtering techniques to effectively mitigate harmonics and improve power quality in electrically distorted networks.

3. Conclusions and Future Perspectives

The accepted papers of this issue collectively advance the theoretical and practical understanding of symmetry and asymmetry in contemporary power systems, delivering comprehensive solutions to critical challenges arising from renewable integration, grid modernization, and power electronics proliferation. These contributions establish a unified framework for maintaining power symmetry and balance across diverse operational scenarios—from microgrid frequency control and multi-terminal DC system optimization to transformer management, renewable inertia support, and fault resilience. By integrating advanced computational techniques with system-level stability analysis, these papers provide actionable methodologies to enhance grid reliability, operational efficiency, and resilience under increasing asymmetry induced by variable renewable generation and complex grid topologies. This body of work not only deepens the foundational knowledge of symmetry principles in power systems but also offers implementable strategies for transitioning toward sustainable, stable, and intelligent power networks in the energy transition era.

Future research in symmetry/asymmetry studies for modern power systems should focus on three interconnected dimensions: (1) developing adaptive control frameworks that dynamically maintain power symmetry under extreme renewable variability and grid-forming converter interactions, particularly for multi-inverter systems operating in weak grid conditions; (2) integrating physics-informed machine learning with real-time stability assessment to overcome the computational bottlenecks in transient stability analysis during high-asymmetry events; and (3) establishing standardized metrics for quantifying symmetry degradation across diverse grid architectures—from distribution networks with high PV penetration to multi-terminal HVDC systems—enabling the holistic design of resilient, low-carbon power systems. These directions will be critical for enabling seamless grid integration of next-generation renewable resources while ensuring operational stability in the face of increasingly complex, asymmetric power flows.

Author Contributions: Conceptualization, T.Z. and C.W.; writing—original draft preparation, T.Z.; writing—review and editing, C.W.; project administration, T.Z.; funding acquisition, C.W. All authors have read and agreed to the published version of the manuscript.

Funding: This work was supported by the Natural Science Foundation of Jiangsu Province (No. BK20241481) and the Fundamental Research Funds for the Central Universities (No. 30922010709).

Data Availability Statement: Data availability is not applicable to this article as it is an Editorial and no new data were generated or analyzed.

Conflicts of Interest: The authors declare no conflicts of interest.

List of Contributions:

1. Wu, M.; Ma, D.; Xiong, K.; Yuan, L. Deep Reinforcement Learning for Load Frequency Control in Isolated Microgrids: A Knowledge Aggregation Approach with Emphasis on Power Symmetry and Balance. *Symmetry* **2024**, *16*, 322. https://doi.org/10.3390/sym16030322.
2. Zhang, J.; Wang, J.; Cao, Y.; Li, B.; Li, C. A Coordinated Control Strategy of Multi-Type Flexible Resources and Under-Frequency Load Shedding for Active Power Balance. *Symmetry* **2024**, *16*, 479. https://doi.org/10.3390/sym16040479.

3. Zou, D.; Sun, X.; Quan, H.; Yin, J.; Peng, Q.; Wang, S.; Dai, W.; Hong, Z. Power Transformer On-Load Capacity-Regulating Control and Optimization Based on Load Forecasting and Hesitant Fuzzy Control. *Symmetry* **2024**, *16*, 679. https://doi.org/10.3390/sym16060679.
4. Zhou, T.; Zheng, Y.; Wang, C.; Chen, L.; Liu, B.; Chen, Z. Small-Signal Modeling and Frequency Support Capacity Analysis of Power Load Considering Voltage Variation Effect. *Symmetry* **2024**, *16*, 918. https://doi.org/10.3390/sym16070918.
5. Zou, D.; Xu, H.; Quan, H.; Yin, J.; Peng, Q.; Wang, S.; Dai, W.; Hong, Z. Top-Oil Temperature Prediction of Power Transformer Based on Long Short-Term Memory Neural Network with Self-Attention Mechanism Optimized by Improved Whale Optimization Algorithm. *Symmetry* **2024**, *16*, 1382. https://doi.org/10.3390/sym16101382.
6. Chen, Z.; Li, Y.; Zhou, Q. Inertia Support Capability Evaluation for Wind Turbine Generators Based on Symmetrical Operation. *Symmetry* **2025**, *17*, 31. https://doi.org/10.3390/sym17010031.
7. Chen, X.; Lei, J.; Zhang, X. Bi-Level Optimization Scheduling Strategy for PIES Considering Uncertainties of Price-Based Demand Response. *Symmetry* **2025**, *17*, 43. https://doi.org/10.3390/sym17010043.
8. Zhang, Y.; Feng, Y.; Xu, T.; Li, Y.; Du, X.; Yuan, C.; Chen, H. An Adaptive Voltage Reference-Based Multi-Objective Optimal Control Method for the Power Flow Symmetry of Multi-Terminal DC Systems with the Large-Scale Integration of Offshore Wind Farms. *Symmetry* **2025**, *17*, 105. https://doi.org/10.3390/sym17010105.
9. Lv, Z.; Jia, B.; Song, Z.; Li, H.; Zhou, S.; Li, Z. Stability Control Method Utilizing Grid-Forming Converters for Active Symmetry in the Elastic Balance Region of the Distribution Grid. *Symmetry* **2025**, *17*, 263. https://doi.org/10.3390/sym17020263.
10. Li, Y.; Zhang, M.; Zhang, S.; Zhou, Y. Quantitative State Evaluation Method for Relay Protection Equipment Based on Improved Conformer Optimized by Two-Stage APO. *Symmetry* **2025**, *17*, 951. https://doi.org/10.3390/sym17060951.
11. Hoon, Y.; Chew, K.W.; Mohd Radzi, M.A. Integrated I-ADALINE Neural Network and Selective Filtering Techniques for Improved Power Quality in Distorted Electrical Networks. *Symmetry* **2025**, *17*, 1337. https://doi.org/10.3390/sym17081337.
12. Zhou, Y.; Gao, Y.; Tang, Y.; Liu, Y.; Tu, L.; Zhang, Y.; Liu, Y.; Zhang, X.; Yu, J.; Cao, R. Distributed Active Support from Photovoltaics via State–Disturbance Observation and Dynamic Surface Consensus for Dynamic Frequency Stability Under Source–Load Asymmetry. *Symmetry* **2025**, *17*, 1672. https://doi.org/10.3390/sym17101672.

References

1. Hatziargyriou, N.; Milanovic, J.; Rahmann, C.; Ajjarapu, V.; Canizares, C.; Erlich, I.; Hill, D.; Hiskens, I.; Kamwa, I.; Pal, B.; et al. Definition and classification of power system stability—Revisited & extended. *IEEE Trans. Power Syst.* **2021**, *36*, 3271–3281. [CrossRef]
2. Ming, X.; Wang, X.; Liu, F.; Qu, Y.; Zhou, B.; Zhang, S.; Yu, P. Mechanical Parameter Identification of Permanent Magnet Synchronous Motor Based on Symmetry. *Symmetry* **2025**, *17*, 1929. [CrossRef]
3. Chang, M.; Salem, M.; Mohamed, F.A. Adaptive virtual inertia emulation based on policy gradient clipping for low-inertia microgrids with phase-locked loop dynamics. *Comput. Electr. Eng.* **2025**, *125*, 110477. [CrossRef]
4. Steele, A.J.H.; Burnett, J.W.; Bergstrom, J.C. The impact of variable renewable energy resources on power system reliability. *Energy Policy* **2021**, *151*, 111947. [CrossRef]
5. Cai, C.; Wang, Y.; Su, Q.; Li, J. Collaborative Cloud-Controlled Defense Mechanism for Low-Carbon Economic Dispatch in Active Distribution Networks Under Interlayer Attack. *IEEE Trans. Reliab.* **2025**, *74*, 2555–2667. [CrossRef]
6. Mars, P.d.; O'Sullivan, A.; Keppo, I. The impact of variable renewable energy on base-load cycling in the GB power system. *Energy* **2020**, *195*, 117041. [CrossRef]
7. Li, Z.; Chan, K.W.; Hu, J.; Guerrero, J.M. Adaptive droop control using adaptive virtual impedance for microgrids with variable PV outputs and load demands. *IEEE Trans. Ind. Electron.* **2021**, *68*, 9630–9640. [CrossRef]
8. Li, Y.; Cao, J.; Xu, Y.; Zhu, L.; Dong, Z.Y. Deep learning based on Transformer architecture for power system short-term voltage stability assessment with class imbalance. *Renew. Sustain. Energy Rev.* **2024**, *189*, 113913. [CrossRef]
9. Lee, H.-J.; Yoon, S.W.; Yoon, Y.-D. Hybrid distribution transformer based on an existing distribution transformer and a series-connected power converter. *IEEE Trans. Power Deliv.* **2022**, *37*, 4202–4211. [CrossRef]

10. Hu, J.; Lai, J.; Yin, X.; Yin, X.; Xiao, F.; Chen, J. Coordination control of V/V-connected hybrid distribution transformer for comprehensive power quality improvement under grid faults. *IEEE J. Emerg. Sel. Top. Power Electron.* **2025**, *13*, 6430–6445. [CrossRef]
11. Yao, L.; Wang, Y.; Xiao, X. Concentrated solar power plant modeling for power system studies. *IEEE Trans. Power Syst.* **2024**, *39*, 4252–4263. [CrossRef]
12. Sun, J.; Xing, X.; Zhang, C. Modeling, analysis, and mitigation of active power oscillation in parallel VSGs system. *IEEE Trans. Power Electron.* **2025**, *40*, 17295–17308. [CrossRef]
13. Yang, Z.; Wang, H.; Liao, W.; Bak, C.L.; Chen, Z. Protection challenges and solutions for AC systems with renewable energy sources: A review. *Prot. Control Mod. Power Syst.* **2025**, *10*, 18–39. [CrossRef]
14. Adewuyi, O.B.; Aki, H. Optimal planning for high renewable energy integration considering demand response, uncertainties, and operational performance flexibility. *Energy* **2024**, *313*, 134021. [CrossRef]
15. Saxena, A.; Shankar, R. An interactive operating demand response approach for hybrid power systems integrating renewable energy sources. *Prot. Control Mod. Power Syst.* **2024**, *9*, 174–194. [CrossRef]
16. Li, Z.; Li, Y.; Yu, Y.; Wang, P.; Ji, X. An Identification Method for Asymmetric Faults with Line Breaks Based on Low-Voltage Side Data in Distribution Networks. *IEEE Trans. Power Deliv.* **2021**, *36*, 3629–3639. [CrossRef]
17. Yang, Z.; Zhang, Q.; Liao, W.; Bak, C.L.; Chen, Z. Harmonic injection based distance protection for line with converter-interfaced sources. *IEEE Trans. Ind. Electron.* **2023**, *70*, 1553–1564. [CrossRef]

symmetry

MDPI

Article

Deep Reinforcement Learning for Load Frequency Control in Isolated Microgrids: A Knowledge Aggregation Approach with Emphasis on Power Symmetry and Balance

Min Wu [1,*], Dakui Ma [1], Kaiqing Xiong [2] and Linkun Yuan [2]

[1] Guangdong Power Grid Co., Ltd., Guangzhou 510600, China; madakui_csg@gdcsg.com
[2] Yangjiang Power Supply Bureau, Guangdong Power Grid Co., Ltd., Yangjiang 529500, China; xiongkq11@gdcsg.com (K.X.); csglinkuny@gdcsg.com (L.Y.)
* Correspondence: minwu1022@gdcsg.com

Abstract: To address the issues of instability and inefficiency that the fluctuating and uncertain characteristics of renewable energy sources impose on low-carbon microgrids, this research introduces a novel Knowledge-Data-Driven Load Frequency Control (KDD-LFC) approach. This advanced strategy seamlessly combines pre-existing knowledge frameworks with the capabilities of deep learning neural networks, enabling the adaptive management and multi-faceted optimization of microgrid functionalities, with a keen emphasis on the symmetry and equilibrium of active power. Initially, the process involves the cultivation of foundational knowledge through established methodologies to augment the reservoir of experience. Following this, a Knowledge-Aggregation-based Proximal Policy Optimization (KA-PPO) technique is employed, which proficiently acquires an understanding of the microgrid's state representations and operational tactics. This strategy meticulously navigates the delicate balance between the exploration of new strategies and the exploitation of known efficacies, ensuring the harmonization of frequency stability, precision in tracking, and the optimization of control expenditures through the strategic formulation of the reward function. The empirical validation of the KDD-LFC method's effectiveness and its superiority are demonstrated via simulation tests conducted on the load frequency control (LFC) framework of the Sansha isolated island microgrid, which is under the administration of the China Southern Grid.

Keywords: knowledge-data-driven load frequency control; power symmetry; isolated microgrid; deep reinforcement learning; knowledge aggregation

1. Introduction

Active power symmetry and power balance refer to the balance between the power provided by the power supplier and the power consumed by the load in the power system. This is the key to the stable operation of a power system. Unbalanced power will lead to problems such as deviations in frequency from normal values, affecting power quality. Therefore, it is very important to maintain the active power symmetry of the microgrid and power system through load frequency control (LFC) methods. The need for renewable energy sources such as wind and solar power is accentuated by the ongoing environmental impacts of fossil fuels and their greenhouse gas emissions. Distributed generation (DG) is increasingly being recognized as a vital approach for utilizing these cleaner energy sources. The efficient operation of DG systems facilitates the use of diverse renewable energies, although their intermittent nature and the integration of auxiliary storage devices introduce operational challenges to the grid [1]. The evolution of microgrid technology, operating in both islanded and grid-connected modes, addresses these issues by harmoniously combining DGs, storage devices, and converters to enhance power reliability and quality [2]. This technology significantly contributes to energy sustainability, grid stability, and the reduction in environmental pollution. Furthermore, advancements in load frequency control, crucial for maintaining grid stability, have paralleled the development of advanced control theory [3]. Recent research has led to the development of adaptive control methods, capable of responding to changing system conditions, thus enhancing frequency regulation performance in power systems. This includes the implementation of model reference adaptive PI controllers [4], adaptive fuzzy logic control [5], and adaptive neuro-fuzzy inference systems [6], each offering unique advantages in optimizing grid operations.

Sliding mode variable structure control represents a nonlinear control technique, distinguished by its adaptability to system uncertainties. This method, proficient in handling system uncertainties, robustly adjusts to parameter changes and external disturbances, ensuring system stability. Particularly in power systems, where disturbances can affect frequency and other states, this control strategy excels by adjusting controller gain in response to frequency shifts and generator outputs, effectively mitigating load variations [7]. However, challenges like system chattering arise due to the oscillatory nature of the frequency response trajectory around the sliding mode surface. Innovations in decentralized sliding mode control for multi-area power systems have been developed, utilizing local state information to stabilize system jitter [8]. Additionally, the integration of disturbance observers in controller designs enhances the prediction accuracy for uncertain disturbances, reducing controller conservatism, diminishing system jitter, and improving response speed [9].

Robust control methods in power systems are designed using the boundary information of system uncertainties, allowing them to effectively address power system uncertainties without pre-identifying the characteristics of perturbations [10]. Recent literature has discussed the development of a robust load frequency controller based on Riccati's equation, which effectively stabilizes power systems against parameter variations and external disturbances [11]. Another study proposed a robust gate-adaptive load frequency control method, combining robust and eye-adaptive control for different ranges of parameter variations [12]. Model predictive control (MPC) is increasingly utilized for managing power system constraints due to its direct handling capabilities and wide industrial applications [13–17]. These studies have predominantly focused on traditional energy storage systems, with recent shifts towards incorporating renewable energy sources like wind and photovoltaic power in system frequency regulation. The stochastic nature and diversity of new power generation units, including those with low inertia, present new challenges for system frequency controllers.

Artificial intelligence methods have made remarkable progress in recent years, enabling their application in various domains due to their high adaptability. However, a power system is subject to uncertainty and stochastic disturbances, such as fluctuations in renewable energy sources, load variations, and network failures, which pose challenges for traditional LFC methods to adapt to complex operating environments, resulting in frequency deviations and increased control costs. To address this issue, some researchers have attempted to apply deep reinforcement learning to LFC, using deep neural networks to learn the system's dynamic model and control strategy to achieve adaptive control and multi-objective optimization of the system.

Reinforcement learning has been applied in all control layers of microgrids, but most of them are focused on energy management and optimal economic allocation for tertiary control. Liu et al. [18] used reinforcement learning for domestic residential heating and hot water installations to effectively control the energy costs of building energy systems. In [19], a microgrid model of wind power generation and battery storage was established, and Q-learning algorithm was used to predict the environment of wind power generation, achieve the optimal scheduling of energy storage, and improve the utilization rate of wind power generation. Dai et al. [20] combined a reinforcement learning algorithm with a distributed optimization method based on multiplier splitting to achieve the optimal power output of different DGs at each moment without knowing the actual power generation cost function. A microgrid with multiple distributed power sources can be regarded as a multi-intelligent system, and although the model-free reinforcement learning approach has a natural advantage, its application in the control field is still very limited, and most applications tend to be converted to a single-intelligence problem to solve. Therefore, the application of reinforcement learning in the secondary control of microgrids still has a great potential. Esmaeili et al. [21] used a reinforcement learning algorithm to improve the PID controller and realized the adaptive control of the microgrid frequency. Adibi et al. [22] applied reinforcement learning to the secondary frequency control of lossy microgrids, effectively dealing with time-varying loads, load impedance, and other common general interference situations. Yu et al. [23] proposed an optimal CPS control strategy based on a Q-learning algorithm under CPS, which was able to optimize the output action commands of the CPS control system online to improve the control system's anti-interference abilities. Bhongade et al. [24] applied a three-layer feedforward neural network to the design of an automatic power generation controller and adopted the back propagation algorithm for training. The feasibility of the proposed strategy was verified through simulation in a multi-area AGC system considering superconducting magnetic energy storage units. In [25], a deep reinforcement learning algorithm with action-weighted optimization and state-of-the-art experience replay was proposed for the stochastic perturbation problem brought by large-scale clean energy.

The efficacy of deep reinforcement learning in power system control, particularly in load frequency control (LFC), is impeded by its reliance on extensive datasets, which are often constrained by the limited, incomplete, or unreliable nature of power system data. This limitation impacts the performance and generalization capacity of deep reinforcement learning, leading to increased frequency bias and higher generation costs in LFC applications. Proximal policy optimization (PPO), a policy-gradient-based deep reinforcement learning algorithm, addresses these issues. It introduces an effective clipping objective function that ensures policy updates are both monotonic and constrained, enhancing efficiency and scalability. The application of the PPO algorithm in LFC optimizes controller parameters, facilitating adaptive control and multi-objective optimization in power systems. PPO, a policy-gradient-based deep reinforcement learning algorithm, addresses the computational complexity and scalability challenges of the trust region policy optimization (TRPO) algorithm. It offers a simple yet effective clipping objective function to ensure policy update monotonicity and constraints, enhancing algorithmic efficiency and scalability. Utilized in load frequency control, PPO optimizes controller parameters for adaptive control and multi-objective optimization in power systems. This approach is particularly suited for islanded microgrid load frequency control but faces challenges in generalization and robustness. The paper introduces a Knowledge Aggregation Proximal Policy Optimization method, combining a priori knowledge and deep neural networks. This method enhances the reinforcement learning process by balancing exploration and exploitation and considering factors like frequency stability and control cost. The effectiveness of this approach is demonstrated through simulations on a four-unit islanded microgrid system, highlighting its potential in the adaptive control and multi-objective optimization of microgrids. This paper introduces two novel contributions to the field of microgrid control:

(1) A Knowledge-Data-Driven Load Frequency Control (KDD-LFC) method is developed. In traditional LFC, it is difficult to achieve adaptive and multi-objective frequency control for complex islanded microgrids [7–17]. This method combines the application of prior knowledge models with the capabilities of deep neural networks to achieve adaptive control and multi-objective optimization in microgrid systems.

(2) A new algorithm named Knowledge Aggregation Proximal Policy Optimization (KA-PPO) is created. Traditional DRL lacks robustness and is difficult to adapt to complex islanded microgrid environments [18–25]. This KA-PPO leverages traditional methods for generating prior knowledge, which is then incorporated into the experience pool. Subsequently, it employs a near-end strategy optimization technique to learn the state representation and behavioral strategies of the microgrid.

The structure of this paper is organized as follows: Section 2 details the model of the islanded microgrid system. Section 3 introduces a novel method, outlining its comprehensive framework. In Section 4, case studies are conducted to evaluate the effectiveness of the proposed method. Finally, Section 5 concludes the paper, summarizing key insights and discussing the primary findings of the research.

2. Islanded Microgrids

2.1. KDD-LFC Model for Islanded Microgrids

An islanded microgrid is defined as a small-scale power system, capable of autonomous operation when disconnected from the main power grid. It encompasses multiple distributed power sources, loads, and energy storage devices and is designed to deliver reliable, high-quality electrical energy services. This paper focuses on load frequency control in isolated microgrids, a critical aspect involving the regulation of active output from each distributed power source to maintain grid frequency stability and power balance. It introduces a distributed control strategy based on the PPO algorithm. PPO, a reinforcement learning algorithm, efficiently optimizes stochastic policies in various environments, simplifying complex gradient calculations and constraints. The proposed islanded microgrid model incorporates a detailed description using mathematical equations or simulation software, encompassing specific components outlined in the paper:

The distributed power model in this paper details generation characteristics, output power, and control strategies of various distributed power sources, such as wind turbines, solar panels, micro gas turbines, and fuel cells. It also includes load modeling, which describes the electricity demand, characteristics, and profiles of different loads, including lamps, air-conditioners, and electric vehicles. The energy storage device model covers the storage characteristics, as well as the charging and discharging behaviors of devices like batteries, supercapacitors, and flywheels. Additionally, the grid model encompasses parameters like voltage, current, power, impedance at each grid node, and details on grid topology and operational mode. The LFC model integrates these components, which is further illustrated in Figure 1 [19].

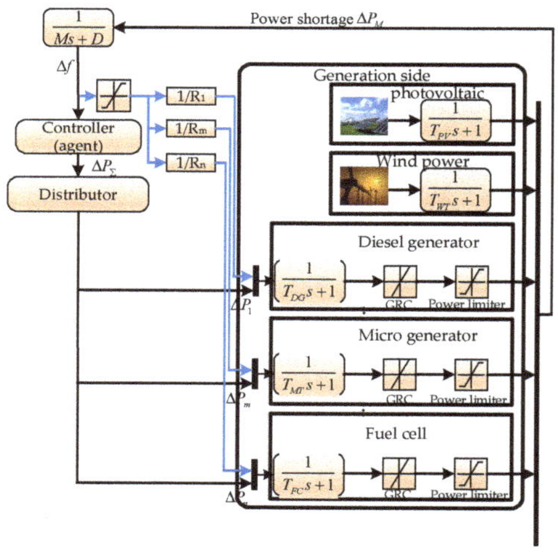

Figure 1. KDD-LFC model.

2.2. Generation Costs

The calculation of generation cost is delineated as follows:

$$C_i(P_{Gi}) = a_i P_{Gi}^2 + b_i P_{Gi} + c_i \tag{1}$$

where P_{Gi} is the output of the ith unit; a_i, b_i, and c_i are constants; and C_i is the cost of the ith unit.

$$C_i(P_{Gi,\text{ actual}}) = C_i\left(P_{Gi,\text{plan}} + \Delta P_{Gi}\right) = \alpha_i \Delta P_{Gi}^2 + \beta_i \Delta P_{Gi} + \gamma_i \tag{2}$$

$$\begin{cases} \alpha_i = a_i \\ \beta_i = 2a_i P_{Gi,\text{plan}} + b_i \\ \gamma_i = a_i P_{Gi,\text{plan}}^2 + b_i P_{Gi,\text{plan}} + c_i \end{cases} \tag{3}$$

where $P_{Gi,\text{plan}}$ is the planned output of unit i, ΔP_{Gi} is the regulation output of the ith unit, $P_{Gi,\text{ actual}}$ is the output of unit I; and α_i, β_i, and γ_i are coefficients.

2.3. Objective Functions and Constraints

Traditional LFC methods in microgrids often prioritize frequency stabilization while neglecting the aspect of cost efficiency. This paper introduces a KDD-LFC method, which effectively addresses both reducing frequency variations and minimizing power production costs in isolated microgrids. The KDD-LFC method employs a multi-objective optimization approach, aiming to minimize the combined effect of frequency variation and the power production cost. This approach balances the dual objectives of maintaining grid stability and enhancing cost efficiency in power generation within the microgrid.

$$\min \sum_{t=1}^{T} |\Delta f| + \sum_{t=1}^{T} \sum_{i=1}^{n} \left(\alpha_i \Delta P_{Gi}^2 + \beta_i \Delta P_{Gi} + \gamma_i\right) \tag{4}$$

$$\begin{cases} \sum\limits_{i=1}^{n} \Delta P_i^{\text{in}} = \Delta P_{\text{order-}\Sigma} \\ \Delta P_{\text{order-}\Sigma} \times \Delta P_i^{\text{in}} \geq 0 \\ \Delta P_i^{\text{min}} \leq \Delta P_i^{\text{in}} \leq \Delta P_i^{\text{max}} \\ |\Delta P_{Gi}(t) - \Delta P_{Gi}(t+1)| \leq \Delta P_i^{\text{rate}} \end{cases} \tag{5}$$

where $\Delta P_{\text{order-}\Sigma}$ is the total command, ΔP_i^{max} and ΔP_i^{min} are the limits of the ith unit, ΔP_i^{rate} is the ramp rate of the ith unit, and ΔP_i^{in} is the command of the ith unit.

2.4. MDP Modelling of KDD-LFCs

The LFC problem of an islanded microgrid is solved using a deep reinforcement learning algorithm. Firstly, the load frequency control of the islanded microgrid is re-described as an MDP model, and then the KA-PPO algorithm is used to solve the constructed model. The MDP model can be represented by the tuple M = (S, A, π, R, γ), i.e., the state space S, the action space A, the state transfer probability π, the reward function R, and the discount factor γ. In the context of an islanded microgrid, the agent represents the decision-making entity. The environment space encompasses elements beyond the agent's control, such as frequency information and the total output of units within the microgrid. The action space chosen for this study includes the unit output as the control variable. During a control period, the agent's objective in the islanded microgrid is to maximize future expected returns. This involves optimizing the sum of discounted rewards to achieve efficient and effective control of the microgrid's operations. The action value function can be expressed as Q_π (s,a). In order to obtain the action value function, the state transfer probabilities need to be known, but due to the existence of a large number of disturbances in reality, it is not possible to obtain the state transfer probabilities. The goal of reinforcement learning is to learn to obtain a policy that maximizes the expected return π_θ, and the objective function of the agent is defined by means of discounting the return as in Equation (2):

$$J^R(\theta) = \mathop{\mathbb{E}}_{(s_t,a_t)\sim\rho_{\pi_\theta}} \left[\sum_t \gamma^t r(s_t, a_t)\right] \tag{6}$$

where $\rho_{\pi\theta}$ is the distribution of trajectories determined by the strategy π_θ and $\gamma \subset [0, 1]$ is the discount rate, denoting the weight of the long-run returns.

In the islanded microgrid LFC problem, a strategy used to maximize the cumulative return of the team, π_{θ^*}, is obtained, and in this paper, the objective of the islanded microgrid LFC problem is defined as in Equation (7):

$$\theta^* = \underset{\theta_i}{\text{argmax}} \sum_{i}^{N} J^R(\theta_i) \tag{7}$$

where N denotes the number of agents.

(1) Action space

In the system outlined, the total electricity output is determined by a command generated by the agent. The agent's direct influence is restricted to only 10% of this command, signifying its limited yet strategic control over the system's output. This model underscores the agent's role in fine-tuning the system's performance through its selective, albeit restricted, action.

$$\left[\Delta P_{\text{order-}\Sigma}/10\right] \tag{8}$$

where $\Delta P_{\text{order-}\Sigma}$ is the total generating power command.

(2) State Space

The state space of the microgrid system encompasses two critical variables: the frequency deviation and its integral. Frequency deviation measures the discrepancy between the microgrid's actual frequency and its target frequency. The integral component cumulatively tracks this deviation over time, offering a comprehensive view of frequency stability. The output variable in this context is the total power output, generated by the distributed energy resources within the microgrid. This setup provides a clear framework for monitoring and adjusting the microgrid's operational parameters.

$$[\Delta f \int_0^t \Delta f dt \; \Delta P_G^{total}] \tag{9}$$

where ΔP_G^{total} is the total power output of the generation.

(3) Reward Functions

The controller in this system is designed with the primary goal of minimizing both the frequency fluctuation and the total cost of production. To encourage the agent towards identifying the optimal policy, the reward function incorporates a cost element for control actions. This reward function is constructed to reflect the dual objectives of the system, balancing frequency stability with cost efficiency. The structure of this function is critical in guiding the agent's actions towards the most effective and economical operational strategy for the microgrid.

$$\text{r} = -\mu_2|\Delta f| + \mu_3 \sum_{i=1}^{n} C_i + P_P \tag{10}$$

$$P_P = \begin{cases} 0 & |\Delta f| < 0.05\text{ Hz} \\ -3 & |\Delta f| \geq 0.05\text{ Hz} \end{cases} \tag{11}$$

where r is the reward and P_P is the punishment function, Δf is the frequency error, C_i is the power generation cost for the ith unit, and μ_1 and μ_2 are the weight coefficients, respectively.

3. Knowledge-Aggregation-Based Proximal Policy Optimization Method

Deep reinforcement learning (DRL) is a powerful, model-free, adaptive control method ideal for load frequency control (LFC) of islanded microgrids. It excels at handling complex, high-dimensional, nonlinear continuous action spaces, enhancing control accuracy and responsiveness. DRL's ability to directly learn from data reduces modeling difficulties and errors. Furthermore, it adapts to dynamic changes and uncertainties within system operations. Knowledge aggregation, which merges a priori and data-driven knowledge, significantly augments DRL's learning efficiency and robustness. It guides DRL's learning process, enhances generalization, and updates knowledge, leading to improved control performance. The proposed KA-PPO algorithm in this paper combines traditional methods for generating prior knowledge with advanced learning strategies, effectively balancing exploration and utilization. This method integrates aspects like frequency stability, tracking performance, and control cost into its reward function, achieving comprehensive, end-to-end control for the microgrid.

3.1. PPO Algorithm in Load Frequency Control

The PPO algorithm effectively overcomes the limitations of previous reinforcement learning algorithms, addressing issues such as low data utilization efficiency and poor robustness in traditional policy gradient methods, as well as the complexity associated with the trust region policy optimization (TRPO) algorithm. PPO's main advantages include its ease of deployment, reduced variance during iterations, user-friendliness, and enhanced robustness in training. It solves the challenge of determining the optimal learning rate in policy gradient methods by using a ratio that limits the update range of new strategies, reducing sensitivity to large training steps. In islanded microgrid load frequency control, the critic's Q-function, which assesses action quality, is modeled in a specific manner to enhance control performance.

The PPO algorithm first adopts the generalized advantage estimation (GAE) as the advantage function to estimate the advantage; then, it selects the loss function of the network parameter θ in the Actor network structure as well as the restriction on the KL scatter term during the process of updating the parameter θ; it also selects the loss function of the network parameter Φ in the criterion network structure. Finally, a new working process of the primary and secondary network structures is proposed.

The proximal policy optimization (PPO) algorithm, while effective, faces a challenge concerning the need for the frequent resampling of data in the environment for each parameter update, leading to a slow update process. This requirement to sample extensive data over extended periods is time-consuming and costly. To enhance training speed and enable data reuse, a strategy involving resampling has been proposed to transition from an on-policy to an off-policy approach. This method, outlined in Equation (12), aims to streamline the learning process, thus addressing the inherent inefficiencies of the PPO algorithm in certain tasks.

$$E_{x\sim p(x)}f(x) = E_{x\sim q(x)}\left(\frac{p(x)}{q(x)}f(x)\right) \tag{12}$$

where $p(x)$ is the sampling function for x^i but is unknown for $p(x)$, and $q(x)$ is the known sampling function.

The PPO algorithm is designed to allow strategies to choose actions with a higher "advantage", i.e., a much higher cumulative reward than predicted by the evaluator. The core purpose of the PPO algorithm is the improvement of the PG algorithm. PPO adds an additional constraint to make $p_\theta(a_t \mid s_t)$ and $p_{\theta'}(a_t \mid s_t)$ similar during training, so that the trained θ and θ' are more similar. Proximal policy optimization (PPO) has two main variants based on constraint methods: PPO-penalty and PPO-clip. PPO-penalty adjusts its penalty value by monitoring the KL's divergence, ensuring the new policy is not too far from the old one. On the other hand, PPO-clip does not directly incorporate KL divergence in the likelihood function but applies a clipping mechanism to the objective function. This clipping limits the range of policy updates, contributing to PPO-clip's effectiveness. PPO-clip is often preferred over PPO-penalty for its superior performance and ease of implementation. The optimization objective function of PPO-clip is designed to balance exploration and exploitation efficiently.

$$J_{PPO-clip}^{\theta^k}(\theta) \approx \sum_{(s_t,a_t)} \min\left(\frac{\pi_\theta(a_t \mid s_t)}{\pi_{\theta^k}(a_t \mid s_t)} A^{\theta^k}(s_t,a_t), \text{clip}\left(\frac{\pi_\theta(a_t \mid s_t)}{\pi_{\theta^k}(a_t \mid s_t)}, 1-\varepsilon, 1+\varepsilon\right) A^{\theta^k}(s_t,a_t)\right) \quad (13)$$

where ε is the hyperparameter, which is usually set to 0.1 or 0.2.

The proximal policy optimization (PPO) algorithm incorporates both on-policy and off-policy approaches. The key distinction lies in the alignment of the policy used for interacting with the environment and the policy used for learning. In the off-policy approach, the model's strategy for action selection in the environment differs from the strategy used during Q-value updates. The latter always chooses actions that maximize state benefits, indicating a divergence between learning and sampling strategies. Conversely, the on-policy approach denotes a scenario where the learning and sampling strategies are identical, ensuring consistency between environmental interaction and policy learning.

The Actor network in the PPO algorithm learns the policy by taking the current state as the input and producing the action probability distribution as the output. The training process of the intelligent agent consists of the following steps: first, the Actor network outputs the action probability distribution; second, the action is sampled from the action probability distribution; third, the environmental state is obtained after executing the action; and finally, the Actor network's parameters are updated using the gradient ascent method.

$$\hat{g} = \hat{E}_t\left[\nabla_\theta \log \pi_\theta(a_t \mid S_t)\hat{A}_t\right] \quad (14)$$

The Actor network loss is shown in Equation (15):

$$L_{PG}(\theta) = \hat{E}_t\left[\log \pi_\theta(a_t \mid S_t)\hat{A}_t\right] \quad (15)$$

where π_θ is the stochastic strategy and $\hat{A}_t$ is the estimate of the dominant $\hat{E}_t$ function at the t moment.

The Critic network's function in the PPO algorithm is to compute $v(s_t)$ and A_t, whose current states are used as inputs to the algorithm, and the output of the algorithm is the predicted state values. The Critic network updates the parameters of the network by minimizing a loss function, which is shown in Equation (14) as follows:

$$L^{VF}(\theta) = \left(v_\theta(s_t) - v_t^{\text{target}}\right)^2 \quad (16)$$

where $v_\theta(s_t)$ is the state value predicted by the Critic network, v_t^{target} is the update target obtained by the GAE algorithm, and v_t^{target} is shown in Equation (17) as follows:

$$v_t^{\text{target}} = v_\theta(s_t) + A_t = v_\theta(s_t) + \sum_{l=t}^{T} (\gamma\lambda)^{l-t+1}\delta_{l-1} \quad (17)$$

where δ_{l-1} denotes the timing difference error.

The general PPO algorithm is not satisfactory at learning efficiency and convergence, and it is difficult to adapt to the complex islanded microgrid load frequency control environment. In this paper, a PPO algorithm based on knowledge aggregation is proposed. The samples obtained from other controllers are input into the PPO network as a priori knowledge information, aiming to reduce the training time and interactive data required by the policy learning algorithm and, at the same time, improve the frequency regulation performance and generalization.

3.2. Knowledge-Aggregation Methods and the KA-PPO Algorithm

In reinforcement learning, various kinds of prior knowledge can facilitate the agent's strategy learning and enhance the final quality of the strategy. However, the methods for the integration and utilization of prior knowledge in reinforcement learning differ depending on their forms of expression. Reinforcement learning relies on the sample data generated by the interaction between the agent and the environment. The agent explores the environment and uses the obtained data to guide its strategy learning, which constitutes the general learning process of reinforcement learning. In this process, enhancing exploration will motivate the agent to try some uncertain actions, so as to gain a more comprehensive understanding of the environment and the task and to prevent the policy from falling into a local optimum prematurely. However, enhancing exploration will also reduce the learning efficiency to some extent, and may incur unnecessary costs and computational resources due to meaningless exploration. Enhancing data utilization will prompt the agent to focus on actions that are likely to bring high cumulative returns, thus maximizing the use of the available data samples and accelerating the convergence process of the policy. The learning process is shown in Figure 2. Nevertheless, such a choice may cause the policy learning to fall into local optima, affecting the final performance of decision-making tasks. Therefore, how to balance the factors of exploration and exploitation according to different task scenarios is an important challenge for reinforcement learning.

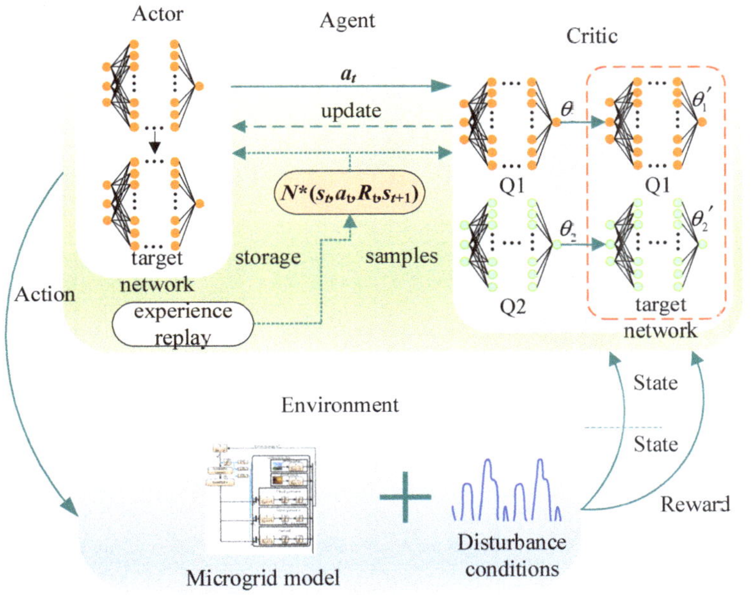

Figure 2. Farmwork of KA-PPO algorithm.

Reinforcement learning with knowledge integration refers to the use of some prior or external knowledge to guide or assist the agent's learning in the process of reinforcement learning, so as to improve the efficiency and performance of learning. This prior knowledge can take various forms, such as the demonstrations of experts, the strategies of scripts, the design of reward functions, and the division of state space. Prior knowledge can help the agent to narrow the search space, reduce the exploration cost, avoid wrong behaviors, accelerate convergence, adapt to changes in the environment, and so on.

Reinforcement learning with knowledge aggregation refers to a combination of imitation learning and reinforcement learning. This type of algorithm stores expert examples in the example replay area and co-trains the behavior policy model with state–action pairs collected from the environment in a certain ratio, so that the agent can both absorb the expert experience and explore the environment beyond the limitations of the expert's experience. The KA-PPO algorithm needs to optimize the TD-error, the loss of supervised learning, and the L2 regular term at the same time, as shown in Equation (18).

$$J(Q) = J_{DQ}(Q) + \lambda_1 J_E(Q) + \lambda_2 J_{L2}(Q) \tag{18}$$

Since the knowledge aggregation reinforcement learning is based on expert example data, the learning efficiency of the agent is significantly improved and the interpretability of the system is enhanced, which has received much attention in the industrial field.

The KA-PPO design employs imitation learning. Each knowledge aggregation module consists of a controller and a distributor. During training, each integrator generates a reasonable result based on its own controller and distributor, converts it to a sample, and adds it to the experience pool. This enriches the public experience pool with valuable samples.

The controller in the knowledge aggregation module uses PI, PSO-PI, FOPI, PSO-tuned fuzzy-PI, and fuzzy-PI algorithms. Due to the frequent occurrence of large disturbances, the controller's objective for the integrators is shown below.

$$\min F_C(t) = \int_0^{\infty} t(e_i^{ACE}(t))^2 \mathrm{d}t \tag{19}$$

The learning process of the KA-PPO algorithm is as follows: First, the information obtained from the knowledge aggregation module is converted into samples according to MDP and used as the a priori knowledge input to the experience pool. Next, the state of the islanded microgrid is inputted to the PPO network, which consists of multiple fully-connected layers. The PPO network then outputs the action distribution for the current state, i.e., the probability of each possible action.

4. Case Studies

To verify the effectiveness of the proposed methodology, we establish an LFC model for the Sansha isolated microgrid of the China Southern Power Grid, which contains both multiple small and medium-sized distributed FM resources and conventional FM units (gas turbines, diesel generators), based on actual data. In the case study, we include not only the KDD-LFC based on the KA-PPO algorithm, but also three deep reinforcement learning algorithm-based KDD-LFCs and three conventional algorithm-based LFCs. The algorithms included for comparison are soft actor–critic (SAC) algorithm-based LFC [23], twin delayed deep deterministic policy gradient (TD3)-based LFC [24], deep deterministic policy gradient (DDPG)-based LFC [23], genetic algorithm fuzzy PI controller (GA-fuzzy-PI) [22], Takagi–Sugeno fuzzy PI controller (TS-fuzzy-PI) [25], and GA optimized PI (GA-PI) [21]. We use a computer with two CPUs with 2.10 GHz Intel Xeon Platinum processors and 16 GB of memory to run the simulation models and procedures that we present in this paper. The simulation software package that we employ is MATLAB/Simulink version 9.8.0 (R2020a).

4.1. Case 1: Step Disturbance and Renewable Disturbance

In Case 1, we add three large-scale load step perturbations and persistent stochastic perturbations in PV and WT energy, with a total perturbation time of 7200 s.

As shown in Table 1, KA-PPO can reduce the deviation in frequency by 12.30–82.83% and the generation cost by 0.0028–0.072% compared with other algorithms. The KDD-LFC with deterministic optimal control policies learned from pre-training can be deployed for online operation to achieve intelligent control of the power system. To simulate the load surge that often occurs in power systems and evaluate the control performance of different controllers, continuous step disturbances with amplitudes of 500 MW, 1000 MW, and 1500 MW (with an evaluation period of 7200 s) are applied as test signals and compared with six other controllers that have also been pre-trained: the proportional–integral controller (PI), proportional–integral–derivative (PID) controller, fuzzy controller (FC), neural network controller (NN), deep deterministic policy gradient algorithm (DDPG), and dual deep q network algorithm (DDQN). In this paper, we use the Knowledge-Aggregation-based Proximal Policy Optimization (KA-PPO) algorithm as the main controller with the following features.

Table 1. Statistical results.

Control Algorithm	Average Frequency Deviation (HZ)	Power Generation Cost (USD)
	$\mid \Delta f \mid_{avg}$	C^{total}
KA-PPO	0.00431	2139.39
SAC	0.00484	2140.67
TD3	0.00493	2140.62
DDPG	0.00656	2139.94
GA-fuzzy-PI	0.00803	2140.94
TS-fuzzy-PI	0.00788	2139.45
GA-PI	0.00699	2139.73

Regarding the output accuracy of the action values, the KA-PPO controller enhances the algorithm's convergence speed and stability by using knowledge aggregation to increase its prior knowledge and applying more prior knowledge to guide its learning process. This algorithm stores expert examples in the replay buffer and samples them with state–action pairs collected from the environment at a certain ratio to co-train the behavior policy model. Thus, the agent can not only learn from the expert experience, but also explore the environment and overcome the limitations of the expert's experience, achieving a better control effect. Other deep reinforcement learning algorithms lack this function and can only obtain more samples for learning through trial and error. Such samples are not diverse, which lowers the agent's learning efficiency, and the algorithms are prone to fall into local optima and struggle to adapt to complex environmental changes. Therefore, as shown in Figure 3, the KA-PPO algorithm achieves a better frequency control effect, and its frequency deviation does not exhibit large overshooting and oscillation, and it returns to stability after only a small amount of overshooting. However, the other deep reinforcement learning algorithms have difficulty in obtaining good performance due to their poor performance. Their frequency deviation oscillates and overshoots, and this overshooting and oscillation severely affect the algorithm's power generation cost. Moreover, repeated adjustments increase the power generation cost of the other deep reinforcement learning algorithms.

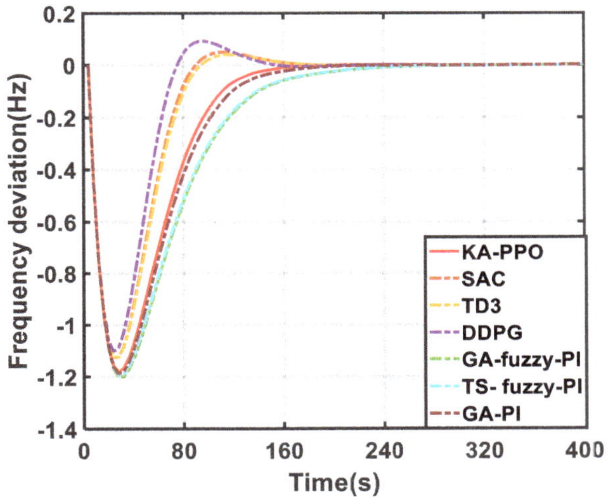

Figure 3. Frequency deviation.

Among other conventional control algorithms, fuzzy rule-based controllers can adaptively regulate the controller's output and adjust and recover in time after the output overshoots. However, since the fuzzy rules are manually formulated, their control accuracy is very low, which causes their frequency to oscillate still. Other optimization-based controllers lack adaptive tuning and thus suffer from different frequency fluctuations and performance under different load perturbations, which severely impairs the frequency control performance, resulting in frequency overshooting and difficulty in controlling it.

4.2. Case 2: Step Disturbance and Renewable Disturbance

This paper presents a smart distribution network model that addresses the challenge of integrating new and distributed energy sources into the grid on a large scale, while maintaining the stability of large grid systems. The proposed model integrates a diverse array of alternative energy sources, including wind power, small-scale hydroelectric systems, micro gas turbines, fuel cells, solar power, and biomass energy. This multifaceted approach to energy generation leverages the unique benefits of each source, aiming to enhance efficiency, reduce environmental impact, and promote sustainability. By diversifying the energy portfolio, the model addresses the growing demand for clean energy solutions and mitigates reliance on traditional fossil fuels. This paper investigates the control performance of KA-PPO in a highly stochastic environment based on this model. Due to the high uncertainty in the output of electric vehicles and wind and solar energy, they are treated as stochastic load disturbances in this paper and are excluded from control. This paper employs finite bandwidth white noise to simulate random wind speeds as an input to the wind turbine and obtain its output. Similarly, this paper uses the simulated variation in solar irradiance throughout the day to obtain the output of the solar power generation model.

Our objective is to investigate how the power system can handle the stochastic fluctuations in load when a large number of new energy sources are integrated into the grid. For this purpose, we introduce random white noise as a load disturbance in the smart distribution network model to evaluate the control effect of the KA-PPO strategy in this complex environment. The KA-PPO algorithm can deal with the random disturbances effectively and produce accurate tracking results. Table 2 presents the statistics of the simulation experiment, where the generation cost indicates the total regulation cost of all generating units in 24 h. The distribution network data reveal that the deviation in frequency of the other algorithms is 11.61–80.64% higher than that of the KA-PPO algorithm, while the generation cost of the KA-PPO algorithm is reduced by 0.067–0.085%. The analysis of control performance metrics demonstrates that the KA-PPO algorithm surpasses other intelligent algorithms in terms of economy, adaptability, and coordinated optimal control performance. We also performed experimental verification of various disturbances such as step waves, square waves, and random waves. The experimental results indicate that KA-PPO has strong convergence ability and high-speed learning efficiency. Particularly in random environments, it exhibits excellent adaptability. It not only mitigates random disturbances, but also improves the dynamic control performance in the interconnected grid environment. Under the control of the total power command, the complementary and synergistic optimal operation of multiple energy sources is accomplished in each time period.

Table 2. Data of Case 2.

Control Algorithms	Average Frequency Error (Hz)	Generation Cost (USD)
	$\lvert \Delta f \rvert_{avg}$	C^{total}
KA-PPO	0.0155	5668.61
SAC	0.0173	5672.65
TD3	0.0195	5672.09
DDPG	0.0242	5670.70
GA-fuzzy-PI	0.0273	5670.95
TS-fuzzy-PI	0.0268	5669.10
GA-PI	0.0280	5669.50

4.3. Case 3: Large-Scale Renewable Disturbance

In this study, Case 3 is introduced, incorporating large-scale renewable energy disturbances, to rigorously evaluate the robustness of the proposed algorithm. This addition aims to simulate realistic conditions under which the algorithm's performance can be assessed against significant fluctuations in renewable energy output, reflecting scenarios such as sudden changes in wind speed or solar irradiance. Through this strategic inclusion, the research endeavors to provide a comprehensive analysis of the algorithm's capability to maintain stability and efficiency in the face of dynamic and unpredictable renewable energy patterns. This approach not only enhances the validity of the algorithm's application in real-world settings but also substantiates its resilience and adaptability to accommodate the inherent variability in renewable energy sources, thereby contributing valuable insights into its potential for optimizing energy distribution and grid stability.

This study aims to explore the capacity of the power system to manage stochastic load fluctuations as a consequence of integrating a substantial number of novel energy sources into the electrical grid. To achieve this, we have introduced random white noise to act as a load disturbance within the smart distribution network model. This methodology is deployed to assess the efficacy of the KA-PPO strategy under such complex conditions. The KA-PPO algorithm demonstrates a proficient capability in addressing random disturbances, delivering precise tracking outcomes. The simulation experiment's results are summarized in Table 3, showcasing the total regulation costs incurred by all generating units over a 24 h period as a measure of generation cost. Data on the distribution network elucidate that the deviation in frequency with other algorithms is 17.11–87.22% higher than that achieved using the KA-PPO algorithm. Simultaneously, the generation cost associated with the KA-PPO algorithm shows a reduction of 0.066–0.083%. This critical analysis of control performance metrics highlights the KA-PPO algorithm's superiority over other intelligent algorithms in delivering economic efficiency, adaptability, and coordinated optimal control performance.

Table 3. Data of Case 3.

Control Algorithms	Average Frequency Error (Hz)	Generation Cost (USD)
	$\lvert \Delta f \rvert_{avg}$	C^{total}
KA-PPO	0.011686	8246.813
SAC	0.012364	8252.201
TD3	0.012858	8251.709
DDPG	0.014776	8249.376
GA-fuzzy-PI	0.016278	8251.551
TS-fuzzy-PI	0.016088	8247.276
GA-PI	0.015794	8248.216

Further experimental validation was conducted to assess the algorithm's response to various disturbances, including step waves, square waves, and random waves. These experimental findings underscore the KA-PPO's robust convergence capabilities and expedited learning efficiency, which is particularly pronounced in environments characterized by randomness. The algorithm's exceptional adaptability is not only pivotal in counteracting random disturbances but also in enhancing the dynamic control performance within the interconnected grid. The KA-PPO algorithm facilitates the complementary and synergistic optimal operation of multiple energy sources across different time intervals, under the directive of total power command. This orchestration underscores the potential of the KA-PPO algorithm to significantly contribute to the stability and efficiency of power systems amidst the increasing integration of renewable energy sources.

5. Conclusions

This manuscript delineates several significant contributions to the realm of microgrid management and optimization. Initially, it introduces the Knowledge-Driven Deep Learning for Load Frequency Control (KDD-LFC) methodology. This sophisticated approach amalgamates prior knowledge models with advanced deep neural network architectures to facilitate adaptive control and achieve multi-objective optimization within microgrid systems. The incorporation of established knowledge into deep learning frameworks allows for a more nuanced and effective control strategy, catering to the dynamic and complex nature of microgrid operations.

Furthermore, the study pioneers a cutting-edge algorithm designated as Knowledge-Augmented Proximal Policy Optimization (KA-PPO). This algorithm ingeniously integrates conventional methodologies to harvest prior knowledge as a form of experiential data, which is subsequently assimilated into the experience pool. Utilizing this enriched dataset, KA-PPO employs an end-to-end policy optimization technique to intricately learn the state representation and behavioral strategies pertinent to microgrid management. This dual-phase approach not only capitalizes on the strengths of traditional and deep learning methodologies but also enhances the algorithm's capability to navigate and optimize the multifaceted microgrid environment.

Empirical validation of the proposed KDD-LFC method and KA-PPO algorithm is conducted through rigorous simulation experiments. These experiments utilize the load frequency control (LFC) model of the Sansha isolated microgrid, operated by the China Southern Power Grid. The outcomes of these simulations unequivocally demonstrate the superiority and efficacy of the proposed solutions. When benchmarked against a suite of prevalent deep reinforcement learning algorithms (such as soft actor–critic (SAC), twin delayed dDPG (TD3), and deep deterministic policy gradient (DDPG)) and conventional control strategies (including genetic algorithm-enhanced fuzzy-PI (GA-fuzzy-PI), Takagi–Sugeno (TS) fuzzy-PI, and GA-PI algorithms), KA-PPO exhibits remarkable performance in minimizing deviations in frequency and reducing generation costs.

Looking ahead, the trajectory of future work is set to pivot towards the practical application and real-world implementation of the proposed algorithm. This endeavor will aim to transcend theoretical validation and simulation experiments, focusing on the deployment and operational efficacy of KA-PPO within live microgrid environments. Such applied research will not only underscore the practical viability of the algorithm but also contribute to its refinement and optimization for broader utility in the field of microgrid management.

Author Contributions: Conceptualization, M.W., D.M. and K.X.; methodology, M.W., D.M. and K.X.; software, M.W., D.M. and K.X.; validation, M.W., D.M. and K.X.; formal analysis, M.W., D.M. and K.X.; investigation, M.W., D.M. and K.X.; resources, M.W., D.M. and K.X.; data curation, M.W., D.M. and K.X.; writing—original draft preparation, M.W., D.M. and K.X.; writing—review and editing, M.W., D.M. and K.X.; visualization, M.W., D.M. and K.X.; supervision, M.W., D.M. and K.X.; project administration, L.Y.; funding acquisition, M.W., D.M. and K.X. All authors have read and agreed to the published version of the manuscript.

Funding: This work was supported by the Science and Technology Project of China Southern Power Grid Corporation, under Grant No. 030000WX24210010.

Data Availability Statement: Data are contained with the article.

Acknowledgments: The authors gratefully acknowledge the support of the National Natural Science Foundation of China.

Conflicts of Interest: Authors were employed by the Guangdong Power Grid Corporation. The authors declare that the research was conducted in the absence of any commercial or financial relationships that could be construed as a potential conflict of interest. .

References

1. Pachaiyappan, R.; Arasan, E.; Chandrasekaran, K. Improved Gorilla Troops Optimizer-Based Fuzzy PD-(1+PI) Controller for Frequency Regulation of Smart Grid under Symmetry and Cyber Attacks. *Symmetry* **2023**, *15*, 2013. [CrossRef]
2. Kumar, A.; Anwar, M.N.; Huba, M. Load Frequency Controller Design Based on the Direct Synthesis Approach Using a 2DoF-IMC Scheme for a Multi-Area Power System. *Symmetry* **2022**, *14*, 1994. [CrossRef]

3. Srikanth, M.; Kumar, Y.V.P. A State Machine-Based Droop Control Method Aided with Droop Coefficients Tuning through In-Feasible Range Detection for Improved Transient Performance of Microgrids. *Symmetry* **2023**, *15*, 1. [CrossRef]
4. Pan, C.T.; Liaw, C.M. An Adaptive Controller for Power System and Load Frequency Control. *IEEE Trans. Power Syst.* **1989**, *4*, 122–128. [CrossRef]
5. Yousef, H.A.; AL-Kharusi, K.; Albadi, M.H.; Al-Badi, A.H. Load Frequency Control of a Multi-Area Power System: An Adaptive Fuzzy Logic Approach. *IEEE Trans. Power Syst.* **2014**, *29*, 1822–1830. [CrossRef]
6. Hosseini, S.H.; Etemadi, A.H. Adaptive Neuro-Fuzzy Inference System Based Automatic Generation Control. *Electr. Power Syst. Res.* **2008**, *78*, 1230–1239. [CrossRef]
7. Bengiamin, N.N.; Chan, W.C. Variable Structure Control of Electric Power Generation. *IEEE Trans. Power Appar. Syst.* **1982**, *PAS-101*, 24–30. [CrossRef]
8. Mi, Y.; Fu, Y.; Wang, C.; Zhang, X.; Zhao, J. Decentralized Sliding Mode Load Frequency Control for Multi-Area Power Systems. *IEEE Trans. Power Syst.* **2013**, *28*, 4301–4309. [CrossRef]
9. Mi, Y.; Fu, Y.; Li, D.; Zhang, X.; Zhao, J. The Sliding Mode Load Frequency Control for Hybrid Power System Based on Disturbance Observer. *Int. J. Electr. Power Energy Syst.* **2016**, *74*, 446–452. [CrossRef]
10. Chen, Y.H.; Leitmann, G.; Kai, X.Z. Robust Control Design for Interconnected Systems with Time-Varying Uncertainties. *Int. J. Control* **1991**, *54*, 1119–1142. [CrossRef]
11. Wang, Y.; Zhou, R.; Wen, C. Robust Load-Frequency Controller Design for Power Systems. *IEE Proc. C-Gener. Transm. Distrib.* **1993**, *140*, 111–116. [CrossRef]
12. Wang, Y.; Zhou, R.; Wen, C. New Robust Adaptive Load Frequency Control with System Parameter Uncertainties. *IEE Proc.-Gener. Transm. Distrib.* **1994**, *141*, 184–190. [CrossRef]
13. Bevrani, H.; Hiyama, T. Robust Decentralised PI Based LFC Design for Time Delay Power Systems. *Energy Convers. Manag.* **2008**, *49*, 193–204. [CrossRef]
14. Xin, H.; Liu, Y.; Wang, Z.; Zhang, B.; Wang, C. A New Frequency Regulation Strategy for Photovoltaic Systems without Energy Storage. *IEEE Trans. Sustain. Energy* **2013**, *4*, 985–993. [CrossRef]
15. Nanou, S.I.; Papakonstantinou, A.G.; Papathanassiou, S.A. A Generic Model of Two-Stage Gridconnected PV Systems with Primary Frequency Response and Inertia Emulation. *Electr. Power Syst. Res.* **2015**, *127*, 186–196. [CrossRef]
16. Liu, Y.; Xin, H.; Wang, Z.; Zhang, B.; Wang, C. Power Control Strategy for Photovoltaic System Based on the Newton Quadratic Interpolation. *IET Renew. Power Gener.* **2014**, *8*, 611–620. [CrossRef]
17. Long, Y.; Liao, K.; Chong, T.; Zhang, Y.; Li, Y. Enhancement of Frequency Regulation in AC Microgrid: A Fuzzy-MPC Controlled Virtual Synchronous Generator. *IEEE Trans. Smart Grid* **2021**, *12*, 3138–3149. [CrossRef]
18. Liu, S.; Henze, G.P. Experimental Analysis of Simulated Reinforcement Learning Control for Active and Passive Building Thermal Storage Inventory: Part 1. Theoretical Foundation. *Energy Build.* **2006**, *38*, 142–147. [CrossRef]
19. Kuznetsova, E.; Li, Y.F.; Ruiz, C.; Zio, E. Reinforcement Learning for Microgrid Energy Management. *Energy* **2013**, *59*, 133–146. [CrossRef]
20. Dai, P.; Yu, W.; Wen, G.; Baldi, S. Distributed Reinforcement Learning Algorithm for Dynamic Economic Dispatch with Unknown Generation Cost Functions. *IEEE Trans. Ind. Inform.* **2020**, *16*, 2258–2267. [CrossRef]
21. Esmaeili, M.; Shayeghi, H.; Nejad, H.M.; Younesi, A. Reinforcement Learning Based PID Controller Design for LFC in a Microgrid. *Int. J. Comput. Math. Electr. Electron. Eng.* **2015**, *34*, 1450–1466. [CrossRef]
22. Adibi, M.; Van der Woude, J. Secondary Frequency Control of Microgrids: An Online Reinforcement Learning Approach. *IEEE Trans. Autom. Control* **2022**, *67*, 4824–4831. [CrossRef]
23. Yu, T.; Zhou, B.; Chan, K.W. Q-learning based dynamic optimal CPS control methodology for interconnected power systems. In Proceedings of the Chinese Society of Electrical Engineering, Beijing, China, 21–23 October 2009; Chinese Society of Electrical Engineering: Beijing, China, 2009; pp. 13–19.
24. Bhongade, S.; Gupta, H.O.; Tyagi, B. Artificial neural network based automatic generation control scheme for deregulated electricity market. In Proceedings of the 2010 Conference Proceedings IPEC, Singapore, 27–29 October 2010; IEEE: Piscataway, NJ, USA, 2010; pp. 1158–1163. [CrossRef]
25. Xi, L.; Sun, M.; Zhou, H.; Li, Y.; Wang, Z.; Zhang, J. Multi-agent deep reinforcement learning strategy for distributed energy. *Measurement* **2021**, *185*, 109955. [CrossRef]

symmetry

Article

A Coordinated Control Strategy of Multi-Type Flexible Resources and Under-Frequency Load Shedding for Active Power Balance

Jian Zhang [1], Jiaying Wang [2], Yongji Cao [3,*], Baoliang Li [1] and Changgang Li [1]

[1] Key Laboratory of Power System Intelligent Dispatch and Control of the Ministry of Education, Shandong University, Jinan 250061, China; 202214664@mail.sdu.edu.cn (J.Z.); 202220677@mail.sdu.edu.cn (B.L.); lichgang@sdu.edu.cn (C.L.)
[2] State Grid Zhejiang Marketing Service Center, Hangzhou 310007, China; wangjy_ee@163.com
[3] Academy of Intelligent Innovation, Shandong University, Jinan 250101, China
* Correspondence: yongji@sdu.edu.cn

Abstract: With the increasing expansion of power systems, there is a growing trend towards active distribution networks for decentralized power generation and energy management. However, the instability of distributed renewable energy introduces complexity to power system operation. The active symmetry and balance of power systems are becoming increasingly important. This paper focuses on the characteristics of distributed resources and under-frequency load shedding, and a coordinated operation and control strategy based on the rapid adjustment of energy storage power is proposed. The characteristics of various controllable resources are analyzed to explore the rapid response capabilities of energy storage. The energy storage types are categorized based on the support time, and the final decision is achieved with power allocation and adjustment control of the energy storage system. Additionally, a comprehensive control strategy for under-frequency load shedding and hierarchical systems is provided for scenarios with insufficient active support. The feasibility of the proposed model and methods is verified via a multi-energy system case.

Keywords: active power balance; active power symmetry; coordinated control; flexible resource; frequency stability; under-frequency load shedding

1. Introduction

Active symmetry and balance refer to the active power balance of a power system between the supply and demand sides [1]. With the rapid development of distributed renewable energy, active distribution networks have become a trend in the evolution of power systems [2]. However, the uncertainty and volatility of distributed renewable energy make the operation of power systems more complex and unstable [3,4]. Concurrently, with the integration of clean energy and distributed controllable loads, the inertia resources and regulation capabilities in the power system are weakened, resulting in a decrease in the system's frequency response capability and an increased risk of frequency decline [5,6]. Hierarchical coordinated control of controllable resources is crucial for achieving the stable operation and decarbonization of high-penetration renewable energy grids [7]. Research on decentralized resource active support and automatic load-shedding coordination control methods in response to the risk of frequency decline holds significant importance [8,9].

Distributed generation, energy storage systems, controllable loads, and other decentralized resources have distinct characteristics and control capabilities [10,11]. They provide frequency support measures to the power system, effectively reducing the risk of frequency decline and enhancing the stability and reliability of the power system [12,13]. Meanwhile, different types of controllable resources exhibit varying response speeds and control capabilities. The energy storage system has the advantage of flexible output control [14,15]. It is both a key part of the energy Internet and a key support for dealing with large-scale renewable energy grid integration problems and increasing the economy and reliability of the power grid [16,17]. Based on this, the study investigates a hierarchical coordinated control approach for controllable resources in the context of active distribution networks [18,19]. It fully utilizes controllable resources to enhance the flexibility and response speed of the power system [20,21].

Coordination control primarily involves the coordinated control of distributed energy resources, energy storage, loads, and under-frequency load shedding within microgrids. Its essence lies in considering the output characteristics of different types of resources, coordinating output control to achieve power balance. Ref. [22] proposes the use of constant power and constant voltage–frequency control strategies. Ref. [23] achieves coordinated control of the microgrid by monitoring the operating state of the microgrid side and the main power grid through upper-level central control. Ref. [24] sets distributed energy source units in maximum power tracking mode, engages energy storage units in peak shaving during load peaks, and charges them at a constant power to the upper limit of the state of charge during non-peak periods. While various energy source units are coordinated to fully leverage their respective advantages, they can also flexibly coordinate with under-frequency load shedding. Ref. [25] optimizes the allocation of automatic generation control (AGC) commands based on the prediction of perturbation signals as well as information about the state of charge (SOC) of energy storage in combination with a grid adaptive search algorithm on the basis of subdividing the continuous control period. Ref. [26] takes flywheel energy storage as the research object and ensures that the remaining capacity is always in the optimal output state while considering its SOC recovery. Ref. [27] proposes control strategies for two cases of adaptive frequency regulation and self-restoration of charge state for energy storage to realize the coordination between battery storage and thermal power units under the premise of energy storage battery charge state management.

There are many studies on load-shedding emergency control systems. However, separate emergency control tends to over-cut or under-cut phenomena triggering negative effects. Additionally, the current coordination control framework often directs energy storage systems to maintain maximum power output during emergency situations. If the storage SOC is insufficient, it would be difficult for its output power to reach the desired value. This is equivalent to a new power deficit under an emergency situation. At this time, it is difficult to maintain the stability of the system with the coordination system of energy storage and load shedding. Therefore, this paper proposes a coordinated control strategy based on the flexible output of energy storage. Based on the flexible output characteristics of the energy storage unit, the fluctuation of renewable energy is suppressed, and the means of low-frequency load shedding is combined to accomplish the coordinated control of the system as a whole. The main contribution of this paper has three main aspects. Firstly, an energy storage output control measure based on different SOC states is designed to provide leaner control of energy storage in the face of different demands. Secondly, a mathematical model of load slicing based on differential evolutionary algorithm is proposed to compress the amount of load slicing under the premise of satisfying the power balance. Thirdly, a complete coordinated control framework is proposed, which takes into account the characteristics of multiple flexible resources, achieves the minimum fluctuation curve in the steady state, and cooperates with the load shedding to complete the emergency control in the emergency state.

The rest of this paper is organized as follows. Section 2.1 considers the characteristics of the different types of controllable resources involved in the frequency response. Additionally, the energy storage control strategies with emergency load-shedding optimization models for different SOC states are presented in Section 2.2. Section 2.3 presents the hierarchical coordinated control approach for systems containing multiple types of controllable resources. The case study is carried out in Section 3. Finally, conclusions and discussion are drawn in Section 4.

2. Materials and Methods

2.1. Controllable Resources Participating in Frequency Response

2.1.1. Inertia Analysis

First, the difference between the synchronous generator in traditional thermal power unit and frequency regulation capabilities of energy storage was analyzed. The total inertia of the generator and prime mover in traditional thermal power units accelerates through the imbalance of applied torque, and the motion equation and inertia expression are expressed by (1) and (2):

$$J\frac{\mathrm{d}\omega}{\mathrm{d}t} = T_{\mathrm{m}} - T_{\mathrm{c}} \tag{1}$$

$$J = \int r^2 dm \tag{2}$$

where J is the moment of inertia, ω is the rotor angular velocity, r is the rotational radius, and m is the mass of the rigid body.

The inertia time constant of the synchronous machine is defined as the ratio of the rotor kinetic energy to the rated capacity under the unit's rated speed operation, which is expressed by (3).

$$H = \frac{E_K}{S_B} = \frac{J\omega_n^2}{2S_B} \tag{3}$$

From Equations (1)–(3), it can be observed that the synchronous machine inertia J is related to the generator rotor mass m. The inertia of each generator's synchronous machine is fixed and cannot be changed. Synchronous machines can only provide inertia during operation, meaning inertia is provided only when the system is in operation. The inertia value remains fixed and unchangeable, resulting in a discrete characteristic of system inertia due to start–stop states, as shown in Table 1.

Table 1. Comparison between synchronous inertia and virtual inertia.

Aspect	Synchronous Machine Inertia	Energy Storage Virtual Inertia
Inertia Continuity	Fixed and discrete	Continuous and designable
Inertial Response Time	Instantaneous action	Some inherent delay
Source of Response Energy	Mechanical energy from synchronous machines	Electrical energy from energy storage

Due to the power shortage causing a decline in system frequency, energy storage employs droop control to mimic the output characteristics of synchronous machines. When the grid frequency deviates from the rated frequency, the energy storage output is determined based on the difference between the measured frequency and the reference frequency. The mathematical expression for the increase in energy storage output, ΔP_{ESS1}, is expressed by (4):

$$\Delta P_{\mathrm{ESS1}} = -K_{\mathrm{ep}}\Delta f \tag{4}$$

where K_{ep} is the proportionality control coefficient set for the energy storage system.

In addition to droop control, energy storage enhances inertial response capability through additional virtual inertia control, adjusting the active power output to mitigate the system's power imbalance. In this case, the increase in active power output from energy storage, ΔP_{ESS2}, can be expressed as (5):

$$\Delta P_{\mathrm{ESS2}} = -K_{\mathrm{ed}}\frac{\mathrm{d}\Delta f}{\mathrm{d}t} \tag{5}$$

where K_{ed} is the virtual inertia control coefficient set for the energy storage system.

The virtual inertia capability of energy storage is related to the coefficient and time constant size and is a designable parameter, thus exhibiting a continuity different from synchronous machine inertia. In summary, the comparison between synchronous machine inertia and virtual inertia is shown in Table 1.

2.1.2. Comparison of Frequency Regulation Capability

To maintain system frequency stability, synchronous generators need to reserve net space for frequency regulation backup. Refs. [28,29] indicate that thermal power units require less than 3 s. In addition to the action delay, the complete response time for synchronous machine frequency regulation is about 10 to 20 s [30,31], no longer able to meet the frequency regulation requirements of current low-inertia systems.

Compared to synchronous generators, energy storage exhibits faster regulation speed, and its output is theoretically adjustable across the full power range, albeit typically with certain amplitude limitations. During the frequency response process, energy storage devices also experience some delay. The comparison of their frequency regulation capabilities is illustrated in Figure 1. T_1 and T_2 are the times used to increase the output of the storage unit and the thermal unit, respectively; T_{del1} and T_{del2} are the control delays of the storage unit and the thermal unit, respectively.

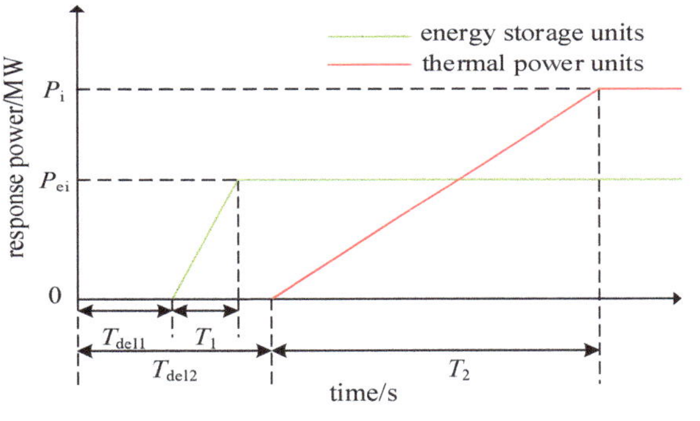

Figure 1. Comparison of frequency regulation capability between energy storage units and thermal power units.

According to [32], energy storage can output active power to the grid within 140 ms after triggering, and the UK specifies that the response delay of energy storage under significant disturbances is generally not more than 0.5 s. Following the triggering action, the complete response time of energy storage does not exceed 2 s [33]. Therefore, energy storage possesses a rapid frequency regulation characteristic, enabling it to provide stable and swift support in low-inertia systems. A comparison of the frequency regulation-related characteristics between synchronous generators and energy storage is presented in Table 2.

Table 2. Comparison of synchronous machines and energy storage frequency regulation capabilities.

Aspect	Synchronous Machine Frequency Regulation	Energy Storage Frequency Regulation
Response Delay	Not more than 3 s	Not more than 0.5 s
Complete Response Time	10~20 s	Not more than 2 s
Output Power Coupling	Mechanical energy from synchronous machines should not exceed governor response limits	Theoretically full power range adjustment, generally with certain limits

Based on the aforementioned characteristic analysis, the proposed controllable resource hierarchical coordinated control strategy in this study utilizes energy storage as the primary support for rapid power adjustment, supplemented by thermal power support, and derives relevant constraints.

2.1.3. Emergency Load Shedding Based on the Differential Evolution Algorithm

In emergency conditions, when other emergency control resources act rapidly but cannot prevent a rapid frequency decline, load-shedding measures are needed for emergency power support in the system [34].

With the objective of minimizing the cost of load shedding, the cost factor of the load is set as a function that increases with the load-shedding ratio. The final objective function for emergency load shedding is expressed as (6):

$$\begin{cases} \min F = \sum\limits_{j=1}^{N_L} c_j \rho_j P_{L_j,0} \\ c_j = \varphi(\rho_j) = k_j \rho_j + b_j \end{cases} \tag{6}$$

where F is the total cost of load shedding; N_L is the number of load-shedding stations; c_j is the load-shedding cost factor for each station; φ is the function representing the variation of the cost factor with the load-shedding ratio, approximating the cost factor as a linear function; k_j is the growth coefficient of the cost factor; b_j is the base cost factor; p_j is the shedding ratio; and $P_{Lj,0}$ is the active load of the shedding station at steady state before direct current (DC) blocking.

The research on the emergency load-shedding model is relatively mature, and its constraint conditions can be summarized by (7):

$$\begin{cases} \eta_f \geq \zeta_f \\ \eta_v \geq \xi_v \\ \eta_I \leq \zeta_I \\ \eta_\delta \leq \xi_\delta \\ 0 \leq \rho_j \leq \rho_{j,max} \end{cases} \tag{7}$$

where η_f, η_v, η_I, and η_δ are transient frequency, voltage security index, line current security index, and rotor angle security index; ξ_f, ξ_v, ξ_I, and ξ_δ are the set limits for transient frequency, voltage, current, and rotor angle constraints; and $\rho_{j,max}$ is the maximum load-shedding ratio.

The differential evolution algorithm (DE) is more suitable for real-valued optimization problems. The main operations of the DE algorithm include the generation of the initial population, mutation, crossover, and selection, with the main parameters being the population size N, evolution generation T, mutation factor K, and crossover factor C. The optimization process is described as follows.

Generate a population randomly within the range of control variables, as shown in (8):

$$\begin{aligned} & X^0 = \{x_1, x_2, x_3, \cdots, x_N\} \\ & x_i = (x_{i,1}, x_{i,2}, x_{i,3}, \cdots x_{i,L}), x_{i,j} \in [0, \rho_{j,\max}] \end{aligned} \tag{8}$$

where N is the number of individuals within the population, L is the number of genes, i.e., the number of cut-loading control variables, and each individual x within the population corresponds to a cut-loading scheme.

In the mutation stage, the parent individuals generate mutated individuals through a mutation strategy, as shown in (9):

$$v_i^{g+1} = x_{r1}^g + K(x_{r2}^g - x_{r3}^g) \tag{9}$$

where g is the current generation; i, r_1, r_2, and r_3 are distinct random integers within the $[1, N]$ interval; v_i is the generated mutated individual; and K is the mutation factor. Equation (9) is commonly referred to as random mutation, and another faster converging optimal mutation method is expressed by (10):

$$v_i^{g+1} = x_{best}^g + K(x_{r1}^g - x_{r2}^g) \tag{10}$$

where x_{best}^g is the optimal individual of the g-th generation population.

The crossover operation involves pairwise crossover between the generated mutated individuals and the parent individuals, as shown in (11):

$$u_{i,j}^{g+1} = \begin{cases} v^{g+1}, \text{ if } rand_j \leq C_R \text{ or } j = rand(j) \\ x^g, \quad \text{otherwise} \end{cases} \tag{11}$$

where C_R is the crossover factor, ranging between [0, 1]; u_i is the individual after crossover; $rand_j$ is a random number within [0, 1]; and $rand(j)$ is a random integer between 1 and L, ensuring that at least one control variable is updated.

The selection operation involves pairwise comparison between the individuals generated after crossover and the parent individuals. The more optimal individual is selected to proceed to the next generation of evolution, as shown in (12):

$$x_i^{g+1} = \begin{cases} u_i^{g+1}, \ if \ f(u_i^{g+1}) \leq f(x_i^g) \\ x_i^g, \quad \text{otherwise} \end{cases} \tag{12}$$

where f is the individual evaluation function, i.e., the final established objective function for emergency load-shedding Equation (6).

2.2. Considering Different Operating States of Energy Storage Control Strategies

2.2.1. Energy Output Control Strategies

Currently, commonly used methods for energy storage participating in power system frequency control include virtual inertia control and virtual droop control. Both control methods simulate the role played by traditional generators in the grid. In the event of a sudden change in system frequency, increasing system inertia and adjusting active power in proportion to frequency deviation can achieve the goal of frequency control. The energy storage unit adopts constant-power double-loop control, which achieves the effect of quickly adjusting the charging and discharging power by adjusting the reference value. The two controls with transfer functions are shown in Figure 2.

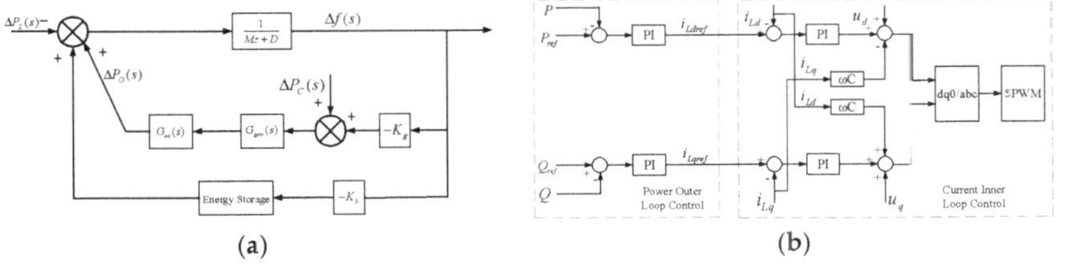

Figure 2. The control scheme diagram of energy storage: (**a**) Virtual droop control; (**b**) PQ double-loop control.

The transfer functions of traditional frequency control units' active power output $\Delta P_G(s)$ in relation to frequency deviation $\Delta f(s)$, energy storage system's active power output $\Delta P_b(s)$ in relation to frequency deviation $\Delta f(s)$, and the disturbance of load $\Delta P_L(s)$, $\Delta P_G(s)$, $\Delta P_b(s)$, $\Delta f(s)$ are expressed by (13).

$$\begin{cases} \Delta P_G(s) = -K_g G_{en}(s) G_{gov}(s) \Delta f(s) \\ \Delta P_b(s) = -K_b G_b(s) \Delta f(s) \\ \Delta f(s) = \frac{\Delta P_G(s) + \Delta P_b(s) - \Delta P_L(s)}{Ms+D} \end{cases} \tag{13}$$

The relationship between load and frequency fluctuations can be summarized as (14).

$$\Delta f(s) = \frac{-\Delta P_L(s)}{Ms + D + K_b G_b(s) + K_g G_{en}(s) G_{gov}(s)} \tag{14}$$

In the novel power system with coordinated control of multiple resources, the advantage of rapid charging and discharging of energy storage is utilized to mitigate the fluctuations of wind and photovoltaic fields, as shown in (15):

$$P_{ess} = P_{ref} - P_w - P_{PV} \tag{15}$$

where P_{ref} is the expected grid-connected power of the renewable energy generation system, and P_w and P_{PV} are the uncontrollable original power of the wind and photovoltaic fields. When $P_{ref} > P_w + P_{PV}$, the energy storage system releases its stored energy to make up the difference between them. When $P_{ref} < P_w + P_{PV}$, the energy storage system absorbs excess power. The principle is illustrated in Figure 3.

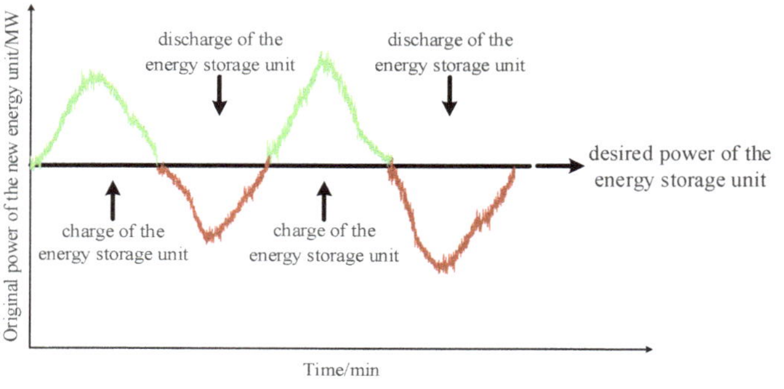

Figure 3. The schematic diagram of minimum output fluctuation based on energy storage.

With the increasing proportion of renewable energy, the system's inertia level decreases, and there is a higher probability of major power deficits or faults such as DC blocking. In the face of a rapid decline in frequency during major fault events, it is necessary for energy storage to transition to operating modes that rapidly increase output for power support.

In emergency situations, it is common practice to immediately adjust energy storage to its maximum output for power support. While this control strategy supports the system's power deficit in milliseconds, it does not consider the state of each energy storage unit. This oversight can lead to a decrease in power due to insufficient SOC after the unit's action, resulting in new power deficits and worsening the system's frequency stability.

To address this issue, we propose a classification control strategy for energy storage under emergency conditions, categorizing energy storage into four classes based on its state, as illustrated in Figure 4.

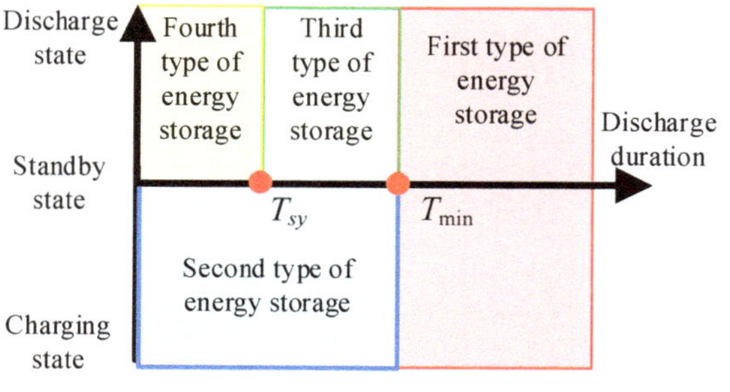

Figure 4. The classification of energy storage based on different discharge times.

First, set the indices for the shortest discharge time (T_{min}) and remaining discharge time (T_{sy}). The shortest discharge time (T_{min}) represents the minimum time required for energy storage to provide power support in emergency situations. The remaining discharge time (T_{sy}) indicates the sustainable time that energy storage can output power at the maximum rate.

In emergency situations, grid operators are more concerned about the supporting power and supporting time of the units, hence the establishment of two indicators, T_{min} and T_{sy}. Under this setup, for an energy storage unit corresponding to the shortest discharge time t_{min}, SOC_{min} is shown in Equation (16):

$$SOC_{\min} = \frac{P_{dis,t}t_{\min}}{\eta_{dus}} \tag{16}$$

where $P_{dis,t}$ is the unit output required by the operator and η_{dus} is the discharge efficiency of energy storage. Similarly, corresponding to the remaining discharge time t_{sy}, SOC_{sy} is shown in Equation (17).

$$SOC_{sy} = \frac{P_{dis,t}t_{sy}}{\eta_{dus}} \tag{17}$$

Different degrees of power deficits require energy storage units to provide varying levels of output support. Under different conditions, even for the same energy storage unit at the same SOC state, the corresponding T_{min} and T_{sy} may vary.

If the energy storage capacity is sufficient, and the T_{sy} is longer than T_{min}, it falls into the first category of energy storage. If the T_{sy} is shorter than T_{min}, indicating insufficient SOC, and the system is in a charging state, it is categorized as the second type of energy storage. If the system is in a standby or discharge state and the SOC is too low to provide power support, it falls into the fourth category of energy storage, and charging or discharging is stopped. Otherwise, it is categorized as the third type of energy storage, providing power support by reducing output.

For the first type of energy storage, discharge at the maximum output to support the power deficit is required by the system within a specified time, as is shown in (18):

$$P_{\text{dis},1}(t) = \frac{SOC(t-1) - S_{\min}}{\Delta t}\eta_{dus} \tag{18}$$

where $P_{\text{dis},1}(t)$ is the output power of the first type of energy storage, $SOC(t-1)$ is the remaining energy at time $t-1$, and $S_{\min}$ is the minimum remaining charge allowed when energy storage outputs at the current power.

For the second type of energy storage, when it is in the charging state, it is considered as a load. It should immediately switch to standby mode to reduce power demand, and based on its SOC, it transitions to the third or fourth type of energy storage.

For the third type of energy storage, the power output should be reduced to support until the shortest discharge time. Equation (19) designs the attenuation coefficient based on the SOC state and power support requirements to derive the output power:

$$P_{\text{dis},3}(t) = P_{\text{dis},1}\sqrt{\frac{SOC(t-1) - SOC_{\text{high}}}{SOC_{\max} - SOC_{\text{high}}}} = \lambda P_{\text{dis},1} \tag{19}$$

where $P_{\text{dis},3}(t)$ is the output power of the third type of energy storage, SOC_{high} is the minimum charge corresponding to the allowable power support of energy storage, $SOC_{\max}$ is the maximum charge of energy storage, and λ is the attenuation coefficient.

In summary, the energy storage control strategy is illustrated in Figure 5.

2.2.2. Adaptive Constraint Handling in Under-Frequency Load Shedding Optimization Strategy

The optimal solution to the emergency load-shedding optimization problem lies on the boundary of the feasible region. Evaluating individuals based on feasibility rules may overlook some better individuals near the boundary, hindering the discovery of optimal solutions.

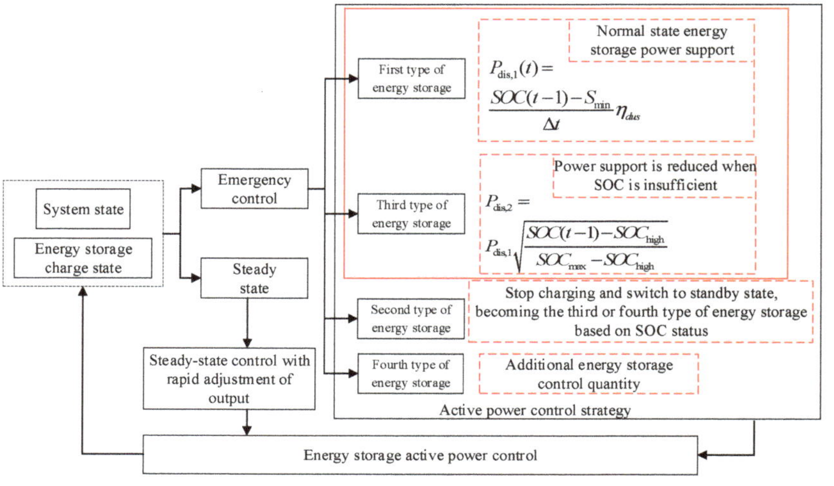

Figure 5. The control strategies for different states of energy storage.

By adopting the ε-adaptive constraint-handling method, we consider superior but infeasible solutions in the optimization process, building upon the feasibility rules. Let x_i and x_j be two individuals under consideration, with $x_i > x_j$ indicating that x_i is superior to x_j. Setting ε as the threshold for constraint violation, the specific comparison criteria are as follows:

Criterion 1: If both solutions meet the constraint requirements, select the one with lower control cost, as shown in (20).

$$x_i > x_j \Leftrightarrow F(x_i) \leq F(x_j) \tag{20}$$

Criterion 2: If one solution satisfies the constraints while the other does not, first calculate the constraint violation degree $G(x)$ for each solution. If the constraint violation degree is less than ε, choose the load-shedding plan with lower control cost. If the violation degree exceeds ε, select the solution that satisfies the constraint requirements according to (21).

$$x_i > x_j \Leftrightarrow \begin{cases} G(x_i) < \varepsilon, G(x_j) = 0, F(x_i) \leq F(x_j) \\ G(x_i) = 0, G(x_j) > \varepsilon \end{cases} \tag{21}$$

The degree of violation of the constraint is defined as (22):

$$G(x) = \sum \frac{\Delta g}{g_\xi} \tag{22}$$

where Δg is the constraint violation value of the cut-load scheme and g_ζ is the constraint limit value.

Criterion 3: If both solutions fail to satisfy the constraints, and the constraint violation degrees for both are less than ε, choose the individual with lower cost. In other cases, select the individual with a smaller constraint violation degree, as shown in (23).

$$x_1 > x_2 \Leftrightarrow \begin{cases} G(x_1) \leq \varepsilon, G(x_2) \leq \varepsilon, F(x_1) \leq F(x_2) \\ G(x_1) \leq \varepsilon, G(x_2) > \varepsilon \\ \varepsilon < G(x_1) < G(x_2) \end{cases} \tag{23}$$

Compared to the feasibility rule, criteria 2 and 3 retain the load-shedding plans with small constraint violation degrees and low control costs, thereby increasing the exploration of the feasible domain boundaries. ε is adaptively adjusted based on the overall violation of constraints, incorporating infeasible individuals in the early stages and setting ε to 0 in the later stages of evolution, as shown in (24):

$$\mathcal{E}(t) = \begin{cases} \mathcal{E}(0)e^{-\alpha t/T_e}, t \leq T_e \\ 0, t > T_e \end{cases} \tag{24}$$

where α is the decreasing coefficient and T_e is the truncation algebra, which is transformed to the feasibility law when taking 0.

In summary, the load-shedding optimization process based on the improved differential evolution algorithm is outlined as follows:

(1) Generate a large number of load-shedding plans randomly within the $[0, \rho_{max}]$ space, serving as the initial set S.
(2) Filter samples that satisfy frequency constraints. Utilize a uniform removal approach to explore and determine approximate lower (P_{low}) and upper (P_{up}) limits of load-shedding amounts. Choose plans in S that meet $P_{low} < P_{up}$ to compose the sample set S_f.
(3) Filter samples that satisfy current or voltage constraints. Calculate the sensitivity of all load-shedding stations and identify the station L_s with the maximum sensitivity. Explore and determine its approximate lower limit with the maximum sensitivity. Explore and determine its approximate lower limit ($P_{low,L}$). Choose plans in the sample set S_f that meet the requirements to form the sample set S_1.
(4) Randomly select N plans from S_1 to form the initial population X_0.

The diagram of adaptive constraint handling in under-frequency load-shedding optimization strategy is shown in Figure 6.

2.3. Hierarchical Coordinated Control Method for Systems with a High Proportion of Renewable Energy Sources

After the large-scale integration of renewable energy sources into the power grid, it is necessary to conduct research on the hierarchical coordinated control of various controllable resources for more intelligent control. The goal is to fully leverage the advantages of different types of resources. During normal operation, when a certain type of resource is affected, other resources can quickly coordinate and complement, thereby improving the economic efficiency and reliability of the power system.

Compared to traditional systems, the coordinated control problem in multi-energy systems exhibits complex nonlinear characteristics. To fully utilize the power support from thermal power generators and the flexibility of energy storage systems, a coordinated control scheme is proposed. This scheme divides the system into two hierarchical levels, each containing different energy resources.

Figure 7 illustrates the coordinated control block diagram after the introduction of load shedding. In each unit, wind and photovoltaic units, unable to support reactive power, adopt fixed DC voltage and reactive power control to maintain zero reactive power output. Active power support is adjusted based on the daily wind and solar conditions.

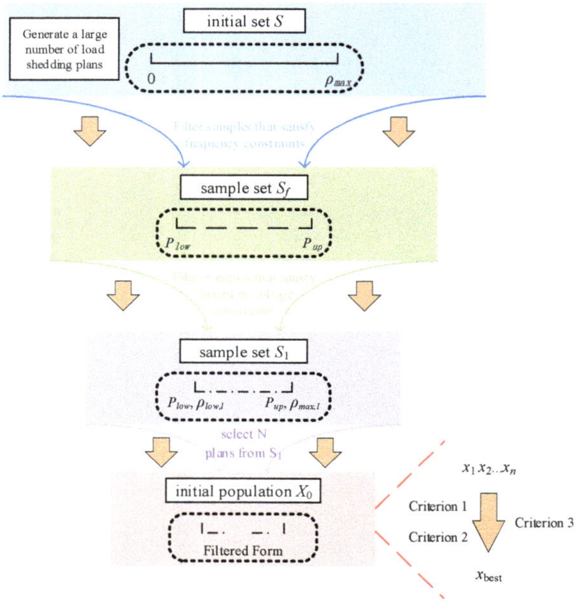

Figure 6. The diagram of adaptive constraint handling in under-frequency load-shedding optimization strategy.

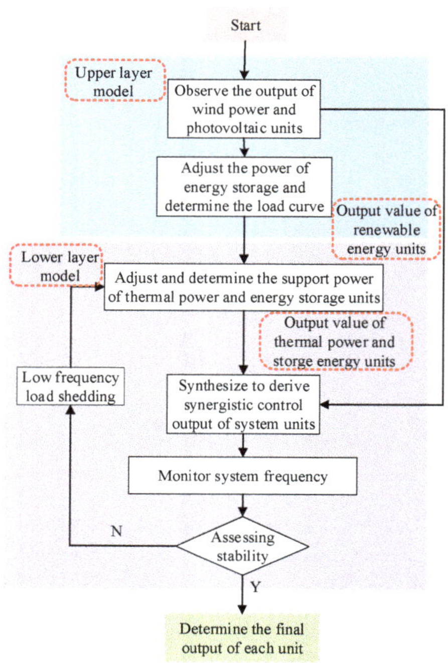

Figure 7. The hierarchical coordinated control system framework for systems with a high proportion of renewable energy sources.

Currently, widely used wind and photovoltaic generation models are expressed by (25) and (26) [35]:

$$P_{pv} = P_{STC}\frac{G_C}{G_{STC}}[1 + k(T_C - T_{STC})] \tag{25}$$

$$P_W(t) = \begin{cases} 0, & v(t) \leq v_{in}, v(t) \geq v_{out} \\ P_r\frac{v(t)-v_{in}}{v_r-v_m}, & v_{in} \leq v(t) \leq v_r \\ P_r, & v_r \leq v(t) \leq v_{out} \end{cases} \tag{26}$$

where P_{PV} is the actual output power of the photovoltaic panels; k is the power temperature coefficient, commonly set to $-0.3\%/^\circ C$ [36]; G_C and T_C denote real-time solar radiation intensity and surface temperature of the photovoltaic modules, respectively; $P_W(t)$ is the actual output power of the wind turbine at time t; v_r is the rated wind speed; v_{in} and v_{out} are the cut-in and cut-out wind speeds of the wind turbine, respectively; and P_r is the rated power of the wind turbine.

The core of the upper-level model is to determine the wind and solar output. The energy storage unit adopts constant-power dual-loop control to achieve rapid adjustment of charging and discharging power by adjusting reference values. Under steady-state conditions, the rapid response capability of energy storage is utilized to follow the fluctuations of wind, solar, and load, with the goal of optimizing the output of the wind–solar–storage integrated system to minimize net load fluctuations. Based on this, the output of the energy storage unit is set to (27):

$$P_{BESS} = P_{ess} = P_{ref} - P_w - P_{PV} \tag{27}$$

where P_{ess} is the power deviation value, P_{ref} is the expected grid-connected power of the renewable energy unit system, and P_w and P_{PV} are the uncontrollable original power of the wind and photovoltaic fields. When $P_{ref} > P_w + P_{PV}$, indicating insufficient output from the renewable energy unit, P_{BESS} takes a positive value, representing discharging power, releasing stored energy to compensate for the gap between them. When $P_{ref} < P_w + P_{PV}$, indicating excess output from the renewable energy unit, P_{BESS} takes a negative value, representing charging power, absorbing the surplus power.

The lower-level system is primarily responsible for the power support of thermal power and energy storage units. Under steady-state conditions, it provides output support for thermal power units based on the equivalent load curve transmitted from the upper-level system. The output of participating controlled thermal power units is determined, yielding the final results.

In emergency situations, the power deficit is determined based on frequency response. Utilizing the emergency output control strategy of energy storage under different classifications as described in Section 2.2.1, the output of the energy storage system is determined. This involves a rapid response to adjust charging and discharging power to achieve successful power support, serving the purpose of swiftly providing power assistance. Thermal power, constrained by its response time, participates in primary frequency control power restoration.

If, after reaching maximum output from the energy storage system, a significant power deficit persists, load shedding is implemented. Simultaneously, the output of the energy storage is adjusted, allowing the system to continue stabilizing. The amount of load shedding is expressed by (28):

$$\Delta P_{sho.real}(t) = \sum_{t=1}^{n}\left[P_L(t) - P_G(t) - P_{p,w}(t) - P_{BESS}(t)\right] \tag{28}$$

where $P_L(t)$, $P_G(t)$, $P_{p,w}(t)$, and $P_{BESS}(t)$ are the power at time t for load, thermal power, renewable energy, and the energy storage system, respectively. The power supported by the energy storage is expressed by (29):

$$P_{BESS}(t) = P_{\text{dis},1}(t) + P_{\text{dis},3}(t) \tag{29}$$

where $P_{\text{dis},1}(t)$ and $P_{\text{dis},3}(t)$ are the output of the energy storage units for the first and third categories.

The system proposed coordinates control based on the flexible output of energy storage. The performance is closely related to the SOC of the energy storage. Using the emergency load-shedding amount to measure the performance of the system, let the index be denoted as D, which can be expressed as (30):

$$\begin{aligned} D &= \frac{\Delta P_{sho,real}(t)}{\Delta P_{\max}} = \frac{\Delta P_e - P_{\text{dis},1}(t) - P_{\text{dis},3}(t)}{\Delta P_{\max}} \\ &== \frac{\Delta P_e - (1+\lambda)\frac{SOC(t-1)-S_{\min}}{\Delta t}}{\Delta P_{\max}} \eta_{dus} \end{aligned} \tag{30}$$

where $\Delta P_{\max}$ is the maximum load-shedding capacity, ΔP_e is the power deficit of the entire system excluding energy storage units, and λ is the attenuation coefficient. It can be observed that as the SOC decreases, the value of D increases. When D is negative, the system quickly adjusts output through energy storage to achieve stability control without the need for load shedding, thus maintaining system stability. When D is a positive number less than 1, the system initiates emergency load shedding for coordinated control to achieve stability. When D is greater than 1, indicating that the power deficit exceeds the maximum load-shedding capacity and SOC is too low to support stable control, the system becomes unstable.

3. Case Studies

3.1. Introduction to the Algorithm

To validate the proposed strategy, a simulation model is constructed as illustrated in Figure 8, where L_1 and L_2 represent active loads, and the distribution network operates at a voltage of 10 kV and a frequency of 50 Hz, consisting of wind, solar, storage, and synchronous generator units.

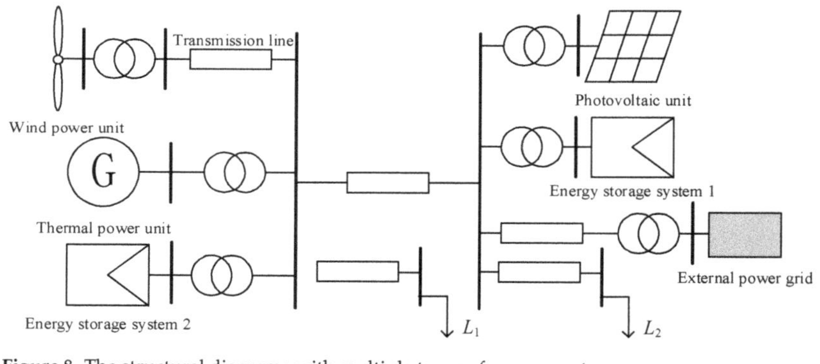

Figure 8. The structural diagrams with multiple types of energy systems.

Multiple scenarios were set up for comparative analysis to validate the effectiveness of the proposed method.

Scenario 1: Load surge. Set load 2 to increase by 600 kW at 4 s into the simulation.

Scenario 2: Illumination perturbation. Introduce irradiance disturbance by setting irradiance to decrease from 1000 to 500 at 4 s into the simulation, and then increase from 500 to 800 at 6 s.

Scenario 3: Wind speed disturbance. Introduce wind speed disturbance by setting wind turbine speed to decrease from 11 m/s to 8 m/s at 4 s into the simulation.

Scenario 4: Insufficient storage. Set the initial SOC of the storage system to 30%. When the simulation reaches 4 s, decrease SOC to below 26%, simulating insufficient energy storage and reducing output power.

Scenario 5: Continuous load increase, insufficient power support. Simulate a continuous increase in load, resulting in insufficient power support and the system frequency dropping below the lower limit.

Scenario 6: Load fluctuation (similar to Scenario 5), with coordinated control of controllable loads.

3.2. Analysis of Different Scenarios

The frequency fluctuations of the system under the first four disturbance scenarios are depicted in Figure 9. It can be observed that, in the face of different fluctuation scenarios, the proposed strategy can effectively achieve coordination among various units in a short period. This involves controlling energy storage tracking and thermal power support, enhancing the integration of renewable energy sources, and stabilizing the system frequency. Taking the scenarios of irradiance fluctuation and insufficient SOC in energy storage as examples, the control strategies of different units under layered coordinated control are analyzed based on their respective power outputs.

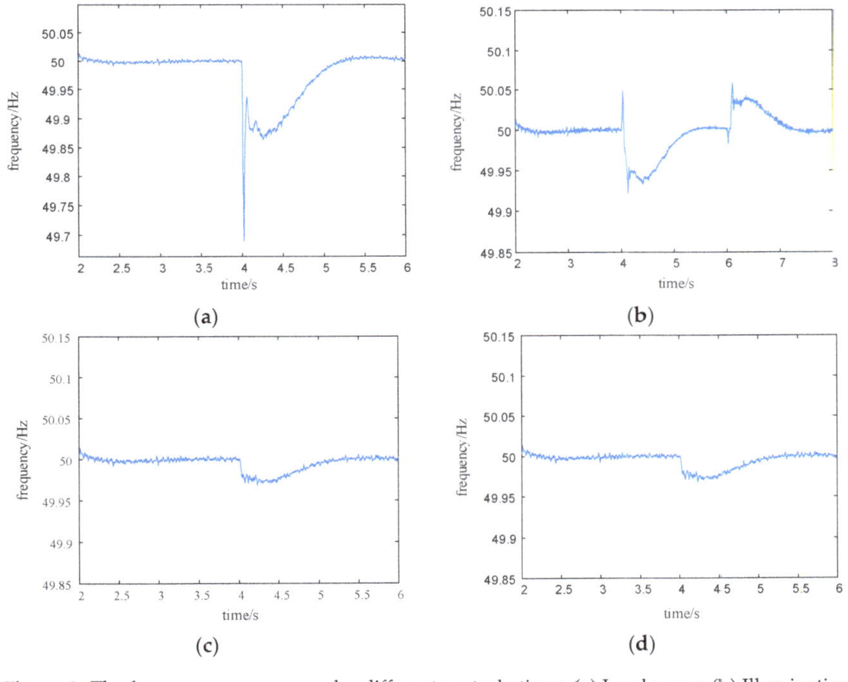

Figure 9. The frequency response under different perturbations: (**a**) Load surge; (**b**) Illumination perturbation; (**c**) Wind speed disturbance; (**d**) Insufficient storage.

The output variations of each unit are shown in Figure 10. It can be observed that, under the scenario of fluctuating sunlight, the output curve of the photovoltaic unit changes with the sunlight variation. The dual-loop controlled energy storage unit adjusts the constant power output by changing the reference value, while the wind power unit shows no fluctuations. The upper-level model optimizes the entire system to achieve the minimum net load. In this case, the lower-level model does not require adjustments to the thermal power unit.

In the scenario of insufficient energy storage SOC, the energy storage unit is forced to reduce its output power and cannot support the power deficit in the system. In this case, the thermal power unit in the lower-level model receives instructions to increase power support, with a response time slightly slower than that of the energy storage unit, but still maintaining system frequency stability.

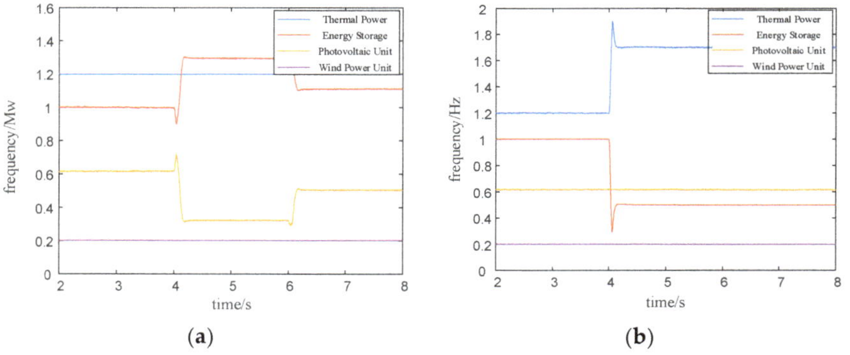

Figure 10. The change in output of each unit: (**a**) Illumination perturbation; (**b**) Insufficient storage.

Figure 11 illustrates the system response during an emergency situation when the power support of units is insufficient due to a sudden load increase. In this scenario, for rapid support, priority is given to the energy storage units to act first, slowing down the frequency drop and raising the minimum frequency point. The wind and solar units remain unchanged, followed by the thermal power units providing power support for primary frequency regulation. From the power output variations of each unit Figure 11a, it can be observed that during the three load spikes at 4 s, 5 s, and 6 s, the energy storage units can quickly increase their output power to achieve the expected results. Energy storage unit 1, with a higher SOC, rapidly increases power support during each disturbance. Energy storage unit 2 operates in the third-class state, only increasing power support when the system power support is insufficient, ensuring the supply time. This validates the effectiveness of the proposed strategy. The thermal power unit responds slightly slower than the energy storage unit when participating in power support. At 6 s, facing another load increase, both types of energy storage cannot further increase power support as they have reached their maximum discharge power. At this point, when the frequency drops below 49.5 Hz, load-shedding measures are implemented, and the thermal power unit subsequently provides power support. The load-shedding value is determined according to Section 2.2, and in this study, nearly 0.4 MW is shed. Compared with the control without load cutting measures, the stability of system is maintained at the cost of 0.4 MW.

Considering the scenario where the overall maximum output power of the units cannot support the power deficit, load shedding is a crucial control measure for system stability. From the frequency comparison chart Figure 11b, it can be observed that after considering load shedding, the original control method provides more stable system support in the emergency situation after 6 s, highlighting the necessity of load shedding as a coordinated control strategy.

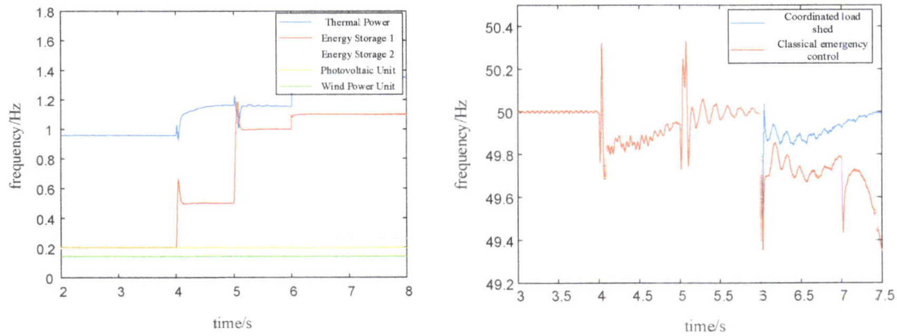

Figure 11. System response under power deficit: (**a**) The change in output of each unit; (**b**) The comparison of frequency curve.

4. Discussion

In the trend of growing active distribution networks, this paper presents an innovative study aimed at solving the layered coordinated control of controllable resources with automatic load shedding in power systems. The characteristics of various controllable resources are analyzed in order to employ adaptive control methods, and a layered coordinated control strategy is proposed. Simulation results show that the controllable resources are effectively coordinated and controlled in various scenarios.

Significantly, this paper highlights the critical role of energy storage systems in active distribution networks. Classification of energy storage based on power support time and the corresponding control strategy are developed, which finally realizes the output control of the energy storage unit. The proposed method exploits the rapid response characteristics of energy storage systems, enhancing the stability of the overall control strategy.

Lastly, this study introduces coordinated control with load shedding on the basis of coordinated control, providing a feasible solution for frequency dropping again. This comprehensive research framework not only proposes effective solutions for the coordinated control of power systems but also provides innovative insights to cope with challenges like recurrent frequency drops. These research findings are promising to provide robust support for the stability and reliability of power systems.

Author Contributions: Conceptualization, J.Z. and Y.C.; methodology, J.W.; software, J.Z.; writing—original draft preparation, J.Z. and B.L.; writing—review and editing, Y.C. and C.L.; supervision, Y.C. and C.L. All authors have read and agreed to the published version of the manuscript.

Funding: This research was funded by Shandong Provincial Natural Science Foundation, grant number No. ZR2021QE133.

Data Availability Statement: Data are available upon reasonable request to the corresponding author.

Conflicts of Interest: Author Jiaying Wang was employed by the company State Grid Zhejiang Marketing Service Center. The remaining authors declare that the research was conducted in the absence of any commercia or financial relationships that could be construed as a potential conflict of interest.

Nomenclature

The main symbols appearing in this paper are defined below.

SOC	State of charge
AGC	Automatic generation control
ES	Energy storage
BESS	Battery energy storage system

PV	Photovoltaic system
DE	Differential evolution
DC	Direct current
T_{min}	Shortest discharge time
T_{sy}	Remaining discharge time

References

1. Wu, M.; Ma, D.; Xiong, K.; Yuan, L. Deep Reinforcement Learning for Load Frequency Control in Isolated Microgrids: A Knowledge Aggregation Approach with Emphasis on Power Symmetry and Balance. *Symmetry* **2024**, *16*, 322. [CrossRef]
2. Cao, Y.; Zhang, H.; Shi, X.; Xu, Q.W.; Li, C.G.; Li, W. Preliminary Study on Participation Mechanism of Large-scale Distributed Energy Resource in Security and Stability Control of Large Power Grid. *Autom. Electr. Power Syst.* **2021**, *18*, 1–8.
3. Wu, D.; Wu, L.; Wen, T.; Li, L. Microgrid Operation Optimization Method Considering Power-to-Gas Equipment: An Improved Gazelle Optimization Algorithm. *Symmetry* **2024**, *1*, 83. [CrossRef]
4. Cao, Y.; Wu, Q.; Zhang, H.; Li, C. Optimal Sizing of Hybrid Energy Storage System Considering Power Smoothing and Transient Frequency Regulation. *Int. J. Electr. Power Energy Syst.* **2022**, *142*, 108227. [CrossRef]
5. Li, Q.; Wang, D.; Liang, X.; Qin, Y.; Zhang, Y.; Ma, C. An Iteratively-Nested Modeling Method for Stability Analysis of the Multirate Power System. *IEEE Trans. Ind. Inform.* **2024**, *3*, 3070–3081. [CrossRef]
6. Zhou, Y.; Zhao, J.; Wu, Z. A Review of Symmetry-Based Open-Circuit Fault Diagnostic Methods for Power Converters. *Symmetry* **2024**, *2*, 204. [CrossRef]
7. Cao, Y.; Wu, Q.; Zhang, H.; Li, C. Multi-Objective Optimal Siting and Sizing of BESS Considering Transient Frequency Deviation and Post-Disturbance Line Overload. *Int. J. Electr. Power Energy Syst.* **2023**, *144*, 108575. [CrossRef]
8. Li, B.; Sun, Y.; Li, Y.; Lu, T.; Yu, T.; Pang, Y. Research on coordinated control of AC/DC system considering energy storage power limitation. *Energy Rep.* **2022**, *8*, 1168–1175. [CrossRef]
9. Nimalsiri, N.; Ratnam, E.; Smith, D.; Mediwaththe, C.; Halgamuge, S. A Distributed Coordination Approach for the Charge and Discharge of Electric Vehicles in Unbalanced Distribution Grids. *IEEE Trans. Ind. Inform.* **2024**, *3*, 3551–3562. [CrossRef]
10. Alpízar-Castillo, J.; Ramirez-Elizondo, L.; Bauer, P. Assessing the Role of Energy Storage in Multiple Energy Carriers toward Providing Ancillary Services: A Review. *Energies* **2023**, *1*, 379. [CrossRef]
11. Tushar, M.; Zeineddine, A.; Assi, C. Demand-Side Management by Regulating Charging and Discharging of the EV, ESS, and Utilizing Renewable Energy. *IEEE Trans. Ind. Inform.* **2018**, *1*, 117–126. [CrossRef]
12. Petrollese, M.; Cascetta, M.; Tola, V.; Cocco, D.; Cau, G. Pumped thermal energy storage systems integrated with a concentrating solar power section: Conceptual design and performance evaluation. *Energy* **2022**, *247*, 123516. [CrossRef]
13. Hunt, J.D.; Nascimento, A.; Zakeri, B.; Jurasz, J.; Dąbek, P.B.; Barbosa, P.S.F.; Brandão, R.; de Castro, N.J.; Filho, W.L.; Riahi, K. Lift Energy Storage Technology: A solution for decentralized urban energy storage. *Energy* **2022**, *254*, 124102. [CrossRef]
14. Rekioua, D. Energy Storage Systems for Photovoltaic and Wind Systems: A Review. *Energies* **2023**, *9*, 3893. [CrossRef]
15. Zhang, H.; Li, Y.; Gao, D.W.; Zhou, J. Distributed Optimal Energy Management for Energy Internet. *IEEE Trans. Ind. Inform.* **2017**, *6*, 3081–3097. [CrossRef]
16. Dowling, J.-A.; Lewis, N.-S. Long-duration energy storage for reliable renewable electricity: The realistic possibilities. *Bull. At. Sci.* **2021**, *6*, 281–284. [CrossRef]
17. Sarker, M.R.; Saad, M.H.M.; Riaz, A.; Lipu, M.S.H.; Olazagoitia, J.L. Micro Energy Storage Systems in Energy Harvesting Applications: Analytical Evaluation towards Future Research Improvement. *Micromachines* **2022**, *4*, 512. [CrossRef] [PubMed]
18. Zhao, J.; Wen, F.; Dong, Z.Y.; Xue, Y.; Wong, K.P. Optimal Dispatch of Electric Vehicles and Wind Power Using Enhanced Particle Swarm Optimization. *IEEE Trans. Ind. Inform.* **2012**, *8*, 889–899. [CrossRef]
19. Lin, H.; Liang, T.; Chen, S.M. Estimation of Battery State of Health Using Probabilistic Neural Network. *IEEE Trans. Ind. Inform.* **2013**, *2*, 679–685. [CrossRef]
20. Cao, Y.; Wu, Q.; Zhang, H.; Li, C.; Zhang, X. Chance-Constrained Optimal Configuration of BESS Considering Uncertain Power Fluctuation and Frequency Deviation Under Contingency. *IEEE Trans. Sustain. Energy* **2022**, *13*, 2291–2303. [CrossRef]
21. Khan, M.-I.; Asfand, F.; Al-Ghamdi, S.-G. Progress in research and technological advancements of thermal energy storage systems for concentrated solar power. *J. Energy Storage* **2022**, *55*, 105860. [CrossRef]
22. Majidi, M.; Parvania, M.; Byrne, R. Risk-based stochastic scheduling of centralised and distributed energy storage systems. *IET Smart Grid* **2023**, *6*, 596–608. [CrossRef]
23. Adewumi, O.B.; Fotis, G.; Vita, V.; Nankoo, D.; Ekonomou, L. The Impact of Distributed Energy Storage on Distribution and Transmission Networks' Power Quality. *Appl. Sci.* **2022**, *13*, 6466. [CrossRef]
24. Serra, F.; Lucariello, M.; Petrollese, M.; Cau, G. Optimal Integration of Hydrogen-Based Energy Storage Systems in Photovoltaic Microgrids: A Techno-Economic Assessment. *Energies* **2020**, *16*, 4149. [CrossRef]
25. Kubankova, E.-A.; Arkharov, I.-A. A Review of Energy Storage Systems. *Chem. Pet. Eng.* **2023**, *11*, 936–942. [CrossRef]
26. Trahey, L.; Brushett, F.R.; Balsara, N.P.; Ceder, G.; Cheng, L.; Chiang, Y.-M.; Hahn, N.T.; Ingram, B.J.; Minteer, S.D.; Moore, J.S.; et al. Energy storage emerging: A perspective from the Joint Center for Energy Storage Research. *Proc. Natl. Acad. Sci. USA* **2020**, *23*, 12550–12557. [CrossRef] [PubMed]

27. Liu, T.; Yang, Z.; Duan, Y. Short- and long-duration cooperative energy storage system: Optimizing sizing and comparing rule-based strategies. *Energy* **2023**, *281*, 128273. [CrossRef]
28. Luerssen, C.; Verbois, H.; Gandhi, O.; Reindl, T.; Sekhar, C.; Cheong, D. Global sensitivity and uncertainty analysis of the levelised cost of storage (LCOS) for solar-PV-powered cooling. *Appl. Energy* **2021**, *286*, 116533. [CrossRef]
29. Dai, R.; Esmaeilbeigi, R.; Charkhgard, H. The Utilization of Shared Energy Storage in Energy Systems: A Comprehensive Review. *IEEE Trans. Smart Grid* **2021**, *4*, 3163–3174. [CrossRef]
30. Gabbar, H.-A.; Othman, A.-M.; Abdussami, M.R. Review of Battery Management Systems (BMS) Development and Industrial Standards. *Technologies* **2021**, *2*, 28. [CrossRef]
31. Lai, C.-S.; Locatelli, G. Economic and financial appraisal of novel large-scale energy storage technologies. *Energy* **2021**, *214*, 118954. [CrossRef]
32. Kandari, R.; Neeraj, N.; Micallef, A. Review on Recent Strategies for Integrating Energy Storage Systems in Microgrids. *Energies* **2023**, *1*, 317. [CrossRef]
33. Lin, Y.; Lin, W.; Wu, W.; Zhu, Z. Optimal Scheduling of Power Systems with High Proportions of Renewable Energy Accounting for Operational Flexibility. *Energies* **2023**, *14*, 5537. [CrossRef]
34. Ai, C.; Zhang, L.; Gao, W.; Yang, G.; Wu, D.; Chen, L.; Chen, W.; Plummer, A. A review of energy storage technologies in hydraulic wind turbines. *Energy Convers. Manag.* **2022**, *264*, 115584. [CrossRef]
35. Bakirci, K.; Yuksel, B. Simulation study of solar-source heat pump system with sensible energy storage. *J. Therm. Anal. Calorim.* **2021**, *4*, 1853–1864. [CrossRef]
36. Guo, J.; Chen, T.; Chaudhuri, B.; Hui, S.Y.R. Stability of Isolated Microgrids with Renewable Generation and Smart Loads. *IEEE Trans. Sustain. Energy* **2020**, *11*, 2845–2854. [CrossRef]

symmetry

MDPI

Article

Power Transformer On-Load Capacity-Regulating Control and Optimization Based on Load Forecasting and Hesitant Fuzzy Control

Dexu Zou [1,2], Xinyu Sun [3], Hao Quan [3,*], Jianhua Yin [3], Qingjun Peng [2], Shan Wang [2], Weiju Dai [2] and Zhihu Hong [2]

1 School of Electrical Engineering, Chongqing University, Chongqing 400044, China; zoudexu@csg.cn
2 Electric Power Research Institute, China Southern Power Grid Yunnan Power Grid Co., Ltd., Kunming 650217, China; pengqingjun@csg.cn (Q.P.); wangshan@csg.cn (S.W.); daiweiju@csg.cn (W.D.); hongzhihu@csg.cn (Z.H.)
3 School of Automation, Nanjing University of Science and Technology, Nanjing 210094, China; sunxinyu1@njust.edu.cn (X.S.); yinjhhust@163.com (J.Y.)
* Correspondence: quanhao@njust.edu.cn

Abstract: The operational stability of a power transformer exerts an extremely important impact on the power symmetry, balance, and security of power systems. When the grid load fluctuates greatly, if the load factor of the transformer cannot be maintained within a reasonable range, it leads to increased instability in grid operation. Adjusting the transformer capacity based on load changes is of great significance. The existing control methods for on-load capacity-regulating (OLCR) transformers have low timeliness, and the daily switching frequency of the capacity-regulating switch is not controlled. To ensure the safe and stable operation of transformers, this paper proposes a control method for OLCR transformers based on load prediction and fuzzy control. Firstly, the operating principle of OLCR transformers is analyzed, and a multi-strategy enhanced dung beetle optimizer (MSDBO) combined with a CNN−LSTM model is proposed for load forecasting. On this basis, the daily switching frequency of the capacity-regulating transformer is introduced, and hesitant fuzzy control is used to select the optimal capacity-regulating strategy relying on three factors: loss, economy, and switching frequency. Finally, simulation models are constructed using the MATLAB/SIMULINK platform and simulation analysis is conducted to verify the effectiveness and superiority of the proposed control method. For the three scenarios in this paper, the method reduces daily power loss by 28.5% to 56.3% and daily operating costs by 25.4% to 50.8%. The method used in this paper can sacrifice 3.5% to 9.2% of the loss reduction capability in exchange for reducing the number of switch operations by 28.6% to 57.1%, significantly extending the lifespan of the switches and thereby increasing the operational lifespan of the transformer.

Keywords: on-load capacity-regulating transformer; power system symmetry; fuzzy control; load forecasting; CNN−LSTM

1. Introduction

Power transformers play a crucial role in ensuring the three-phase symmetry and stability of power transmission in the power system [1]. Serving as the infrastructure for power transmission and distribution, transformers are widely used in industries, agriculture, transportation, and other fields [2]. With the development of various sectors, the scale of power generation and consumption is continuously expanding, leading to a gradual increase in the capacity and voltage levels of power transformers [3]. As the scale of transformers expands, their ability to operate stably has become one of the key factors determining whether the power system can maintain symmetry and balance [4,5].

Excessive or insufficient load on transformers can present a potential hazard to the functioning of the electrical grid [6]. Prolonged operation of transformers under overload conditions may result in significant temperature rises and even lead to safety incidents, such as damage [7,8]. On the other hand, extended periods of light or no load can cause substantial power losses and economic damage [9,10]. Therefore, it is of paramount importance to judiciously adjust the rated capacity of transformers in engineering applications based on actual circumstances.

The use of on-load capacity-regulating transformers has proven to be effective in addressing the above issues. On-load capacity-regulating (OLCR) transformers are divided into two modes: high capacity and low capacity. When the grid load is high, exceeding the transformer's critical load, the transformer adjusts to the high-capacity mode, effectively avoiding overload issues. Conversely, when the grid load is low, below the transformer's critical load, the transformer switches to the low-capacity mode, effectively preventing significant power losses in light or no-load conditions [11,12].

Scholars from both domestic and international backgrounds have extensively researched the topic of the control of OLCR transformers. This includes analyses of the principles behind OLCR transformers, the selection of OLCR schemes, and the application of neural network methods to the control of on-load capacity regulation.

Reference [13] has demonstrated the feasibility of two methods for capacity regulation: no-load capacity regulation with automatic voltage regulation and on-load capacity regulation. Reference [14] conducts an analysis of the structural characteristics of OLCR transformers, elucidating the principles behind capacity regulation. Additionally, it provides OLCR schemes for such transformers from an economic perspective. In reference [15], a capacity regulation scheme for on-load transformers with composite switches is proposed. The study demonstrates that the parallel connection of buffering resistors at both ends of the capacity-regulating switch can effectively suppress overcurrent and overvoltage generated during capacity regulating. Based on the principles of capacity-regulating transformers, reference [16] conducts an analysis of the changes in losses, magnetic flux density, and impedance before and after capacity regulating in on-load capacity-regulating transformers. The study proposes a method for calculating the optimal capacity-regulating selection node. Reference [17] explores and designs strategies for the economic and safe operation of on-load capacity-regulating transformers. Reference [18] provides a method for calculating critical economic capacity. The study includes the design of a control system for no-load capacity-regulating transformers and proposes a switching strategy based on a gray theory load forecasting model. The above methods focus on selecting the optimal capacity-regulating node based on the economic considerations of the transformer but do not address the control of the daily switch operation frequency. In response to the issue of the irrational setting of the optimal capacity-regulating node, reference [19] proposes a method for selecting the optimal capacity-tap node for OLCR transformers based on fuzzy control. Reference [20] introduces an improved model that combines the fuzzy consistent matrix control method with the fuzzy preference method. This model applies fuzzy matrix conversion relationships to determine the weights of various indicators. Comparative analysis of the comprehensive evaluation results for a case with the analytic hierarchy process, evidence theory, and super-efficiency data envelopment analysis further demonstrates the applicability and credibility of this evaluation method. Reference [21] introduces a control strategy for on-load capacity-regulating transformers based on load forecasting. The effectiveness of this strategy is validated through experimentation. The mentioned studies primarily focus on transformer power losses and economic factors. They achieve capacity-regulating control by periodically monitoring actual values or using load forecasting. However, there is a lack of comprehensive consideration for factors such as transformer power losses, economic aspects, and the daily frequency of capacity-regulating switch operations.

This paper proposes a power transformer OLCR control and optimization method based on both load forecasting and fuzzy control, which comprehensively consider power losses, economic aspects, and the daily switching frequency all together. To address the issue of insufficient capacity switching for power transformers during frequent grid fluctuations, a method using load forecasting is proposed. This method involves training neural network models on the daily load data of transformers to construct a load forecasting model that enables real-time load prediction and capacity regulation of transformers. Furthermore, hesitant fuzzy control is employed to optimize the capacity-regulating scheme. This approach aims to proactively understand and perceive transformer loads, ultimately achieving rapid and accurate regulation of transformer capacity. The main contributions of this paper are summarized as follows:

- A multi-strategy enhanced dung beetle optimizer (MSDBO), named the MSDBO–CNN–LSTM model for load forecasting, is proposed to adjust the hyperparameters of a CNN−LSTM. This adjustment enhances the model's accuracy. Additionally, different optimization algorithms are compared to validate that the proposed MSDBO algorithm achieves the highest accuracy and the fewest iterations.
- Due to the difficulty of accurately analyzing the importance of various factors for transformer capacity-regulating control and the lack of weights on different factors, conventional fuzzy control methods perform poorly. For multifactor problems with unknown weights, hesitant fuzzy control combines the different opinions of multiple decision-makers for a more comprehensive analysis, resulting in decisions that are more relevant to the integrated opinions of multiple control. Compared to conventional fuzzy control, hesitant fuzzy control provides more accurate and comprehensive results.
- The effectiveness of the proposed capacity control method is assessed through simulation using the MATLAB/SIMULINK platform. The simulation results show that transformers using load forecasting models can accurately and promptly switch capacities. Transformers that combine load forecasting with hesitant fuzzy control can reduce the number of daily capacity switch operations while ensuring low power and economic losses.

The rest of this paper is organized as follows: Section 2 discusses the capacity-regulating principle of on-load capacity-regulating transformers. Section 3 introduces the CNN−LSTM model and the multi-strategy dung beetle optimization (MSDBO) algorithm. Through testing, it demonstrates the superiority of the MSDBO algorithm for optimization. Section 4 determines the regulating schemes for capacity nodes of transformers, introduces the use of hesitant fuzzy control, and presents an overall control strategy for on-load capacity regulation based on a combination of neural networks and fuzzy control. The case study is presented in Section 5. Finally, the conclusions and discussion are presented in Section 6.

2. On-Load Capacity-Regulating Transformer Operating Principles

The study in this paper primarily focuses on on-load capacity-regulating transformers that employ a high-voltage winding star–delta transformation and a low-voltage winding series–parallel transformation, as illustrated in Figure 1. This capacity-regulating method is characterized by its effective loss reduction and widespread applicability. During the transformation of an OLCR transformer from low to high capacity, the number of turns in the low-voltage winding decreases. Simultaneously, the high-voltage winding changes to a Y connection, resulting in an increase in the phase voltage on the high-voltage side. The reduction in the number of turns and the increase in voltage are proportional, ensuring that the output voltage remains constant while the capacity is switched.

Figure 1a illustrates the high-voltage winding side. In this diagram, k1~k6 represent capacity-regulating switches. Figure 1b illustrates the low-voltage winding side. In this diagram, k1~k9 represent capacity-regulating switches.

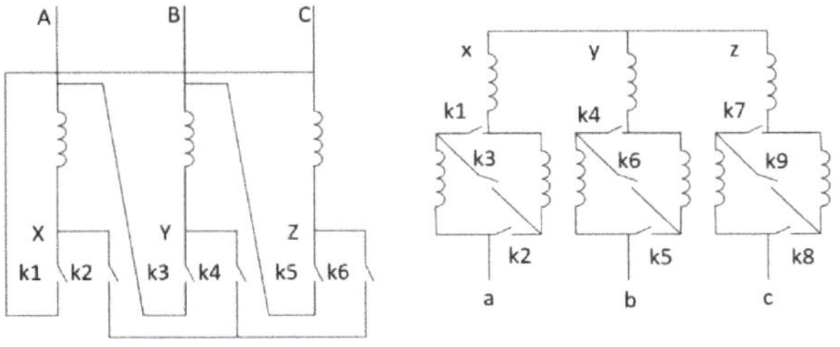

(**a**) High-voltage winding (**b**) Low-voltage winding

Figure 1. High- and low-voltage winding connections.

When the high-voltage winding is connected in a delta configuration, and the low-voltage winding is connected in parallel with the second and third sections of the winding and in series with the first section of the winding, the transformer is in a high-capacity mode. On the other hand, when the high-voltage winding is connected in a star configuration and the low-voltage winding is connected in a series configuration with all three sections, the transformer operates in the low-capacity mode.

3. On-Load Capacity-Regulating Control Based on Load Forecasting

The existing OLCR transformer determines its operational state based on periodic monitoring of the load condition. It compares the determined state with its current state to decide whether the OLCR switch should be activated, thus controlling the OLCR switch operation. However, in situations where the load fluctuates frequently, the issue of delayed switch operation arises.

This chapter proposes an on-load transformer capacity-regulating control method based on MSDBO−CNN−LSTM (multi-strategy enhanced dung beetle optimizer—convolutional neural network—long short-term memory). The MSDBO−CNN−LSTM algorithm is employed for load forecasting and enhances prediction accuracy.

3.1. Load Forecasting Based on CNN−LSTM

The convolutional neural network (CNN) belongs to the category of feedforward neural networks, characterized by a deep structure and incorporating convolutional computations. The CNN comprises convolutional layers and pooling layers, utilizing convolutional computations to extract latent features from the data and employing pooling layers for downsampling and compression of network parameters. A one-dimensional CNN is employed in this study, and the architecture is illustrated in Figure 2.

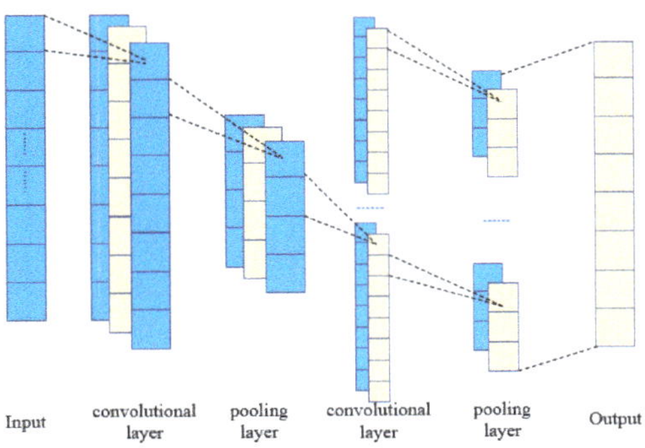

Figure 2. One-dimensional CNN structure.

The long short-term memory (LSTM) network is a variant of the recurrent neural network (RNN), specifically designed to mitigate the challenge of preserving long-term dependencies, a limitation often encountered in conventional RNNs. It enhances the basic RNN structure by introducing forget gates, input gates, and output gates, as shown in Figure 3. The LSTM network, through the incorporation of memory cells and gate mechanisms, is better equipped to capture and handle long-term dependencies within input sequences.

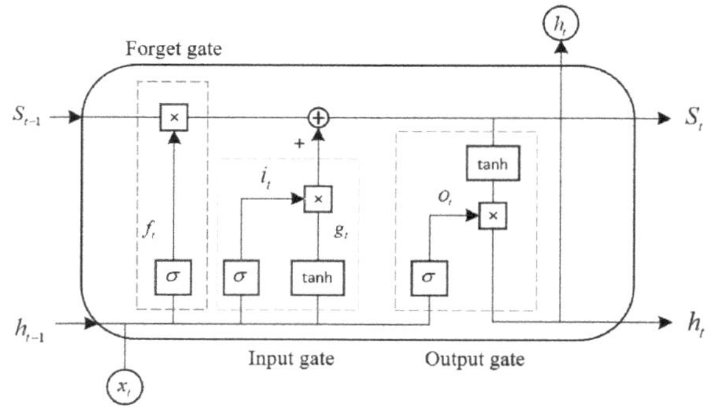

Figure 3. LSTM structure.

The forget gate, utilizing the input, the intermediate state, and the state memory unit, aims to retain useful information while avoiding the transmission of irrelevant information from the previous time step to the next. The roles of the input gate and the output gate include reading data and passing the processed data to the next time step. The computational formulas are presented as Formulas (1)–(6):

$$f_t = \sigma\left(H_{fx}x_t + H_{fh}h_{t-1} + b_f\right) \tag{1}$$

$$i_t = \sigma(H_{ix}x_t + H_{ih}h_{t-1} + b_i) \tag{2}$$

$$n_t = \phi(H_{nx}x_t + H_{nh}h_{t-1} + b_n) \tag{3}$$

$$o_t = \sigma(H_{ox}x_t + H_{oh}h_{t-1} + b_o) \tag{4}$$

$$s_t = g_t \odot i_t + s_{t-1} \odot f_t \tag{5}$$

$$h_t = \phi(s_t) \odot o_t \tag{6}$$

where f_t, i_t, n_t, o_t, s_t, and h_t represent the states of the forget gate, input gate, input node, output gate, state unit, and intermediate output, respectively. H_{fx}, H_{fh}, H_{ix}, H_{ih}, H_{nx}, H_{nh}, H_{ox}, and H_{oh} denote the matrix weights for the corresponding gates, input x_t, and intermediate input h_{t-1}. b_f, b_i, b_n, and b_o are the bias terms for the respective gates. $\odot$ indicates element-wise multiplication in the vector, σ represents the sigmoid function transformation, and ϕ signifies the tanh function transformation.

The CNN–LSTM neural network combines CNN with LSTM. It utilizes input features such as the three-phase current amplitudes, three-phase voltage amplitudes, active power, reactive power, and oil temperature data for the transformer every half hour. The network is employed to predict the load on the transformer.

The predictive method consists of two parts: the CNN part for feature extraction and the LSTM part for load prediction. The CNN in this study comprises two convolutional layers, both of which are one-dimensional (Conv1D). The rectified linear unit (ReLU) is selected as the activation function for both convolutional layers. Following the convolutional layers, max-pooling operations are employed to downsample the features extracted by the convolutional layers, reducing the model parameters. There are also two layers of max-pooling (MaxPooling1D). After the two layers of convolution and pooling, the extracted vector arrays are passed as features to the LSTM network.

Through LSTM, the extracted features are thoroughly learned to capture the intrinsic relationships and periodic patterns among the data, enabling the prediction of future data. The LSTM network consists of one layer of LSTM with 48 neurons, using ReLU as the activation function. Finally, the output is passed through a fully connected layer (Dense) to generate a vector in the specified format.

3.2. The Multi-Strategy Enhanced Dung Beetle Optimizer

3.2.1. Dung Beetle Optimizer

Dung beetle optimizer (DBO) is an optimization algorithm grounded in the behavioral traits observed in dung beetles. This algorithm leverages five distinct behavioral characteristics exhibited by dung beetles—ball rolling, dancing, foraging, stealing, and reproduction—to acquire globally optimal weights and thresholds.

(1) When the dung beetle encounters no obstacles, it navigates using the sun. The repositioning of the dung beetle rolling dung balls is influenced by

$$x_i(t+1) = x_i(t) + a \cdot k \cdot x_i(t-1) + b \cdot \left| x_i(t) - X^W \right| \tag{7}$$

where t represents the number of iterations, $x_i(t)$ represents the position of the ith dung beetle, $a \in [0,1]$ represents the degree of offset caused by natural factors, $k \in [0,2]$ represents the perturbation coefficient $b \in [0,1]$, and $|x_i(t) - X^W|$ represents the change in light intensity, where a larger value indicates a lower light intensity and X^W represents the worst position.

(2) When a dung beetle encounters an obstacle and cannot move forward, it repositions itself through a dance, and its new position is updated to

$$x_i(t+1) = x_i(t) + \tan\theta |x_i(t) - x_i(t-1)| \tag{8}$$

where $\theta \in [0, \pi]$ represents the perturbation angle; when θ is 0 or $\frac{\pi}{2}$, the position remains unchanged. $|x_i(t) - x_i(t-1)|$ represents the offset of the ith dung beetle's position.

(3) The breeding area is simulated using a boundary selection strategy, defined as

$$x_i(t+1) = x_{gbest}(t) + g_1(x_i(t) - L_{g*}) + g_2|x_i(t) - U_{g*}| \tag{9}$$

where $x_{gbest}(t)$ represents the global best position; g_1 and g_2 are two independent random vectors of size $1 \times D$, where D represents the dimension of the optimization problem; L_{g*} and U_{g*} represent the lower and upper bounds of the breeding area, respectively.

(4) When dung beetles forage, the formula for position change is

$$x_i(t+1) = x_i(t) + C_1(x_i(t) - L_{g(t)}) + C_2\left|x_i(t) - U_{g(t)}\right| \tag{10}$$

where $L_{g(t)}$ and $U_{g(t)}$ represent the lower and upper bounds of the foraging area, respectively, C_1 represents random numbers drawn from a normal distribution, and C_2 represents random vectors of size $1 \times D$.

(5) When dung beetles steal, the formula for position change is

$$x_i(t+1) = x_{lbest}(t) + Qf(\left|x_i(t) - x_{gbest}(t)\right| + |x_i(t) - x_{lbest}(t)|) \tag{11}$$

where $x_{lbest}(t)$ represents the best food source, Q represents a constant value, and f is a random vector of size $1 \times D$ drawn from a normal distribution.

3.2.2. The Multi-Strategy Enhanced Dung Beetle Optimizer (MSDBO)

DBO is widely utilized for its simple structure and high accuracy. However, it is susceptible to becoming trapped in local optima, resulting in poor global optimization capabilities. To address this issue, this paper proposes a multi-strategy-enhanced DBO.

(1) Bernoulli chaos mapping

The traditional DBO initializes the population positions using random number generation, which may not cover all positions in the environment, thus affecting the optimization effectiveness and convergence rate of the algorithm. Using Bernoulli chaotic mapping to explore as many positions in the environment as possible, the expression is given by

$$Z(t+1) = \begin{cases} Z(t)/(1+\rho) & Z(t) \in (0, 1-\rho] \\ (Z(t-1)+\rho)/\rho & Z(t-1) \in (1-\rho, 1) \end{cases} \tag{12}$$

where $Z(t)$ represents the current value of the chaotic sequence at the tth iteration and ρ is the control coefficient. When $\rho = 0.5$, the algorithm exhibits the best exploratory behavior.

(2) Levy flight strategy

The Levy flight mechanism, which involves long-range, short-distance roaming, is beneficial for enhancing the diversity of the population. Long-range jumps with variable directions ensure detailed exploration of the nearby regions, and the mechanism's mutation provides certain advantages for exploring a large space. By combining short-range and long-range flight modes, it reflects thorough optimization of the search space, thereby improving the algorithm's global search capability.

If a dung beetle exhibiting stealing behavior becomes stuck in a local optimum, and its position update stagnates, applying the Levy flight strategy to update the individual's position can help it escape from the local optimum and diffuse to more distant positions. The position is updated to

$$x_i(t+1) = x_i(t) + \alpha \oplus Levy(\lambda) \tag{13}$$

where α represents a random step length, $\oplus$ represents the dot product, and $Levy$ is a random search path following the Levy distribution, constrained as follows:

$$Levy(\lambda) \sim \frac{\varphi u}{|v|^{1/2}} \tag{14}$$

$$\varphi = \left[\frac{\Gamma(1+\lambda)\sin(\pi\lambda/2)}{\Gamma(1+\lambda/2)\lambda 2^{(\lambda-1)/2}} \right]^{1/\lambda} \tag{15}$$

where u, v follow a normal distribution and $\lambda = 1.5$.

3.2.3. MSDBO Testing

To evaluate the effectiveness of the MSDBO algorithm, we selected test functions for performance testing and compared it with traditional algorithms, such as GWO, SSA, WOA, NGO, and DBO. The test functions included four unimodal functions and four multimodal functions, as shown in Table 1.

Using MATLAB, simulation experiments were conducted on the test functions. To ensure fair comparison, the maximum number of iterations was uniformly fixed at 500, the population size was N = 30, and the dimension was D D = 30. Each algorithm was run 50 times, and the simulation results are shown in Figure 4.

Table 1. Test function.

Test Function	Value Range
$F_1(x) = \sum_{i=1}^{n} x_i^2$	$[-100, 100]$
$F_2(x) = \sum_{i=1}^{n} \lvert x_i \rvert + \prod_{i=1}^{n} \lvert x_i \rvert$	$[-10, 10]$
$F_3(x) = \sum_{i=1}^{n} \left(\sum_{j-1}^{i} x_j \right)^2$	$[-100, 100]$
$F_7(x) = \sum_{i=1}^{n} i x_i^4 + random[0, 1)$	$[-1.28, 1.28]$
$F_9(x) = [x_i^2 - 10\cos(2\pi x_i) + 10]$	$[-5.12, 5.12]$
$F_{10}(x) = -20\exp\left(-0.2\sqrt{\frac{1}{n}\sum_{i=1}^{n} x_i^2}\right) - \exp\left[\frac{1}{n}\sum_{i=1}^{n}\cos(2\pi x_i)\right] + 20 + e$	$[-32, 32]$
$F_{11}(x) = \frac{1}{4000} * \sum (x_i^2) - \prod \left(\cos\left(\frac{x_i}{\sqrt{i}}\right)\right) + 1$	$[-600, 600]$
$F_{12}(x) = \frac{\pi}{n}\left\{10\sin(\pi y_1) + \sum_{i=1}^{n-1}(y_i - 1)^2[1 + 10\sin^2(\pi y_{i+1})] + (y_n - 1)^2\right\} + \sum_{i=1}^{n} u(x_i, 10, 100, 4)$	$[-50, 50]$

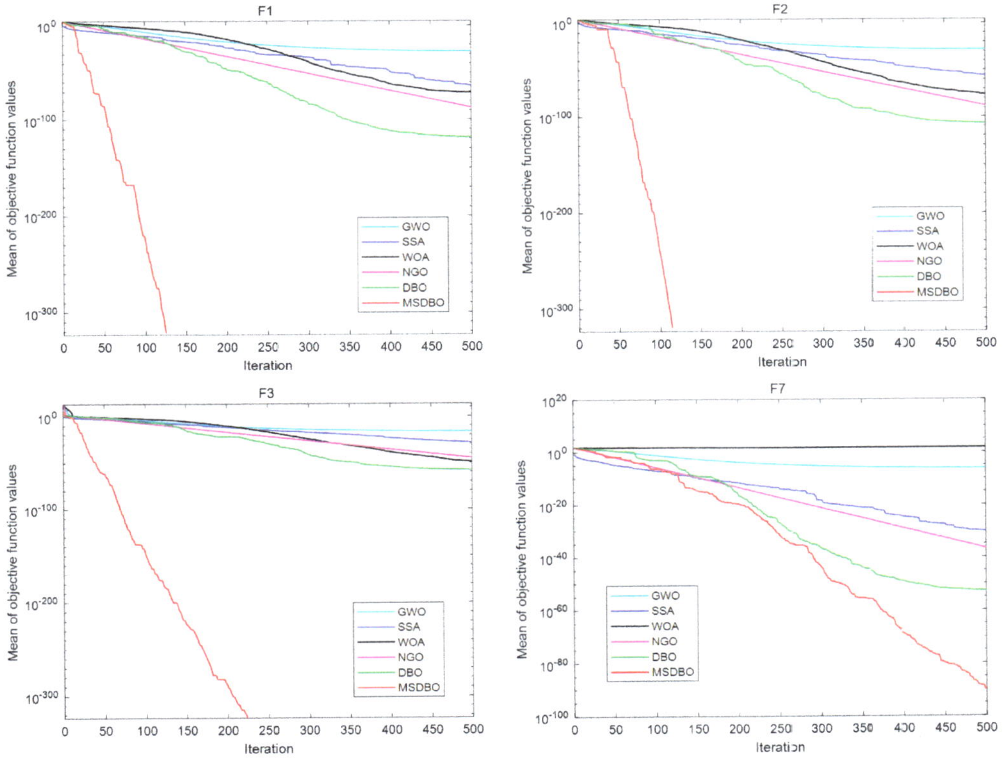

Figure 4. *Cont.*

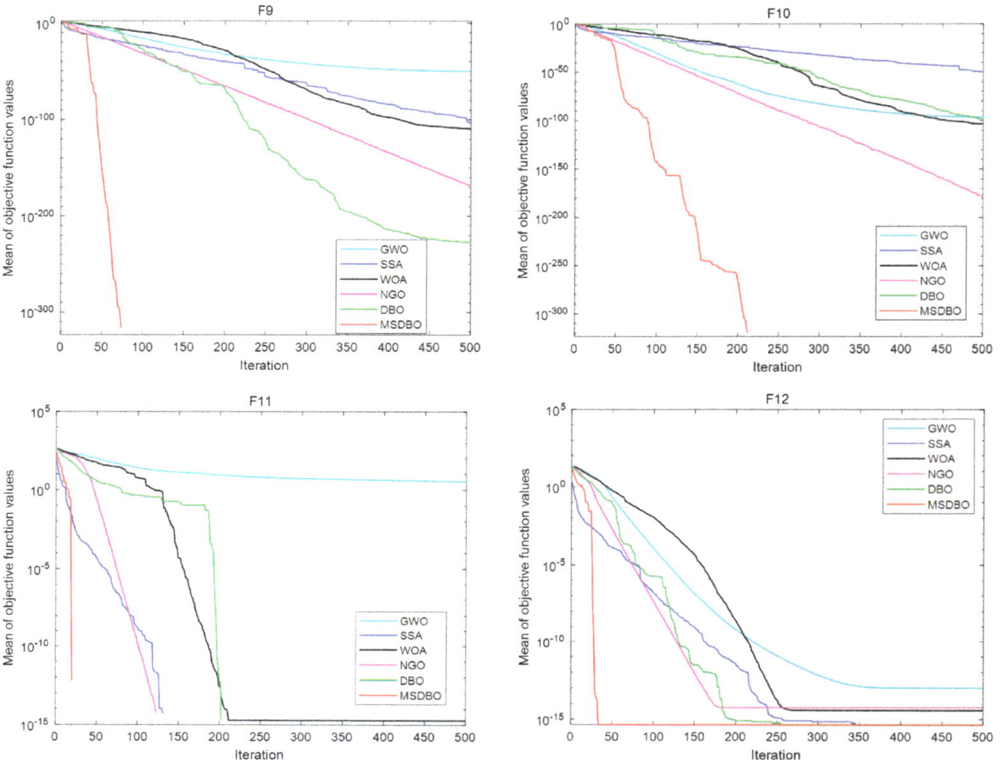

Figure 4. Iteration curves.

From Figure 4, it can be observed that for both unimodal and multimodal functions, the MSDBO algorithm converges faster, achieves higher accuracy, and requires fewer iterations compared to other traditional algorithms. The optimal and average values obtained by the MSDBO algorithm are closer to the optimal value of the function, which demonstrates the significant advantages of the MSDBO algorithm used in this study over the other algorithms.

3.3. MSDBO−CNN−LSTM

To address the issue of local optima and improve the predictive performance of the CNN–LSTM neural network, this study utilizes the MSDBO algorithm to optimize the hyperparameters of the CNN–LSTM, such as the dropout rate and batch size. This optimization algorithm exhibits strong global search capabilities and fast convergence, effectively resolving the inherent problems of the CNN−LSTM. By applying the optimized hyperparameters to the CNN−LSTM, the optimization of the CNN−LSTM is achieved, resulting in more accurate load prediction results.

Figure 5 illustrates the architecture of the MSDBO−CNN−LSTM neural network.

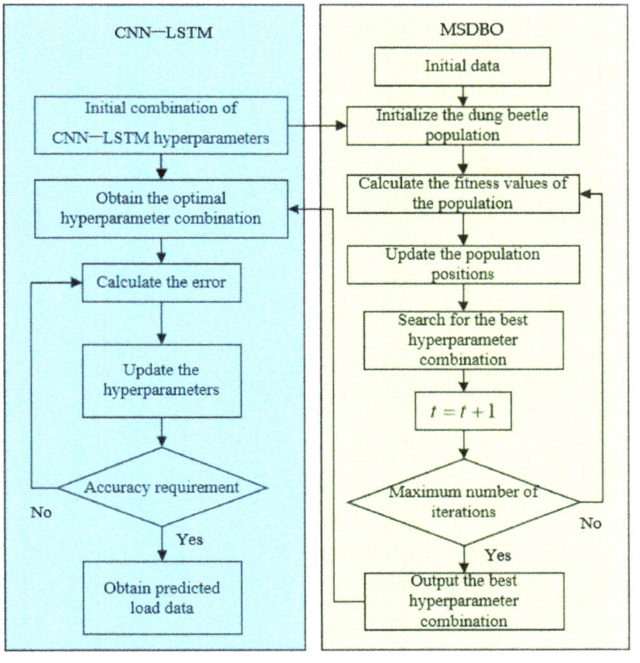

Figure 5. MSDBO−CNN−LSTM structure.

4. On-Load Capacity Regulation and Optimization Based on Fuzzy Control

4.1. Calculation and Economic Analysis of Capacity-Regulating Nodes

The power loss of a transformer with different capacity levels is composed of no-load loss and short-circuit loss, as expressed by the following formula:

$$S_H = S_{OH} + \beta_H^2 S_{XH} \tag{16}$$

$$S_L = S_{OL} + \beta_L^2 S_{XL} \tag{17}$$

where S_H, S_L represent the power loss for a transformer with different capacity levels, S_{OH}, S_{OL} represent the no-load loss for a transformer with different capacity levels, β_H, β_L represent the load factors for a transformer with different capacity levels, and S_{XH}, S_{XL} represent the short-circuit loss for a transformer with different capacity levels.

$$\beta_L = \frac{\beta_H S_{NH}}{S_{NL}} \tag{18}$$

where S_{NH}, S_{NL} represent the rated capacity of the transformer for different capacity levels.

If we assume that $S_H = S_L$, meaning that the power loss is the same for different capacity levels, then, based on Formulas (16) and (17), the expression for the critical load can be obtained as follows:

$$S^{L-H} = \sqrt{\frac{P_{OH} - P_{OL}}{\frac{P_{XL}}{S_{NL}^2} - \frac{P_{XH}}{S_{NH}^2}}} \tag{19}$$

The operational cost calculation formula for the transformer is given by

$$C_y = [T_1 \times (P_0 + 0.05 \times I_0 \times S_N/100) + T_2 \times (P_x + 0.05 \times U_x \times S_N/100)] \times C \tag{20}$$

where T_1, T_2 represent the transformer no-load time and equivalent full-load time, respectively, P_0, P_x represent the transformer no-load loss and load loss, respectively, I_0, U_x represent the transformer no-load current and short-circuit impedance, respectively, and S_N, C represent the transformer rated capacity and electricity price, respectively, where $C = 0.5$ yuan/kWh and is a constant value.

The formula for calculating the power loss of a transformer is given by [16]:

$$\begin{cases} P_{0z} = P_0 + K_Q Q_0 + K_P P_0 \\ P_{kz} = P_k + K_Q Q_k + K_P P_k \\ Q_0 = I_0\% S_N/100 \\ Q_k = U_k\% S_N/100 \end{cases} \tag{21}$$

where P_{0z}, P_{kz} represent no-load comprehensive loss and load comprehensive loss, respectively, Q_0, Q_k represent no-load reactive power loss and short-circuit reactive power loss, respectively, $I_0\%$, $U_k\%$ represent no-load current and short-circuit impedance respectively, and $K_Q = 0.1, K_p = 0.2$.

4.2. Hesitant Fuzzy Control Making for Capacity-Regulating Nodes

The core idea of fuzzy control is to simulate human fuzzy reasoning and decision-making. By translating the expertise or experience of experts into fuzzy rules, it transforms real-time signals from sensors into fuzzy signals. These fuzzy signals are then used as inputs for fuzzy rules, and the output obtained from fuzzy reasoning is added to the actuator to achieve intelligent control of the system. The block diagram illustrating the principles is shown in Figure 6:

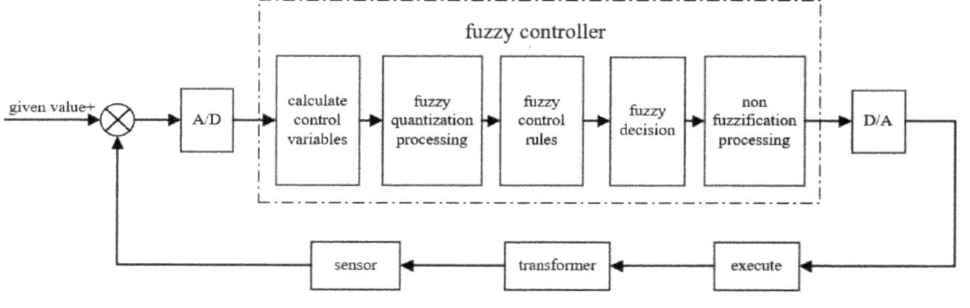

Figure 6. Fuzzy control principle block diagram.

The fuzzy controller has inputs for switch times, power loss, and economic loss, and an output for adjustment node value. The domains and fuzzy subsets for each variable are as follows:

The number of switch changes: domain [0, 10], fuzzy subsets {LS (low), MS (medium), HS (high)};

Power loss: domain [0, 60], fuzzy subsets {LP (low), MP (medium), HP (high)};

Operating costs: domain [0, 30], fuzzy subsets {LC (low), MC (medium), HC (high)};

Switch taps: domain [30, 70], fuzzy subsets {VL (very low), L (low), M (medium), H (high), VH (very high)};

Triangular membership functions are used to describe the membership degrees of each variable, as shown in Figure 7.

Based on the fuzzy set relationships of switch changes, power loss, and operating costs, a more suitable switch node is determined. A fuzzy rule table is established accordingly, as shown in Table 2.

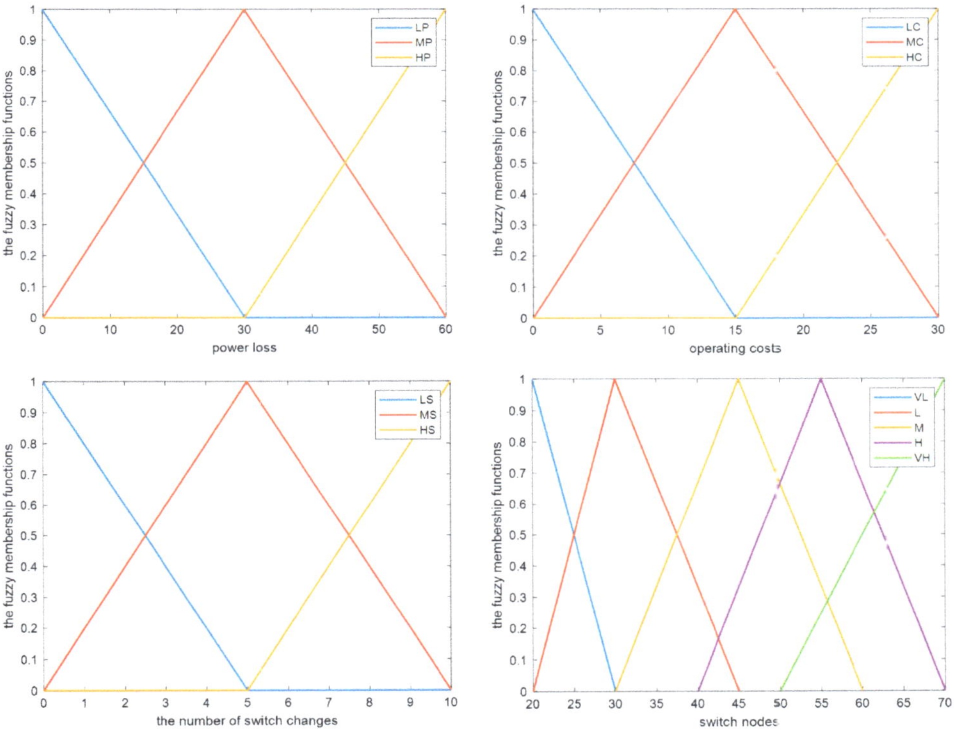

Figure 7. The fuzzy membership functions of input and output.

Table 2. Fuzzy rule base.

			The Number of Switch Changes		
			LS	**MS**	**HS**
	LP	LC	VH	H	H
	LP	MC	H	H	M
	LP	HC	H	M	M
Power loss	MP	LC	H	H	M
and	MP	MC	H	M	M
operating costs	MP	HC	M	M	L
	HP	LC	H	M	M
	HP	MC	M	M	L
	HP	HC	M	L	VL

Due to the difficulty in accurately analyzing the importance of various factors for transformer capacity-regulating control, and the lack of weights on different factors, conventional fuzzy control methods perform poorly. For multifactor problems with unknown weights, hesitant fuzzy control combines the different opinions of multiple decision-makers for a more comprehensive analysis, resulting in decisions that are more relevant to the integrated opinions of multiple decision-makers. Compared to conventional fuzzy control, hesitant fuzzy control provides more accurate and comprehensive results. The hesitant fuzzy control method is described in the following text:

Through factor $G = \{G_1, G_2, \cdots, G_n\}$, decision-makers need to evaluate plan $A = \{A_1, A_2, \cdots, A_m\}$, where n is the number of factors and m is the number of plans.

Assuming h_{ij} represents the evaluation by decision-makers of factor G_j in option A_i, all h_{ij} values constitute the hesitant fuzzy control matrix $H = (h_{ij})_{m\times n}$, $h_{ij} = \bigcup_{t=1}^{l_{ij}} \left\{\gamma_{ij}^{(t)}\right\} = \left\{\gamma_{ij}^{(1)}, \gamma_{ij}^{(2)}, \cdots \gamma_{ij}^{(t)}\right\}$, $\gamma_{ij}^{(t)}$ represents the tth evaluation value among h_{ij}. If the decision-maker's attitude is comparatively optimistic, it will lead to missing values in the matrix. In the case of inconsistent lengths of hesitant fuzzy control elements, the maximum membership value should be added to the fewer elements to obtain the standardized matrix $\widetilde{H} = \left(\widetilde{h}_{ij}\right)_{m\times n}$. The weight vector λ represents the relative importance of the factors, i.e., $\lambda = (\lambda_1, \lambda_2, \cdots, \lambda_n)^T$, satisfying the formula $\lambda_j \in [0,1], \sum_{j=1}^{n} \lambda_j = 1$. The operator-related weight vector $\omega = (\omega_1, \omega_2, \cdots, \omega_n)^T$ satisfies the formula $\omega_j \in [0,1], \sum_{j=1}^{n} \omega_j = 1$.

The factor weight represents the relative importance of the factors and the preferences of the decision-maker. Due to the unknown weight coefficients, the hesitant fuzzy hamming distance is used to extend the maximum deviation method to the hesitant control environment, obtaining the optimal weight vector $\lambda = (\lambda_1, \lambda_2, \cdots, \lambda_n)^T$.

$$\lambda_j = \frac{Y_j}{\sum_{j=1}^{n} Y_j}, j = 1, 2, \cdots, n, \tag{22}$$

where $Y_j = \sum_{i=1}^{m} \sum_{k=1}^{m} \left(\frac{1}{l}\sum_{t=1}^{l}\left|\widetilde{\gamma}_{ij}^{(t)} - \widetilde{\gamma}_{kj}^{(t)}\right|\right), j = 1, 2, \cdots, n.$

The operator-related weight vector is an important part of the generalized hesitant fuzzy mixed-weighted aggregation operator used for information integration. In this paper, the operator-related weight vector $\omega = (\omega_1, \omega_2, \cdots, \omega_n)^T$ is determined using a normal distribution. Table 3 shows the variation of ω.

Table 3. The variation of ω for n ranging from 2 to 5.

n	Operator-Related Weight Vector $\omega=(\omega_1,\omega_2,\cdots,\omega_n)^T$
2	$\omega = (0.5, 0.5)^T$
3	$\omega = (0.2429, 0.5142, 0.2429)^T$
4	$\omega = (0.1550, 0.3450, 0.3450, 0.1550)^T$
5	$\omega = (0.1117, 0.2365, 0.3036, 0.2365, 0.1117)^T$

For the hesitant fuzzy multifactor control method under unknown weights, the algorithm proceeds as follows:

(1) Generate the hesitant fuzzy control matrix $H = (h_{ij})_{m\times n}$ based on the decision-maker's evaluations of the options, and obtain matrix $\widetilde{H} = \left(\widetilde{h}_{ij}\right)_{m\times n}$ by standardizing the matrix.

(2) Calculate the optimal weight vector $\lambda = (\lambda_1, \lambda_2, \cdots, \lambda_n)^T$ based on Formula (12).

(3) Select the operator-related weight vector $\omega = (\omega_1, \omega_2, \cdots, \omega_n)^T$ based on the value of n.

(4) Using the new hesitant fuzzy control weighted aggregation operator (NGHFHWA), integrate $\widetilde{h}_{ij}$ into the hesitant fuzzy element $\widetilde{h}_i$ for each option A_i.

$$\widetilde{h}_i = NGHFHWA(\widetilde{h}_{i1}, \widetilde{h}_{i2}, \cdots \widetilde{h}_{in}) \tag{23}$$

(5) Calculate the score function $S(\widetilde{h}_i)$ based on $\widetilde{h}_i$. If the score functions are equal, differentiate them through the hesitant fuzzy-order central aggregation function $p(\widetilde{h}_i)$ for sorting.

$$s(h) = \frac{1}{l}\sum_{i=1}^{l} \gamma^{(i)} \tag{24}$$

(6) Arrange the options according to the score function $S(\widetilde{h}_i)$ and then choose the one with the highest score as the best option.

The hesitant fuzzy control process diagram is illustrated in Figure 8:

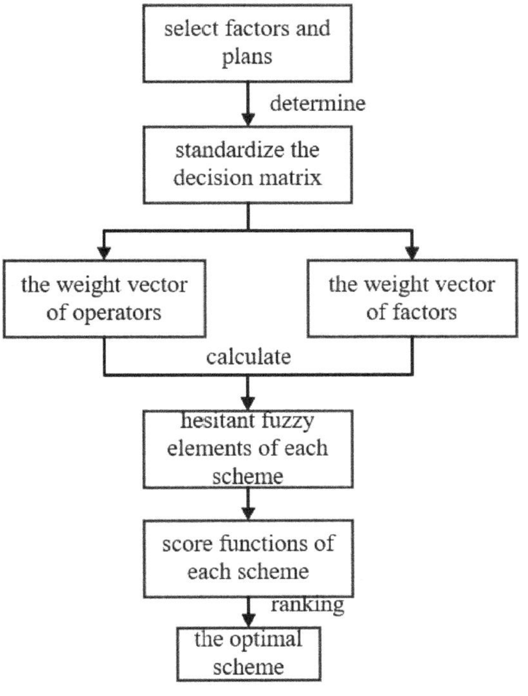

Figure 8. Hesitant fuzzy control flowchart.

5. On-Load Capacity-Regulating Control Strategy and Methodology

The variation in load leads to changes in data, such as current, voltage, and active power. Capacity regulation of the transformer can be achieved through sensor monitoring and feedback. However, this type of capacity-regulating method relies heavily on real-time data, lacks adaptability to learning from data, and cannot provide warnings for future load conditions over a certain period.

To address the above issues, this study employs an MSDBO−CNN−LSTM neural network to predict load data for the next 24 h. Based on these data, suitable capacity-regulating schemes are selected through hesitant fuzzy control. The merit values for switch operation frequency, power losses, and operating costs are calculated, and a weighted sum is used to determine the comprehensive merit value for each scheme. The optimal capacity-regulating scheme is then identified. Control of the capacity-regulating switches is carried out using a step wave signal. Additionally, the load data for this time are added to the database for use in the next round of MSDBO−CNN−LSTM predictions, ensuring the real-time and accurate nature of subsequent forecasting results.

This study proposes a method for the capacity adjustment of OLCR transformers in distribution networks using a combination of load prediction and hesitant fuzzy control, as shown in Figure 9. The methodology consists of the following steps:

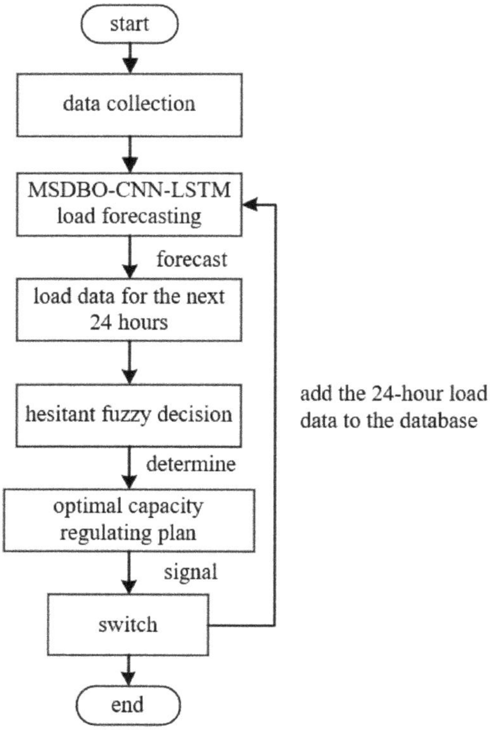

Figure 9. Load forecasting and capacity-regulating control flowchart.

Step 1. Data collection:

Collect historical load data from the distribution network, including daily load profiles, for analysis and prediction. The load data should encompass various periods to capture seasonal variations and different operational scenarios.

Step 2. Load forecasting:

Preprocess the historical load data to ensure its quality and consistency. Utilize the MSDBO−CNN−LSTM model to predict future load demand based on historical data.

Step 3. Hesitant fuzzy control:

(1) Define the evaluation criteria for selecting the optimal adjustment scheme, including the number of switch times, power losses, and operating costs.
(2) Design a fuzzy controller with inputs of the number of switch changes, power loss, and operating costs, and an output of adjustment node value.
(3) Define the fuzzy subsets and membership functions for each input and output variable.
(4) Develop fuzzy rules based on expert knowledge and the fuzzy rule base table to determine the adjustment node value.

Step 4. Optimal capacity-regulating plan:

(1) Develop a detailed capacity adjustment schedule for the 24-h period based on the optimal adjustment nodes obtained through load forecasting and hesitant fuzzy decision-making methods.
(2) Reduce power losses and operating costs through the capacity adjustment schedule while minimizing the number of switch changes as much as possible.

Step 5. Action and analysis:

(1) Control the switching of OLCR transformer capacity by implementing the optimal capacity-regulating plan in the form of signals.

(2) Compare the performance of the conventional OLCR transformer with the proposed method in terms of the number of switching times, power losses, and operating costs.

(3) Analyze the results of different scenarios to demonstrate the effectiveness and superiority of the proposed method in capacity regulation.

6. Case Studies and Results Analysis

6.1. Example Introduction

According to the principles of OLCR transformers, simulation models are constructed using the MATLAB/SIMULINK platform. Figure 10a is a diagram illustrating the regulation of transformer capacity based on real-time load currents. Figure 10b is a diagram of a transformer capacity-regulating control using an MSDBO−CNN−LSTM neural network and hesitant fuzzy control.

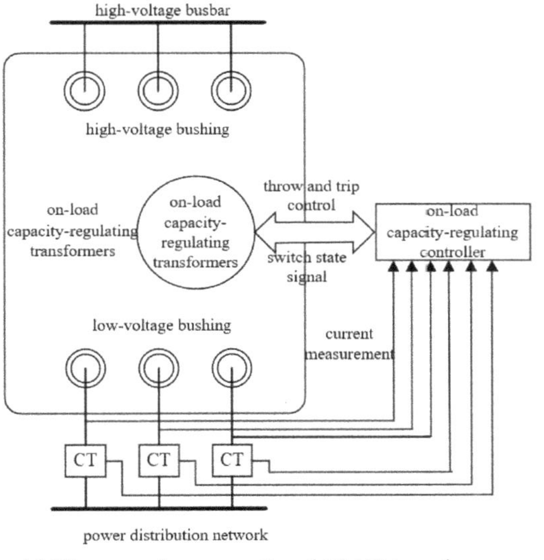

(**a**) Diagram of a conventional OLCR transformer

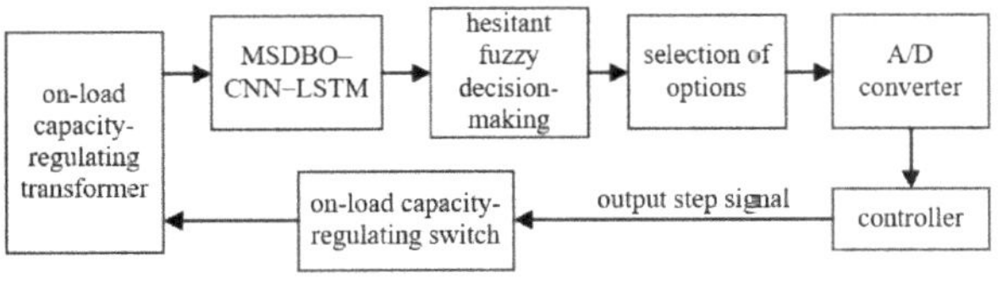

(**b**) Diagram of a load forecasting and hesitant fuzzy control capacity-regulating transformer

Figure 10. Diagrams of capacity-regulating transformers.

6.2. Load Forecasting Based on an MSDBO−CNN−LSTM Model

Using an MSDBO−CNN−LSTM neural network, the next 24 h of data are predicted, and Figure 11 presents a comparative graph between the predicted results and the actual values over the 24-h period.

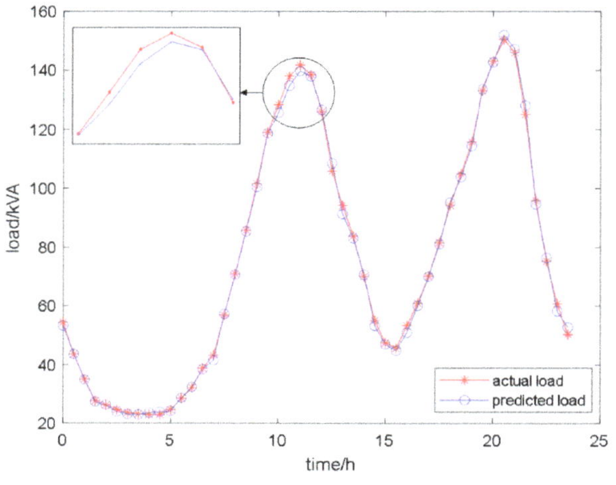

Figure 11. Comparison chart between actual load and predicted load.

The relative error between the actual and predicted results for the 24-h load is presented, as shown in Figure 12. From the graph, it can be observed that the relative errors are all less than 5%.

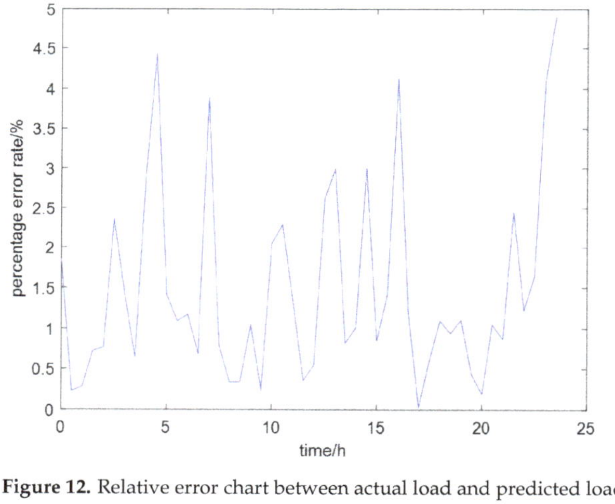

Figure 12. Relative error chart between actual load and predicted load.

R^2 is the coefficient of determination, an accuracy evaluation metric in machine learning. In this case, the obtained R^2 value is 0.9789.

6.3. Capacity-Regulating Control Based on Hesitant Fuzzy Control

Choosing the S11-M series three-phase transformer with a rated capacity of 315/100 kVA, the critical load is calculated by substituting the data from Table 4 into Formula (19).

The calculation results from the table above indicate that 49.73 kVA is the transformer's critical load. The transformer runs in low-capacity mode when the load is less than 49.73 kVA. On the other hand, the transformer goes into high-capacity mode when the load surpasses 49.73 kVA. Power loss is reduced and economic efficiency is increased by switching between these two capacity modes.

Table 4. S11-M three-phase transformer operating parameters.

Rated Capacity (kVA)	No-Load Loss (kW)	Load Loss (kW)	No-Load Current (%)	Short-Circuit Impedance (%)	Critical Value (kVA)
315/100	0.48 0.2	3.65 1.5	1.1 1.6	4 4	49.73

To calculate the daily operational cost through exemplification, the measured load data of a transformer on a specific day are considered. Within the course of that day, the transformer operated at reduced capacity for a total of 8 h and at full capacity for 16 h.

The following results are obtained through the calculation using Formula (20), as shown in Table 5.

Table 5. Daily operating costs of three-phase transformers.

Rated Capacity (kVA)	Operating Cost of the OLCR Transformer (CNY)	Rated Capacity (kVA)	Operating Cost of the Fixed-Tap Transformer (CNY)	Cost Savings with OLCR Transformer (CNY)	Cost Savings with OLCR Transformer (%)
315/100	16.61	315	20.68	4.07	19.68

The load data for a transformer over the course of one year is considered, during which the transformer operates at reduced capacity for a total of 6108 h and at full capacity for 2532 h.

The following results are obtained through the calculation using the above Formula (20), as shown in Table 6.

Table 6. Annual operating cost of three-phase transformers.

Rated Capacity (kVA)	Operating Cost of the OLCR Transformer (CNY)	Rated Capacity (kVA)	Operating Cost of the Fixed-Tap Transformer (CNY)	Cost Savings with OLCR Transformer (CNY)	Cost Savings with OLCR Transformer (%)
315/100	4334.80	315	8240.50	3905.70	47.40

In the provided data, the maximum load over the course of a year is 200 kVA. If a transformer with a rated capacity of 250 kVA, such as the S11-M-250/10 model, were employed, the daily operational cost for the transformer would be 19.025 yuan, resulting in a cost saving of 12.69%. Furthermore, the annual operational cost for the transformer would amount to 6859.3 yuan, leading to a cost saving of 36.8%.

After performing calculations and economic analyses of the switching capacity nodes, the selection of these nodes can be initiated through hesitant fuzzy control. Based on the following three factors, choose the appropriate regulating scheme, where G_1 represents the number of switch changes, G_2 represents power loss, and G_3 represents operating costs. Suppose there are three regulating schemes $A_i(i = 1,2,3)$ to choose from, with capacity nodes of 40 kVA, 50 kVA, and 60 kVA, respectively. Due to the difficulty of precisely analyzing the importance of the three factors, it is challenging to directly decide which scheme to choose. By using hesitant fuzzy sets to represent the evaluation values of each scheme relative to the three factors, the hesitant fuzzy control matrix $H = (h_{ij})_{3\times3}$ can be obtained, as shown in Table 7.

Table 7. Hesitant fuzzy control matrix.

	G1	G2	G3
A_1	{0.7, 0.8, 0.9}	{0.4, 0.6}	{0.5, 0.6, 0.7}
A_2	{0.5, 0.6}	{0.7, 0.8, 0.9}	{0.4, 0.5}
A_3	{0.3, 0.4, 0.5}	{0.6, 0.7}	{0.7, 0.8}

After standardizing the matrix, the matrix $\widetilde{H} = \left(\widetilde{h}_{ij}\right)_{3\times 3}$ is obtained as shown in Table 8.

Table 8. Standardize the hesitant fuzzy control matrix.

	G1	G2	G3
A_1	{0.7, 0.8, 0.9}	{0.4, 0.6, 0.6}	{0.5, 0.6, 0.7}
A_2	{0.5, 0.6, 0.6}	{0.7, 0.8, 0.9}	{0.4, 0.5, 0.5}
A_3	{0.3, 0.4, 0.5}	{0.6, 0.7, 0.7}	{0.7, 0.8, 0.8}

According to Table 2, when n = 3, $\omega = (0.2429, 0.5142, 0.2429)^T$.

Using Formula (22), the factor weight vector can be obtained as $\lambda = (0.4138, 0.2759, 0.3103)^T$.

For the capacity-regulating schemes A_1, A_2, and A_3, the corresponding $\widetilde{h}_i$ can be calculated using the NGHFHWA operator. This paper takes A_1 as an example.

$$\begin{aligned}\widetilde{h}_1 &= NGHFHWA(\widetilde{h}_{11}, \widetilde{h}_{12}, \widetilde{h}_{13}) \\ &= NGHFHWA(\{0.7, 0.8, 0.9\}, \{0.4, 0.6, 0.6\}, \{0.5, 0.6, 0.7\}\end{aligned}$$

According to the formula, it can be concluded that $s(\widetilde{h}_{11}) = \frac{0.7+0.8+0.9}{3} = 0.8$, $s(\widetilde{h}_{12}) = \frac{0.4+0.6+0.6}{3} = 0.5333$, $s(\widetilde{h}_{13}) = \frac{0.5+0.6+0.7}{3} = 0.6$.

At this point, $\widetilde{h}_{11} > \widetilde{h}_{13} > \widetilde{h}_{12}$. According to the normal distribution of ω, it can be inferred that $\omega'_1 = (0.2429, 0.2429, 0.5142)^T$. Using this information, the following calculation can be carried out:

$$\frac{\lambda_1\omega'_{11}}{\sum_{j=1}^{3}\lambda_j\omega'_{1j}} = 0.3073, \frac{\lambda_2\omega'_{12}}{\sum_{j=1}^{3}\lambda_j\omega'_{1j}} = 0.2049, \frac{\lambda_3\omega'_{13}}{\sum_{j=1}^{3}\lambda_j\omega'_{1j}} = 0.4878.$$

When $p = 1$, perform the following calculation.

$$\begin{aligned}\widetilde{h}_1 &= NGHFHWA(\widetilde{h}_{11}, \widetilde{h}_{12}, \widetilde{h}_{13}) \\ &= \bigcup_{i=1}^{3}\left\{\left(1-\left(1-\left(\widetilde{\gamma}_{11}^{(i)}\right)^p\right)^{0.3073}\left(1-\left(\widetilde{\gamma}_{12}^{(i)}\right)^p\right)^{0.2049}\left(1-\left(\widetilde{\gamma}_{13}^{(i)}\right)^p\right)^{0.4878}\right)^{\frac{1}{p}}\right\} \\ &= \{0.5544, 0.6749, 0.7711\}\end{aligned}$$

Using the same method, the hesitant fuzzy elements $\widetilde{h}_2 = \{0.5280, 0.5308, 0.6771\}$ and $\widetilde{h}_3 = \{0.5540, 0.6607, 0.6797\}$ synthesized from A_2 and A_3, respectively, can be calculated.

For the capacity-regulating schemes A_1, A_2, and A_3, the respective score values $s(\widetilde{h}_i)$ are calculated as follows:

$$s(\widetilde{h}_1) = 0.6668, s(\widetilde{h}_2) = 0.5786, s(\widetilde{h}_3) = 0.6315.$$

Based on the score values $s(\widetilde{h}_i)$, the capacity regulating schemes A_1, A_2, and A_3 are ranked in the following order: $A_1 > A_3 > A_2$. Therefore, A_1 is the best capacity-regulating scheme.

6.4. Simulation Results and Analysis

In the experiment of this paper, the three-phase transformer is formed by three multi-winding single-phase transformer models, with parameters set to have one primary winding and three secondary windings (one 27% turn and two 73% turn windings), achieving an S11-M 350/100 kVA capacity-adjustable distribution transformer. All transformer parameter settings are consistent, with $S_N = 100$ kVA, $U_1/U_2 = 5774/220$ V, high and low voltage side 4-section winding resistance per unit values of 0.075, 0.041, 0.041, and 0.028, respectively, leakage reactance per unit values for each section of 0.02, 0.11, 0.11, and 0.08, respectively, and excitation impedance per unit values of 50, 50, respectively. In this study, the voltage source in this simulation is provided by a three-phase generator. The three-phase generator has a rated power of 320 kW for main use and 350 kW for standby, operating at a speed of 1500 rpm. It has a power factor of 0.8 and a rated voltage of 400/230 V. The peak amplitude of an ideal sinusoidal AC voltage source is 5774 V with a frequency of 50 Hz, and each phase is 120 degrees apart.

To simulate the capacity-regulating transformer SIMULINK model, the initial state assumes that the transformer is in a reduced capacity state. At 1.9 s, when the load capacity exceeds 49.73 kVA, the transformer switches to a full-capacity state.

Figure 13 illustrates the voltage and current waveforms for the A−, B−, and C−phases for both the conventional OLCR transformer and the load prediction-controlled OLCR transformer.

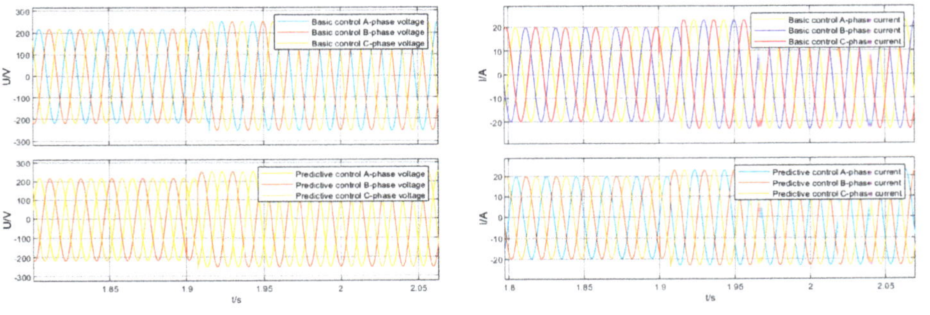

Figure 13. A−, B−, and C−phase voltage and current results of the basic control and predictive control during capacity change.

At 1.9 s, when the capacity state switches, no significant fluctuation is observed in the voltage and current in the A−, B−, and C−phases. This indicates that, after the capacity switch, the output voltage and current of the capacity-regulating transformer remain relatively stable. The capacity-regulating transformer controlled by load prediction demonstrates the ability to switch between the reduced and full capacities accurately and rapidly. As shown in the figure, the load forecasting control switches in a more timely fashion than in a conventional OLCR transformer. This is because the conventional OLCR transformer makes judgments on capacity regulation at fixed intervals, while predictive control involves real-time prediction and judgment for switching, making it capable of faster capacity regulation.

Based on load forecasting, fuzzy control is employed to evaluate the comprehensive merit values of multiple scenarios using switch-switching frequency, power loss, and operational cost as fuzzy control criteria.

The optimal variable-capacity-regulating scheme is selected, and after incorporating it into the simulation, the voltage and current waveforms for the A−, B−, and C−phases are obtained, as shown in Figure 14. With reasonable capacity regulation, the number of switching operations for the transformer is reduced from eight to four, avoiding frequent switching actions and achieving the goal of hesitant fuzzy control.

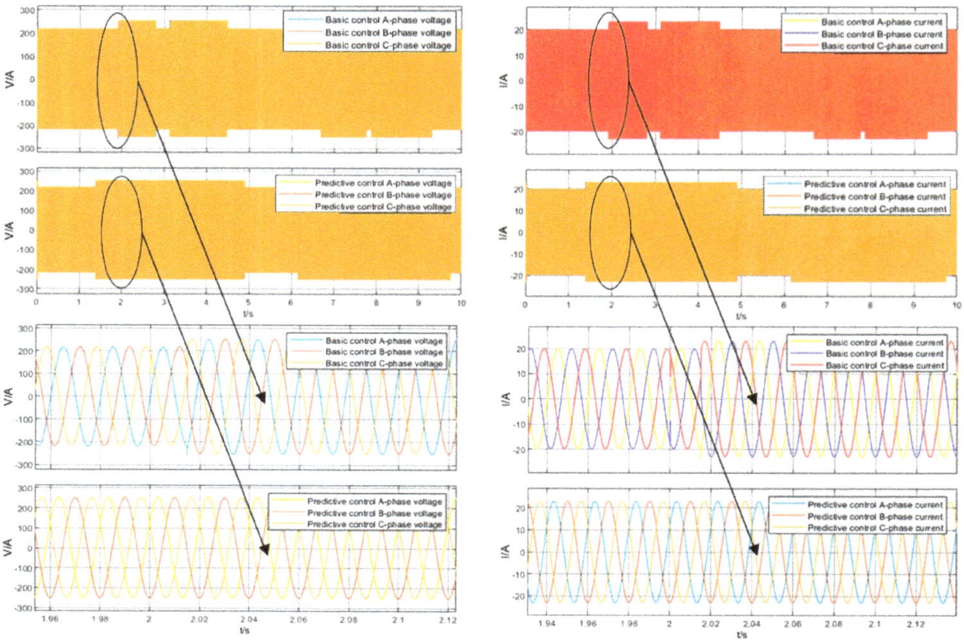

Figure 14. The switching times in A−, B−, and C−phase voltage and current of the basic control (8 times) and predictive control (4 times) under fuzzy control.

6.5. Scene Analysis

To test the effectiveness and reliability of the proposed method, this paper selects three different scenarios for analysis, as detailed below:

(1) Scenario 1: During the busy agricultural season, electricity consumption in rural power grids is concentrated, leading to significant load fluctuations and severe overloading of transformers.
(2) Scenario 2: During the idle agricultural season, electricity consumption in rural power grids is more dispersed. Transformers operate in no-load or light-load conditions for extended periods, leading to higher no-load losses.
(3) Scenario 3: Urban residents' electricity consumption follows daily life rhythms, resulting in large peak–valley differences in electricity usage.

For each of the three scenarios, the MSDBO−CNN−LSTM model is used to make predictions, resulting in 10 sets of 24−h predicted data for this scenario. Ten sets of 24−h actual data are selected to simulate the conventional OLCR transformer, and 10 sets of 24−h predicted data are selected to simulate the OLCR transformer based on the method proposed in this paper. The simulation results are analyzed to validate the superiority of the method proposed in this paper under each operating scenario.

MATLAB/SIMULINK simulations using data from September to October are conducted, corresponding to the busy agricultural season. The simulations are used to calculate the daily operating costs, power loss, and number of switch changes for the transformer. The results from the ten sets of calculations are used to compare the performance of the conventional OLCR transformer and the OLCR transformer based on load prediction and hesitant fuzzy decision-making. A S11−M 315/100kVA OLCR transformer is selected for this scenario. The results are shown in Table 9.

Table 9. Results of 10 cases in the busy agricultural season.

Case	The Number of Switch Changes		Power Loss (kWh)		Operating Costs (CNY)	
	Conventional Method	New Method	Conventional Method	New Method	Conventional Method	New Method
1	2	1	17.6	19.2	10.9	11.8
2	2	1	21.2	23.4	12.6	14.0
3	4	3	31.7	33.6	18.0	19.3
4	1	1	16.7	19.1	10.3	11.7
5	4	3	34.5	36.8	19.3	20.8
6	4	2	30.5	32.1	17.3	18.5
7	3	1	25.6	27.5	14.9	16.2
8	3	1	27.4	29.4	15.8	17.2
9	2	1	20.8	21.8	12.4	13.1
10	3	2	25.0	28.1	14.5	16.4

Calculating the average of the results from the 10 cases can provide a better overall analysis of the effectiveness and superiority of the proposed method in this paper. Table 10 compares the various results of the fixed-tap transformer, the OLCR transformer controlled by the traditional method, and the OLCR transformer controlled by the new method proposed in this paper.

Table 10. Average results for the busy agricultural season.

	The Number of Switch Changes	Power Loss (kWh)	Operating Costs (CNY)
Fixed-tap transformer	/	37.9	21.3
Conventional method	2.8	25.1	14.6
New method	1.6	27.1	15.9

Ten sets of 24−h data from March for the idle agricultural season are used to simulate and verify the conventional OLCR transformer and the OLCR transformer based on the method proposed in this paper, following the same validation method used for the busy agricultural season scenario. A S11−M 315/100kVA OLCR transformer is selected for this scenario. The results are shown in Table 11.

Table 11. Average results for the idle agricultural season.

	The Number of Switch Changes	Power Loss (kWh)	Operating Costs (CNY)
Fixed-tap transformer	/	84.3	42.1
Conventional method	6.3	33.7	19.2
New method	2.7	36.8	20.7

Similarly, for the urban residents' electricity consumption scenario, 10 sets of 24−h data are randomly selected from the entire year for simulation and verification of both the conventional OLCR transformer and the OLCR transformer based on the method proposed in this paper. A S11−M 400/125kVA OLCR transformer is selected for this scenario. The results are shown in Table 12.

Table 12. Average results of urban residents.

	The Number of Switch Changes	Power Loss (kWh)	Operating Costs (CNY)
Fixed-tap transformer	/	58.5	29.2
Conventional method	3.7	29.8	17.2
New method	2.1	31.1	17.8

The following conclusions can be drawn from the above scenarios:

(1) Both the conventional OLCR transformer and the OLCR transformer controlled by the method proposed in this paper can effectively reduce power loss and operating costs, with the reduction effect significantly increasing as the load decreases. For the idle agricultural season in rural power grids, using the method proposed in this paper reduces daily power loss and daily economic loss by an average of 56.3% and 50.8%, respectively. For the busy agricultural season in rural power grids, using the method proposed in this paper reduces daily power loss and daily economic loss by an average of 28.5% and 25.4%, respectively. For urban residents' electricity consumption, using the method proposed in this paper reduces daily power loss and daily economic loss by an average of 46.8% and 39%, respectively.

(2) Compared to a conventional OLCR transformer, an OLCR transformer controlled by load prediction and fuzzy hesitant decision-making can significantly reduce the daily number of switch operations while sacrificing some loss reduction capability. For the three scenarios mentioned above, the method used in this paper can sacrifice 3.5% to 9.2% of the loss reduction capability in exchange for reducing the number of switch operations by 28.6% to 57.1%. This significantly extends the lifespan of the switches, thereby increasing the operational lifespan of the transformer.

7. Conclusions

The control method proposed in this paper enables OLCR transformers to effectively address operating conditions, such as no-load, light-load, overload, and voltage fluctuations caused by load fluctuations, which is of great significance for building energy-efficient and smart grids. This paper utilizes the MSDBO−CNN−LSTM neural network to predict the load situation, calculates the switch node based on transformer parameters, and analyzes its economic feasibility. By employing hesitant fuzzy control to select the optimal switch node, a SIMULINK model for OLCR transformers is constructed and simulated. Experimental validation demonstrates the superiority of this method in different electrical scenarios, leading to the following conclusions:

(1) The utilization of an MSDBO−CNN−LSTM neural network in this study for forecasting the load conditions for the next 24 h enables accurate switching of the transformer capacity mode. This helps avoid prolonged periods of overload or light-load conditions for the transformer. By reducing losses, it extends the operational lifespan of the on-load variable transformer, thereby enhancing the security and stability of the power grid.

(2) In this study, a hesitant fuzzy control approach is employed to select the variable-capacity regulating nodes. Considering the three major factors of loss, economy, and switching frequency, the approach aims to minimize power loss and operating cost while reducing the number of switching operations. This strategy helps avoid a decrease in the operational lifespan of the transformer due to frequent capacity switch operations.

(3) This control method demonstrates good capacity adjustment effects for OLCR transformers in rural distribution networks with seasonal loads and in urban residential networks with symmetric loads. It is applicable in various scenarios and exhibits superior performance. For the three scenarios mentioned above, the method reduces daily power loss by 28.5% to 56.3% and daily operating costs by 25.4% to 50.8%. The

method used in this paper can sacrifice 3.5% to 9.2% of the loss reduction capability in exchange for reducing the number of switch operations by 28.6% to 57.1%. This significantly extends the lifespan of the switches, thereby increasing the operational lifespan of the transformer.

The work presented in this paper focuses on the analysis of the losses and economics of distribution transformers under balanced three-phase scenarios. In future work, the authors plan to conduct further research on the transformer adjustment strategy under three-phase unbalanced scenarios.

Author Contributions: Conceptualization, D.Z.; methodology, D.Z.; software, D.Z.; validation, D.Z., X.S. and H.Q.; formal analysis, X.S. and D.Z.; investigation, H.Q.; resources, Z.H. and W.D.; data curation, S.W.; writing—original draft preparation, D.Z.; writing—review and editing, D.Z., X.S., H.Q., Q.P. and J.Y.; visualization, D.Z., X.S. and J.Y.; supervision, D.Z. and H.Q.; project administration, Q.P. and S.W.; funding acquisition, D.Z. All authors have read and agreed to the published version of the manuscript.

Funding: This work was supported by the Electric Power Research Institute of Yunnan Power Grid Co., Ltd., China (No. YNKJXM20220009).

Data Availability Statement: Data are contained in the article.

Conflicts of Interest: Authors Dexu Zou, Qingjun Peng, Shan Wang, Weiju Dai and Zhihu Hong were employed by the company China Southern Power Grid Yunnan Power Grid Co., Ltd. The remaining authors declare that the research was conducted in the absence of any commercial or financial relationships that could be construed as a potential conflict of interest.

References

1. Zhang, W.; Jiang, J.; Li, B. Residual lifetime evaluation of power transformers based on data fusion and wiener model. *IEEE Trans. Power Deliv.* **2023**, *38*, 4189–4201. [CrossRef]
2. Yao, D.; Li, L.; Zhang, S.; Zhang, D.; Chen, D. The vibroacoustic characteristics analysis of transformer core faults based on multi-physical field coupling. *Symmetry* **2022**, *14*, 544. [CrossRef]
3. Wang, S.; Gao, M.; Zhuo, R. Research on high efficient order reduction algorithm for temperature coupling simulation model of transformer. *High Volt. Appar.* **2023**, *59*, 115–126.
4. Chen, T.; Chen, Y.; Li, X. Prediction for dissolved gas concentration in power transformer oil based on CEEMDAN-SG-BiLSTM. *High Volt. Appar.* **2023**, *59*, 168–175.
5. Zang, C.; Zeng, J.; Li, P. Intelligent diagnosis model of mechanical fault for power transformer based on SVM algorithm. *High Volt. Appar.* **2023**, *59*, 216–222.
6. Xia, Y.; Guo, C.; Chen, Y. Evaluation model of transformer overload capacity considering multi-factor. *J. Mech. Electr. Eng.* **2017**, *34*, 509–514.
7. Zheng, Y.; Hao, Z.; Xue, Z. Exploration of safe operation and active protection technology for large power transformers. *Autom. Electr. Power Syst.* **2023**, *40*, 1–12.
8. Wang, Q.; Guo, Y.; Mo, F. Optimal load distribution of urban high voltage distribution network considering short term emergency load of transformers. *Automat. Electron. Power Syst.* **2023**, *47*, 106–115.
9. Shi, M.; Xu, H.; Li, H. A graded warning model for oil immersed distribution transformers based on multimodal data fusion. *Guangdong Electr. Power* **2022**, *35*, 110–117.
10. Zhang, H.; Chen, X.; Li, Z. Study on thermal circuit model of ONAN transformer with auxiliary heat dissipation equipment. *Guangdong Electr. Power* **2022**, *35*, 95–103.
11. Wang, J.; Sheng, W.; Wang, L.; Yang, H. Study on technical and economical efficiency of amorphous alloy transformer and on-load capacity regulating transformer in distribution network application. In Proceedings of the 2014 China International Conference on Electricity Distribution (CICED), Shenzhen, China, 23–25 September 2014.
12. Yang, S.; Li, K. Study on synthesis loss reduction optimization method considering for rural distribution network planning. *J. Agric. Mech. Res.* **2015**, *37*, 250–253.
13. Wang, J.; Sheng, W. Simulation analysis of capacity regulating transformer. *Transformer* **2009**, *46*, 19–23.
14. Wang, J. Application and integrated economic analysis of on-load capacity regulating transformer. *High Volt. Appar.* **2009**, *45*, 32–35.
15. Yang, T.; Zhao, Y.; Jiang, Z. Study on the technology of non-contact on-load automatic capacity regulating of the distribution transformer in rural power network. *J. Agric. Mech. Res.* **2016**, *38*, 239–244.
16. Han, S.; Guo, X.; Tang, Q. Analysis and calculation of the best capacitive set-point of on-load adjustable capacitive transformer in operation. *Transformer* **2013**, *50*, 1–6.

17. Fan, W.; Han, S. A Control strategy for secure and economic operation of on-load capacity regulating transformer. *Automat. Electron. Power Syst.* **2011**, *35*, 98–102.
18. Meng, Z.; Piao, Z.; Wang, D. Design of switchover strategy and control system for arc-less on-load automatic capacity regulating distribution transformer. *Trans. Chin. Soc. Agric. Eng.* **2013**, *29*, 158–165.
19. Huo, J.; Zhao, Q.; Wang, X. Analysis and selection of optimal capacity adjustment point of on-load capacity regulating transformer. *Transformer* **2018**, *55*, 38–42.
20. Wang, R.; Zhao, W. A Fuzzy decision-based method to evaluate planning scheme for urban high voltage transmission network. *Power Syst. Technol.* **2013**, *37*, 488–492.
21. Wang, Q.; Zhao, Q.; Wang, X. Research on load capacity control strategy based on load forecasting. *Transformer* **2019**, *56*, 23–26.

symmetry

MDPI

Article

Small-Signal Modeling and Frequency Support Capacity Analysis of Power Load Considering Voltage Variation Effect

Tao Zhou [1,2], Yuxin Zheng [1], Cheng Wang [1,*], Lei Chen [2], Bo Liu [3] and Zhong Chen [3]

[1] School of Automation, Nanjing University of Science and Technology, Nanjing 210094, China; zhoutaonjust@njust.edu.cn (T.Z.); zhengyuxin@njust.edu.cn (Y.Z.)
[2] State Key Laboratory of Power System Operation and Control, Tsinghua University, Beijing 100084, China; chenlei08@tsinghua.edu.cn
[3] School of Electrical Engineering, Southeast University, Nanjing 210096, China; 230208789@seu.edu.cn (B.L.); zhongchen@seu.edu.cn (Z.C.)
* Correspondence: chw714@njust.edu.cn

Abstract: The frequency support capacity of power loads is essential for maintaining active power symmetry and balance between the generation and demand sides of power systems. As the proportion of renewable energy sources and power electronic equipment increases, the inertia on the power generation side decreases, highlighting the growing importance of frequency support on the load side. As it is generally believed that the active power balance of power systems determines the frequency stability, few studies have considered the effect of voltage variation on the frequency response dynamics. It is important to note that the node voltage keeps fluctuating throughout the frequency dynamic process, which affects the active power of loads and should not be neglected. Based on the aforementioned rationales, this paper endeavors to investigate the modeling of power load frequency response and analyze its support capability considering the voltage variation effect. This paper initially establishes the small-signal model of dynamic load under frequency dynamics, derives the transfer function relating active power to system frequency deviation, and subsequently develops its frequency response model. Subsequently, commencing with the ZIP model of static load, the power fluctuation of load nodes is derived from the influence of preceding nodes, and the frequency response model of the static load is formulated and its frequency support capacity is scrutinized. Based on this foundation, a comprehensive aggregation model of the complex load is constructed, and its frequency support capability is assessed using actual data. Finally, the proposed model and analysis results are validated through simulation, confirming their correctness and effectiveness.

Keywords: active power symmetry; frequency response modeling; dynamic load; static load; voltage variation effect

1. Introduction

The symmetry between the generation and load sides of power systems plays a crucial role in determining the balance and stability of the overall system [1]. The power load comprises both dynamic and static loads, with their frequency support capabilities playing a pivotal role in maintaining the active power symmetry of power systems [2]. Due to the increasing share of renewable energy and the widespread use of large-scale power electronic equipment, the inertia level in new power systems is on a declining trend, posing a serious threat to their safe operations [3–7]. In comparison with traditional power systems, there has been a gradual decrease in inertia on the generation side and an increase on the load side [8]. Further research is needed to explore the system's frequency support capabilities within the context of decreasing inertia levels [9]. However, in comparison to traditional synchronous systems, this new system encounters two significant challenges [10,11]. Firstly, there is a pronounced fluctuation and intermittency in the output of renewable energy sources. Secondly, a significant number of wind power and photovoltaic systems, which lack rotational inertia, are replacing synchronous generators [12]. As a result, the frequency stability of the new power system is severely compromised. In comparison to the conventional power system, the frequency response capability of the load side should not be overlooked [13,14]. Therefore, it is crucial to investigate the frequency response capability and response principles of the power load to ensure the symmetry and stability of modern power systems [15,16].

Within the framework of load, two frequently used models are the static load model and the dynamic model. The static load model encompasses the constant impedance, constant current, and constant power load (ZIP) model [17]. Previous research has predominantly focused on investigating the correlation between load frequency and active power, with limited attention given to the impact of voltage fluctuation on frequency response dynamics. It is important to acknowledge that the node voltage undergoes fluctuations throughout the dynamic frequency process, thereby influencing the active power of the load and should not be disregarded.

In [18], in order to determine the parameters of the load model, a time-varying comprehensive load model was established. However, this paper does not focus on the relationship between system frequency and load response and does not focus on the coupling relationship between frequency, voltage, and power. In [19], derived from the fifth-order model of the induction motor, the equivalent conduction function of frequency and active power is constructed, which represents the load model's dynamic characteristics. However, this paper assumes that the terminal voltage of the induction machine is constant and does not consider that the voltage change also affects the frequency response of the induction machine load. It is very important to construct a comprehensive mathematical model for the load system frequency response, considering voltage variation. The static load model mainly reflects the frequency characteristics of the load side resistance, inductance, and capacitance of the electrical equipment and depicts the relationship among the power, voltage, and frequency of this type of electrical equipment. The authors in [20] investigated how the static load frequency response model affects the stability of the Shanghai power grid. Meanwhile, the authors in [21] thoroughly analyzed the frequency characteristics of the static load model and developed a simulation model using the PSD-BPA software platform, considering the load composition and proportion of the Yunnan power grid. However, these studies mainly focus on the frequency characteristics of the static load, without incorporating small-signal modeling for the static load during small disturbances. Moreover, the voltage at the load node also changes due to system frequency, intensifying the frequency response of the static load. In contrast, the authors in [22] take into account the voltage fluctuation of the static load and analyze its frequency response under small disturbances. However, it does not consider the perspective of the entire power grid, and the voltage change during frequency disturbances remains uncertain.

To address the above issues, this paper aims to establish a comprehensive load model, including static load and dynamic load considering the voltage variation, and its overall frequency supporting capability is analyzed. The main contributions of this paper are outlined below.

This paper proposes a small-signal model of dynamic load under frequency dynamics. Furthermore, the transfer function is derived for the deviation between active power and system frequency, constructing a dynamic load frequency response model and evaluating its capability to support system frequency.

Based on the ZIP model of static load, a frequency response model is established and tested for its ability to support varying frequencies. The frequency response model of the static load encompasses changes in node voltage, effectively capturing the frequency response capability of the static load.

This paper develops a comprehensive load frequency response model incorporating dynamic and static load. The model considers fluctuations in node voltage to more precisely characterize the load behavior. Simulation results demonstrate that, in comparison with the previous load model, the power curve generated by the proposed model closely aligns with the actual curve.

The remainder of this paper is structured as follows: A frequency response model for dynamic load is established in Section 2. A static load frequency response model is developed in Section 3, taking into account variations in node voltage. A comprehensive load frequency response model is developed in Section 4, which incorporates both dynamic and static loads and considers the fluctuation of node voltage to provide a more precise characterization of load behavior. In Section 5, the proposed comprehensive model considering voltage variation and frequency support capability analysis results are verified through simulations using MATLAB, PSASP, and DIgSILENT. Finally, Section 6 concludes this paper.

2. Dynamic Load Modeling and Analysis

The primary sources of dynamic load in power systems are induction machines, which are extensively utilized in textile, machinery, and other industrial applications. Unlike synchronous machines, induction machines exhibit a slip between their speeds and the grid frequency [23]. Nevertheless, they still contribute to the frequency response and provide support for the power grid frequency, serving as the principal resource for inertia support on the load side. On this basis, this section establishes the small-signal model of the induction machine, which includes both the power-frequency and power-voltage balancing. Utilizing the proposed model, the frequency support capability of the dynamics is analyzed.

2.1. Model Derivation

The power system model is shown in Figure 1, where the induction machine is connected to the generation through transmission lines. To focus on the frequency support capability of the induction machine, the generation is simplified to the infinite power supply. When the system's active power balancing is disturbed, the system frequency will change, which also affects the system power flow [24,25]. Therefore, the terminal voltage of the induction machine also changes, which can be obtained as follows:

$$U_t = \sqrt{(U_n - \frac{P_{em}R_l + Q_{em}X_l}{U_n})^2 + (\frac{P_{em}X_l + Q_{em}R_l}{U_n})^2} \tag{1}$$

where U_t denotes the terminal voltage of the induction machine; P_{em} and Q_{em} are the active power and reactive power of induction machine; the R_l and X_l denote the resistance and reactance of induction machine; and U_n denotes the terminal voltage of the generation.

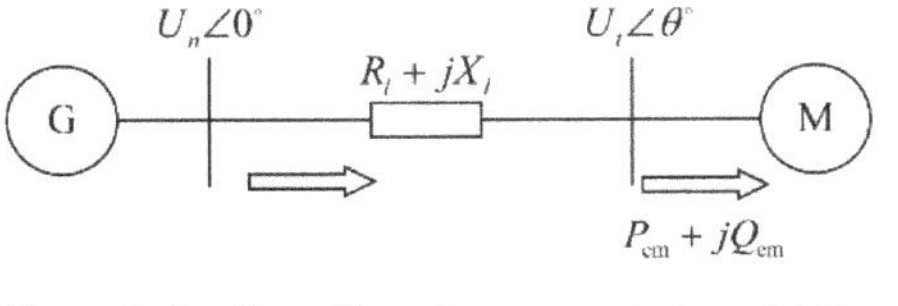

Figure 1. Configuration of power system containing dynamic load.

In the frequency dynamics of the induction machine, the parameters on the rotor side can be transformed to the stator side. The mechanical transient circuit of the stator and rotor windings of the induction machine is illustrated in Figure 2a. Since the mutual inductance x_m of the fixed rotor is much larger than the leakage reactance x_r of the rotor winding in practical applications, the mechanical transient equivalent circuit of the induction machine is depicted in Figure 2b with x_m neglected.

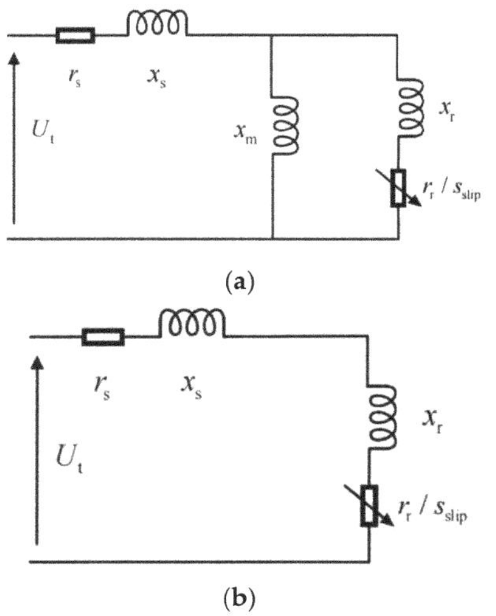

Figure 2. Induction motor mechanical transient time circuit. (**a**) Induction motor mechanical transient equivalent circuit. (**b**) Equivalent circuit of induction motor.

In Figure 2, the parameters r_s and x_s represent the equivalent resistance and leakage reactance of the stator windings, while r_r and x_r denote the equivalent resistance and leakage reactance of the rotor windings. Additionally, x_m signifies the mutual inductance of the stationary rotor, and s_{slip} denotes the slip rate of an induction motor.

The induction machine model under frequency dynamics is widely used in power system modeling, which is shown below [26].

$$\begin{cases} P_{em} = \frac{U_t^2}{(x_s+x_r)^2+(R_s+\frac{R_r}{s_{slip}})^2}(R_s+\frac{R_r}{s_{slip}}) \\ Q_{em} = \frac{U_t^2}{(x_s+x_r)^2+(R_s+\frac{R_r}{s_{slip}})^2}(x_s+x_r) \end{cases} \tag{2}$$

$$P_{mm} = k\omega_r\left[\alpha + (1-\alpha)\left(1-s_{slip}\right)^{\rho}\right] \tag{3}$$

$$2H_{am}\frac{d\omega_r}{dt} = \Delta P_{em} - \Delta P_{mm} \tag{4}$$

$$s_{slip} = \frac{\omega - \omega_r}{\omega_r} \tag{5}$$

where x and R denote the induction machine reactance and resistance; the subscript s and r represent the stator winding and rotor winding of induction machine; P_{mm} represents the mechanical power of an induction machine; S_{slip} and H_{am} denote the slip and inertia of the induction machine; α represents the constant torque component; k represents the load factor; ρ represents the mechanical characteristics of induction machine, which is exponential; and ω and ω_r are the angular speed of power system and induction machine.

In practical power systems, U_n is not an ideal infinite voltage source, but varies due to the active power disturbance. U_n and f are inputs, and P_e, consumed by the induction motor, is output. According to the small-signal method, Equations (1)–(5) can be linearized to obtain the following equations:

$$\Delta U_t = K_1\Delta P_{em} + K_2\Delta Q_{em} + K_v\Delta U_n \tag{6}$$

$$\begin{cases} \Delta P_{em} = K_3 \Delta U_t + K_4 \Delta s_{slip} \\ \Delta Q_{em} = K_5 \Delta U_t + K_6 \Delta s_{slip} \end{cases} \tag{7}$$

$$\Delta P_{mm} = K_7 \Delta s_{slip} + K_8 \Delta \omega_r \tag{8}$$

$$\Delta s_{slip} = K_9 \Delta f + K_{10} \Delta \omega_r \tag{9}$$

$$2H_{am} s \Delta \omega_r = \Delta P_{em} - \Delta P_{mm} \tag{10}$$

$$\begin{cases} K_1 = \dfrac{-R_l(U_n^2 - P_{em0}R_l - Q_{em}X_l) + X_l(P_{em0}X_l + Q_{em}R_l)}{U_n^2\sqrt{\left(U_n - \frac{P_{em0}R_l + Q_{em}X_l}{U_n}\right)^2 + \left(\frac{P_{em0}X_l + Q_{em}R_l}{U_n}\right)^2}} \\ K_2 = \dfrac{-X_l(U_n^2 - P_{em}R_l - Q_{em0}X_l) + R_l(P_{em}X_l + Q_{em0}R_l)}{U_n^2\sqrt{\left(U_n - \frac{P_{em}R_l + Q_{em0}X_l}{U_n}\right)^2 + \left(\frac{P_{em}X_l + Q_{em0}R_l}{U_n}\right)^2}} \\ K_v = \dfrac{U_{n0}^2(P_{em}R_l + Q_{em}X_l - 1)\left[(P_{em}R_l + Q_{em}X_l) - U_{n0}^2\right] - (P_{em}X_l + Q_{em}R_l)^2}{U_{n0}^3\sqrt{(U_{n0} - \frac{P_{em0}R_l + Q_{em}X_l}{U_{n0}})^2 + (\frac{P_{em0}X_l + Q_{em}R_l}{U_{n0}})^2}} \end{cases} \tag{11}$$

$$\begin{cases} K_3 = \dfrac{2U_{t0}}{\left(r_s + \frac{r_r}{s_{slip0}}\right)^2 + (x_s + x_r)^2}\left(\dfrac{r_r}{s_{slip0}} + r_s\right) \\ K_4 = \dfrac{\left[(x_s + x_r)^2 - \left(r_s + \frac{r_r}{s_{slip0}}\right)^2\right]U_t^2\left(-\frac{r_r}{s_{slip0}^2}\right)}{\left[\left(r_s + \frac{r_r}{s_{slip0}}\right)^2 + (x_s + x_r)^2\right]^2} \\ K_5 = \dfrac{2U_{t0}}{\left(r_s + \frac{r_r}{s_{slip0}}\right)^2 + (x_s + x_r)^2}(x_s + x_r) \\ K_6 = \dfrac{(x_s + x_r)\left(r_s + \frac{r_r}{s_{slip0}}\right)U_t^2\left(-\frac{r_r}{s_{slip0}^2}\right)}{\left[\left(r_s + \frac{r_r}{s_{slip0}}\right)^2 + (x_s + x_r)^2\right]^2} \end{cases} \tag{12}$$

$$\begin{cases} K_7 = k\omega_{r0}\rho(1 - \alpha)\left(1 - s_{slip0}\right)^{\rho - 1} \\ K_8 = k\left[\alpha + (1 - \alpha)\left(1 - s_{slip0}\right)^{\rho}\right] \end{cases} \tag{13}$$

$$\begin{cases} K_9 = \frac{1}{\omega_r} \\ K_{10} = \frac{-\omega}{\omega_{r0}^2} \end{cases} \tag{14}$$

where s denotes the differential operator; K_1–K_{10} and K_v denote the coefficient without s; P_{em0} and Q_{em0} represent the initial values of electromagnetic power and reactive power of the induction machine, respectively; and U_{t0} and U_{n0} represent the initial values of the induction machine terminal voltage and the generation terminal voltage, respectively.

In the system frequency response model, it is essential to establish the relationship between active power and system frequency, which is also called the frequency characteristics. According to the Equations (7)–(10), the induction model can be established as shown in Figure 3.

As a typical dynamic load, the dynamic characteristics of an induction machine significantly impact the stability of the power system. As depicted in Figure 3, the grid-connected induction machine exhibits dynamic voltage and frequency balance. During a disturbance in the active power, the system frequency is disrupted and undergoes changes, which further affect the slip rate of the induction machine. As in Equations (8) and (10), the variations in the slip affect both the active and mechanical powers of the induction machine. Equation (5) describes how these changes disrupt the power dynamics equilibrium of the induction machine. Equation (9) further elucidates that alterations in slip also influence reactive power variations. The fluctuations in both active and reactive powers of the induction machine subsequently impact changes in system power flow. Due to the variation in the front-end node voltage, the constant voltage of the induction machine node is no longer maintained, thereby amplifying the frequency response of the induction machine [27].

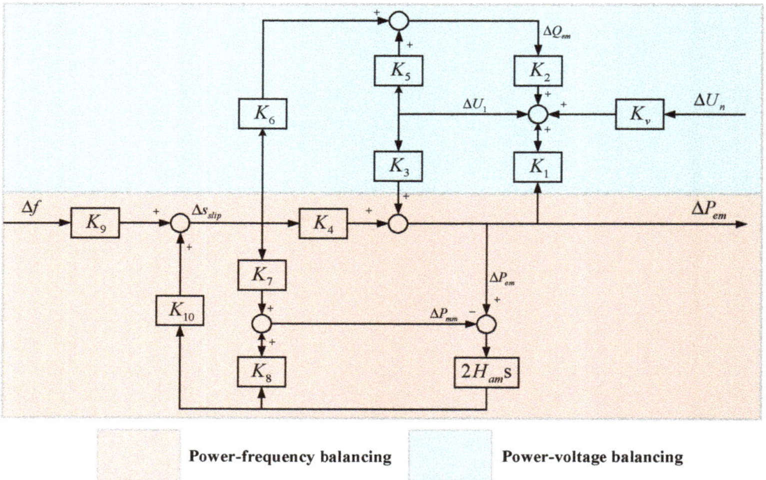

Figure 3. Block diagram illustrating the small-signal model of the induction machine.

2.2. Analysis of Frequency Support Capability of Dynamic Load

In response to system frequency disturbances, the slip rate of the induction machine undergoes instantaneous changes, leading to an immediate adjustment in active power to counteract the differential power of the system. This alteration in active power disrupts the equilibrium of the rotor motion equation of the induction machine [28]. According to the proposed model, both system frequency and node voltage are identified as key factors influencing the active power of the induction machine. The relationship between them and active power is depicted in Figures 4 and 5.

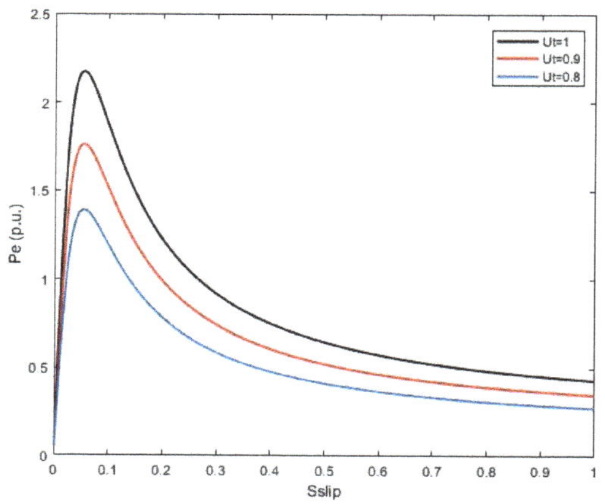

Figure 4. The relation between slip and active power.

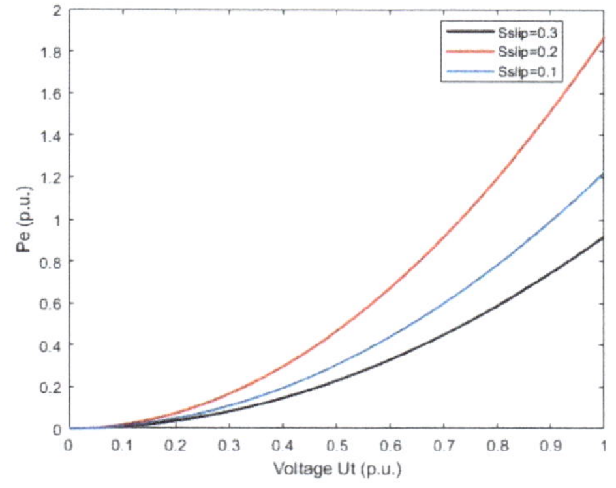

Figure 5. The relationship between voltage and active power at different slip rates.

As the slip of the induction machine increases, the active power initially rises and then declines. The extremum lies within the range of $s_{slip} = [0.055, 0.06]$. Under normal circumstances, the induction machine's steady-state slip is between 0.001 and 0.01. Assuming a positive disturbance in the active power of the system, there is a drop in the system frequency. The speed of the induction machine also decreases, albeit at a slower rate than that of the system frequency. As the power system's frequency decreases, active power from the induction machine is shed, demonstrating its ability to support the system frequency.

The node voltage also affects the active power consumed. According to Equation (2), the relationship between active power and node voltage is expressed as a quadratic function. As the node voltage increases, the active power of the induction machine increases. The reactive power of the induction motor decreases at the same time.

The disruption of the balance of reactive power and voltage in the system leads to a decrease in the system's reactive power level and node voltage. Simultaneously, the reduction in power consumed by the system lines results in a decrease in line voltage drop. Figure 6 illustrates that changes in the reactive power have a greater impact on system voltage than changes in active power during system power flow. Significant variations in the reactive load force cause an increase in the node voltage for induction machines, leading to contradictory processes and minimal voltage fluctuations at the induction machine nodes, ranging from 0 to 0.01 p.u.

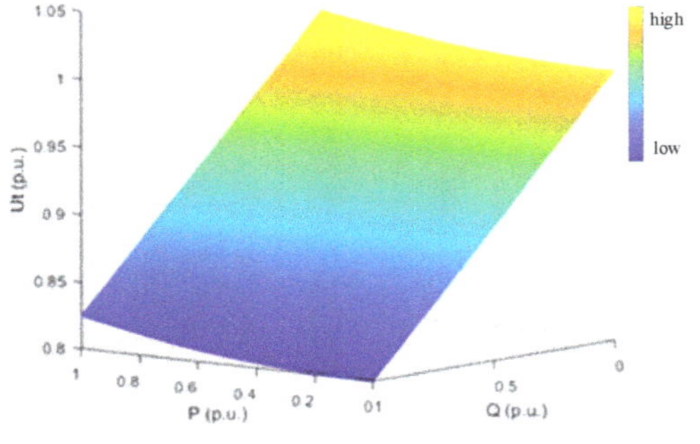

Figure 6. The relationship between reactive power, active power, and node voltage.

3. Static Load Modeling and Analysis

3.1. Model Derivation

Similar to the topological structure in Figure 1, the system, which replaces the dynamic load with static load, is a static load grid-connected model. The ZIP model is always used in simulations of power system loads [29]. The nonlinear mathematical formulas of the ZIP models are shown as follows:

$$\begin{cases} P_{es} = P_n[p_1(\frac{U_t}{U_0})^2 + p_2(\frac{U_t}{U_0}) + p_3](1 + k_{pZIP}\Delta f) \\ Q_{es} = Q_n[q_1(\frac{U_t}{U_0})^2 + q_2(\frac{U_t}{U_0}) + q_3](1 + k_{qZIP}\Delta f) \end{cases} \tag{15}$$

$$\begin{cases} p_1 + p_2 + p_3 = 1 \\ q_1 + q_2 + q_3 = 1 \end{cases} \tag{16}$$

where P_{es} and Q_{es} represent the active power and reactive power; P_n, Q_n and U_0 represent the nominal active and reactive power and the nominal voltage of the load; p_1, p_2, and p_3 represent the participation coefficients of ZIP, respectively, in the active power consumed by the static load; q_1, q_2, and q_3 represent the participation coefficients of ZIP, respectively, in the reactive power consumed by the static load; k_{pZIP} and k_{qZIP} denote the frequency participation for the cases of active and reactive power; and Δf represents the rate of frequency change.

Similarly, in the actual power system, the front-end node voltage of the load will also change under the influence of power disturbance. To investigate the frequency support capability of the static load, the small-signal method is employed to linearize Equations (15) and (16). The linearization results are shown below:

$$\begin{cases} \Delta P_{es} = K_{11}\Delta U_t + K_{12}\Delta f \\ \Delta Q_{es} = K_{13}\Delta U_t + K_{14}\Delta f \end{cases} \tag{17}$$

$$\begin{cases} K_{11} = P_n\left[2p_1\left(\frac{U_{t0}}{U_0}\right) + \frac{p_2}{U_0}\right](1 + k_{pzip}f_0) \\ K_{12} = P_n\left[p_1\left(\frac{U_{t0}}{U_0}\right)^2 + p_2\left(\frac{U_{t0}}{U_0}\right) + p_3\right]\cdot k_{pzip} \\ K_{13} = Q_n\left[2q_1\left(\frac{U_{t0}}{U_0}\right) + \frac{q_2}{U_0}\right](1 + k_{qzip}f_0) \\ K_{14} = Q_n\left[q_1\left(\frac{U_{t0}}{U_0}\right)^2 + q_2\left(\frac{U_{t0}}{U_0}\right) + q_3\right]\cdot k_{qzip} \end{cases} \tag{18}$$

where K_{11}–K_{14} denote the coefficient without s. According to the proposed Equations (17) and (18), the load model can be established as shown in Figure 7.

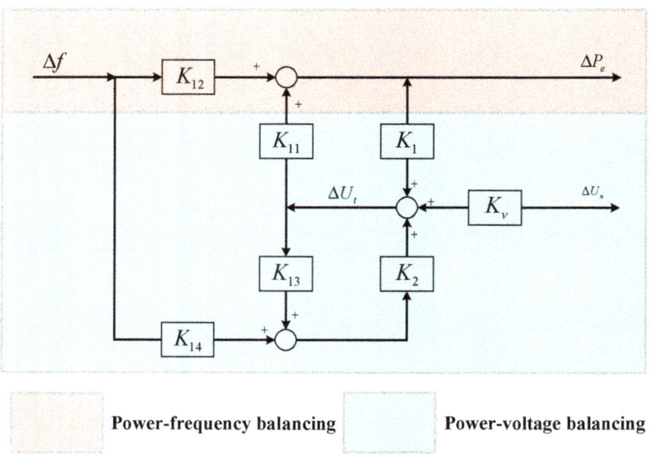

Figure 7. Block diagram illustrating the small-signal model of static load.

As shown in Figure 6, similar to the dynamic machine, the grid-connected static load also includes the dynamic voltage balance and dynamic frequency balance. According to Equations (13) and (14), the variation in system frequency is proportional to the active power consumed by the static load. Upon the occurrence of a frequency disturbance in the system, the frequency drives the static load power change and then causes the system power flow change, which leads to the node voltage change. This dynamic process does not end until a new steady state is reached.

3.2. Analysis of Frequency Support Capability of Static Load

The static load is responsible for adjusting the active power output in response to deviations in system frequency. It exhibits prompt reactivity during frequency dynamics, although it is also subject to significant influence from node voltage. Any changes in voltage at the load node that occur in the opposite direction have the potential to exacerbate the differential power within the system.

According to the proposed model, the system frequency and node voltage are the key factors affecting the active power of the static load. The relationship between them and active power is shown in Figures 8 and 9, with different voltages.

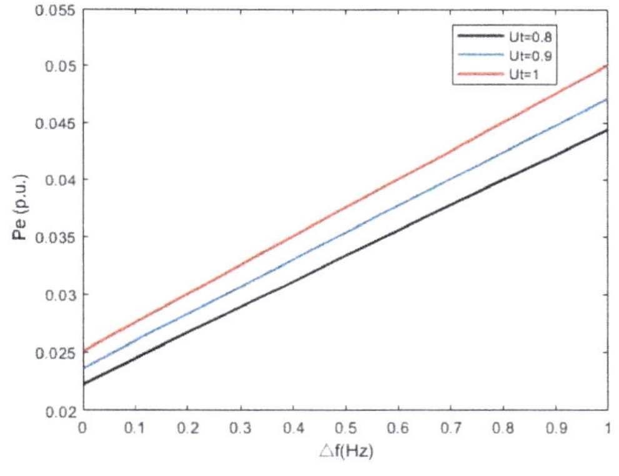

Figure 8. The relation between frequency and active power.

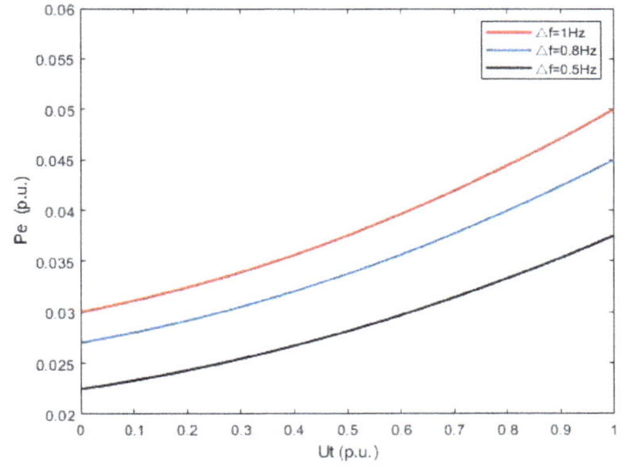

Figure 9. The relationship between voltage and active power under different frequency deviation.

As shown in Figure 8, the proportionality coefficient k_{pZIP} is positive. When the variation in the system frequency is positive and increasing, the active power absorbed by the static load demonstrates a concurrent increase. Assuming the active power disturbance of the system is positive, the system frequency decreases. The active power consumed by the static load will also decrease. However, due to the small value of the proportional coefficient k_{pZIP} and the range of frequency variation in the system is usually between 0 and 0.5 Hz, the range of variation in the active power influenced by the frequency is also very small.

As shown in Figure 9, due to the presence of constant resistance, the voltage has a quadratic relationship with the active power drawn by the static load. As the node voltage rises, the active power is increased by the static load. Similar to the dynamic load, when the system experiences a disturbance in the active power and the system frequency changes, the node voltage of the static load also has a dynamic process. Assuming that the active power disturbance of the system is positive, the system frequency drops. The active and reactive power utilized by the static load will also decrease. The reactive power balance of the system is broken, resulting in a decrease in the total reactive power level and node voltage. At the same time, as the voltage loss on the line decreases, the node voltage level will also increase. The two contradictory dynamic processes continue until a new steady state occurs. Since the proportional coefficient k_{pZIP} of static load is small and the frequency change is small, the power change in the static load is mainly affected by the node voltage. As for the supporting effect of the system frequency, the static load itself has a supporting effect on the frequency stability. But due to the uncertain change in node voltage, its supporting effect is not obvious compared with the dynamic load. In the actual power system, the node voltage has a variety of voltage regulation methods, including: forward and reverse voltage regulation. The voltage regulating operation will also directly affect the node load state.

4. Frequency Support Capacity Analysis of Complex Load

4.1. Comprehensive Modeling

The node load of a power system encompasses both dynamic and static loads. The active power of the complex load represents the aggregate active power, encompassing both dynamic and static loads, as does the reactive power. Its mathematical expression is shown in (19) and (20):

$$\Delta P_e = \sum_{i=1}^{n} \Delta P_{em,i} + \sum_{j=1}^{m} \Delta P_{es,j} \tag{19}$$

$$\Delta Q_e = \sum_{i=1}^{n} \Delta Q_{em,i} + \sum_{j=1}^{m} \Delta Q_{es,j} \tag{20}$$

where ΔP_e and ΔQ_e denote the active and reactive powers of the complex load; n and m denote the total number of dynamic loads and static loads. Based on the same principle as (1), the voltage changes in this node can be determined through the power flow calculation as follows:

$$U_t = \sqrt{(U_n - \frac{P_e R_l + Q_e X_l}{U_n})^2 + (\frac{P_e X_l + Q_e R_l}{U_n})^2} \tag{21}$$

As the calculated values are all nominal unit values, each load needs to have a proportional coefficient to represent the load capacity of the complex load. The proportional coefficient is shown as follows:

$$K_i = \frac{s_i}{\sum_{i=1}^{n} s_i + \sum_{j=1}^{m} s_j} \tag{22}$$

$$K_j = \frac{s_j}{\sum_{i=1}^{n} s_i + \sum_{j=1}^{m} s_j} \tag{23}$$

where K_i and K_j denote the proportional coefficients of the dynamic load i and static load j and s_i and s_j denote the apparent power of the dynamic load i and static load j.

Based on (19) and (21), the configuration of the load model is depicted in Figure 10. Each load has two inputs: node voltage and system frequency. The number of dynamic loads is n, and the number of static loads is m. The output of the complex load comprises both active power and reactive power. It is important to note that the output power must be multiplied by the proportional factor K_i or K_j for each load. The active and reactive powers of the node are the aggregate values of n dynamic loads and m static loads. Since the power sum has been multiplied by the proportional factor K_i or K_j, there is no error in the calculation of the identity value. The sum of the active and reactive powers of the nodes is divided into the node voltage dynamic balance, which makes the nodal voltage calculation more accurate. Finally, the voltage U_t is fed back into the load transfer function, which forms the dynamic balance of power, frequency, and voltage. The complex load has both dynamic and static load characteristics, which can better reflect node load fluctuations. This further facilitates the analysis of the load's frequency support capability.

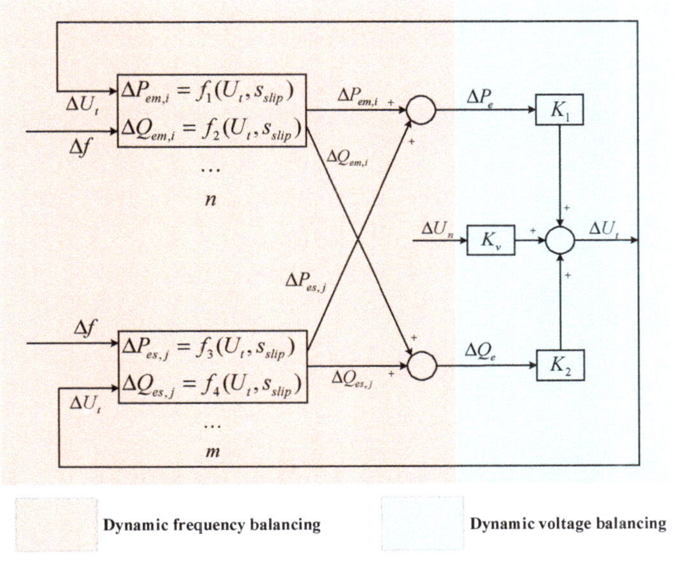

Figure 10. The model of complex load.

4.2. Parameter Influence Analysis

Some parameters indirectly impact the dynamic balance between active power and frequency, influencing the balance process. Taking an induction machine as an example, both its inertia and rotational speed significantly affect the power-frequency dynamic process. The rotational speed of an induction machine directly affects the amount of kinetic energy stored in its rotor, thereby modifying the machine's inertia. This inertia plays a crucial role in preventing sudden changes in system frequency and maintaining system stability. Additionally, it can enhance overall system inertia levels and improve frequency stability [19].

The distribution of each component within the ZIP also impacts the frequency response capability. The constant power component lacks the frequency response capability, as its active power remains unchanged regardless of variations in system frequency and voltage. Consequently, a higher proportion of constant power results in a weaker frequency response capability. Additionally, there exists a quadratic relationship between constant impedance loads and node voltage, leading to the power variation in constant impedance loads being more dependent on changes in node voltage.

Table 1 shows the node load components of the U.S. Eastern Interconnection Grid. As a dynamic load has a better frequency support capability than a static load, the higher the proportion of dynamic loads, the better the frequency response capability of the node.

Table 1. The node load components of the U.S. Eastern Interconnection Grid.

Node	Dynamic Load (%)	Constant Impedance (%)	Constant Current (%)	Constant Power (%)	Remaining Load (%)
MISO Vectren	85.4	0	2.16	11.34	1.1
MISO ATC	61.775	18.6	1	4.425	14.2
West MA	61.46	10.2	0	14.43	13.91

4.3. Overall Framework

Based on the preceding sections, this paper presents a comprehensive load frequency response model that incorporates the impacts of voltage fluctuations and evaluates its frequency support capability. Firstly, the transfer function of power consumption and frequency deviation of the induction machine are derived based on the proposed small-signal model. The frequency support capability of the induction machine is then examined, with an analysis of its response characteristics and parameter influence mechanism at various stages. Furthermore, a static load frequency response model is developed to account for voltage variations, thereby accurately capturing the frequency response capabilities. Based on this foundation, a comprehensive load frequency response model is established to incorporate the impact of voltage fluctuations. This model is integrated into the traditional system frequency response model, and validation and parameter analysis are conducted to assess the load frequency support capacity across different proportions and types, as well as to examine the influence mechanism of parameters. The overall framework of this paper is presented in Figure 11.

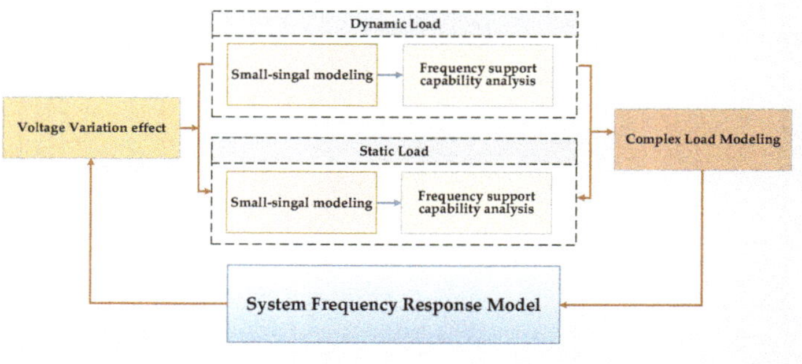

Figure 11. Overall framework.

5. Case Study and Analysis

In this section, the comprehensive load model proposed in this paper is tested and verified through simulations and analysis using MATLAB/SIMULINK and PSASP. The proposed model in the SFR system is implemented and validated using MATLAB, with corresponding practical examples established with PSASP for comparative analysis and verification. To demonstrate the practicality and viability of this model in power systems with actual frequency characteristics, simulations and analyses are conducted using the IEEE 39-node system on the DIgSILENT PowerFactory platform.

5.1. Model Validation

(1) Dynamic load model and static load model

The proposed model is firstly verified using MATLAB/SIMULINK and PSASP using the power system model in Figure 1. Parameters of dynamic load and static load are shown in Tables 2 and 3, respectively. The simulation duration is 50 s, and the system synchronous generator is equipped with primary frequency modulation equipment. The output quantity is system frequency f, node voltage U_t, and dynamic load active power offset P_e. In the steady-state condition, $\Delta f = 0$, $\Delta U_t = 0$, and $\Delta P_e = 0$. At $t = 0$, a disturbance of $\Delta P_d = 0.05$ p.u. is applied, and the changes in system frequency, voltage, and active power of load are shown in Figures 12–14. The dynamic load initial point is $f = 50$ Hz, $U_t = 1.016$ p.u., and $P_{es} = 1.025$ p.u.

Table 2. The parameter of dynamic load.

Dynamic Load Parameter	H_{am}	ω_{r0}	x_s	x_r	r_s
parameter values	0.5	0.973	0.11	0.12	0.011

Table 3. The parameter of static load.

Static Load Parameter	P_{ns}	k_{pZIP}	p_1	p_2	p_3	Q_{ns}	k_{qZIP}	q_1	q_2	q_3
per-unit value	1.025	1	0.2	0.2	0.6	0.94	1	0.2	0.2	0.6

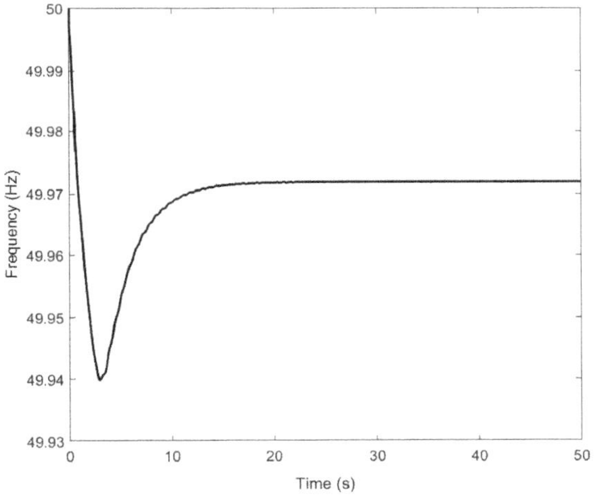

Figure 12. The curve of system frequency considering dynamic load.

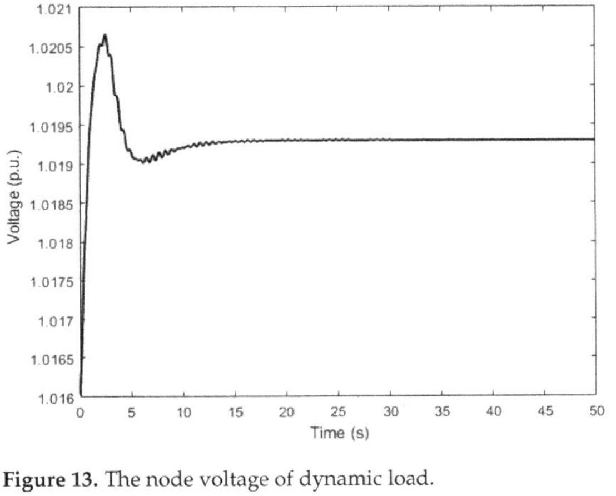

Figure 13. The node voltage of dynamic load.

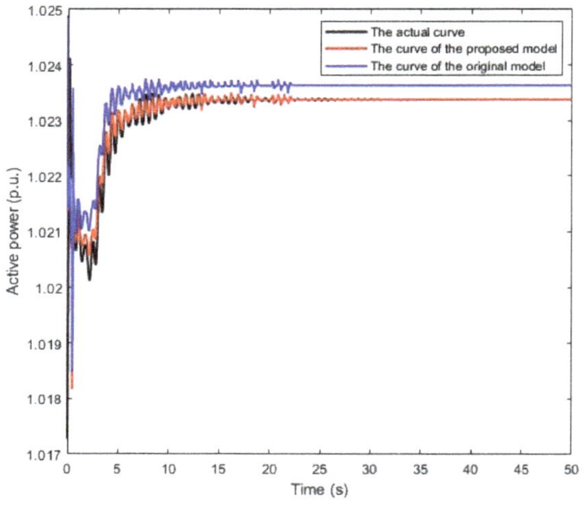

Figure 14. The curves of active power under different models.

Figure 12 shows that when the system experiences a disturbance in the active power, the system frequency first drops to the nadir point and then recovers to a new steady state $f_{nadir} = 49.94$ Hz. As illustrated in Figure 13, the node voltage of the induction machine increases immediately after the frequency changes. The reactive power of the induction machine exhibits a rapid decrease, so that the voltage lost on the line also decreases, leading to the rapid increase in node voltage. The figure illustrates that the node voltage has a small change and the overall change range is within 4.5×10^{-3} p.u.

At the beginning stage of the frequency disturbance, the active power of the induction machine experiences a swift decline, followed by a gradual dissipation of the stored kinetic energy in the rotor, and the active power rises, as shown in Figure 13. This verifies the theory proposed in Section 2.2. Figure 14 has three curves: the actual power curve, the curve under the proposed model, and the curve of the original model. The figure indicates that the curve waveform of the model proposed in this paper aligns with the actual curve and exhibits a higher degree of fitting. This study diverges from traditional approaches that primarily focus on the direct impact of the frequency deviations on the active power balance by incorporating the influence of the voltage variations on the load dynamics. It is observed that voltage fluctuations play a significant role in affecting the active power of loads during frequency dynamics. Therefore, the proposed model in this paper is validated with its precision and reliability. On this basis, a comprehensive analysis of the parameter influence of complex load can be conducted.

Based on the simulation analysis results presented above, it is evident that the proposed model curve aligns well with the actual curve, demonstrating a high degree of fitting in relation to load voltage variations. In contrast to traditional methods, which primarily consider the direct impact of frequency deviation on the active power balance, our proposed model incorporates the influence of voltage fluctuations on load dynamics. It is important to note that in the frequency dynamic process, voltage fluctuations significantly impact the active power of the load.

To verify the frequency response of the static load, the system load is changed to a static load and the same active power disturbance is applied. Figures 15–17 show the system frequency variation, node voltage variation, and static load power variation. The static load initial point is $f = 50$Hz, $U_t = 1.017$ p.u., and $P_{es} = 1.025$ p.u.

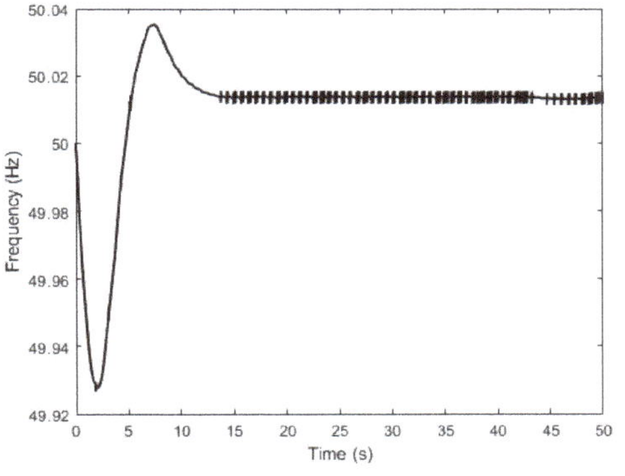

Figure 15. The curve of system frequency considering static load.

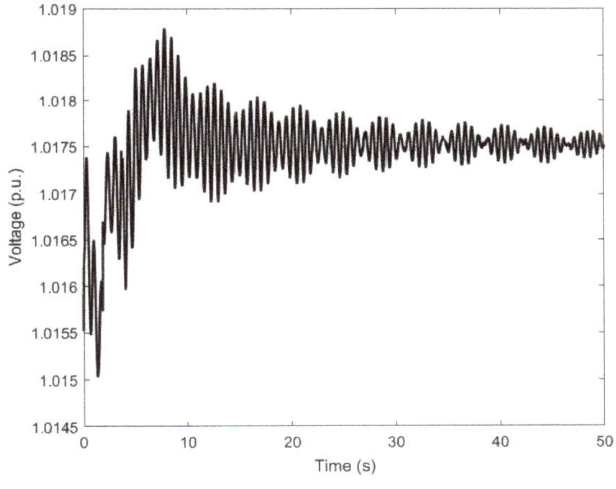

Figure 16. The node voltage of static load.

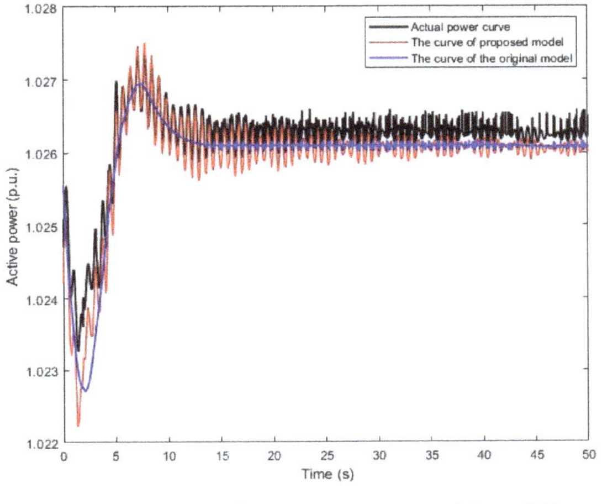

Figure 17. Comparison of active power curves of three different models.

As depicted in Figure 15, the system frequency drops first and then recovers to a new steady state. The nadir point of the system frequency is $f_{nadir} = 49.928$Hz. Hence, when the system frequency alters, the anti-interference power of the system carrying the dynamic load is much greater than that of the static load. Due to the load-shedding device and the primary frequency modulation device, the system frequency settles into a new steady state.

As illustrated in Figure 16, the fluctuation in node voltage is opposite to that of the dynamic load. When the static load voltage is disturbed by frequency, the voltage decreases first and then increases. The general trend of the voltage variation is consistent with that of the frequency. The voltage variation is not only due to the reduction in the reactive power level of the system but also to the action of the synchronous generator itself. Under the dual action of frequency and voltage, the active power of the static load decreases first and then increases, as shown in Figure 17. The frequency response speed of the static load is the same as that of the dynamic load, but it is mainly passive change according to the change in frequency and voltage. It cannot provide stored kinetic energy to prevent a frequency disturbance as dynamic loads do, and it does not have the active support ability of the frequency response. However, due to voltage and frequency changes, active power changes in the static loads can prevent the aggravation of system frequency deterioration.

Figure 17 contains three curves: the actual power curve, the power curve of the proposed model, and the power curve of the original model. The figure demonstrates that the power curve of the model proposed in this paper aligns closely with the actual power curve waveform, indicating a higher degree of fitting. The power curve of the original model, which is proportional to the frequency change, cannot reflect the actual change in the static load more effectively. Therefore, Figure 17 verifies the accuracy of the model proposed in Section 3.1.

(2) The complex model

For consideration of complex models containing dynamic loads and static loads, the analysis focuses on the node MISO ATC in the U.S. Eastern Interconnection Grid, as shown in Table 1. The simulation duration is 100 s. At $t = 25$ s, an active power disturbance of $\Delta P_d = 0.025$ p.u. is applied to the system. The complex load initial point is $f = 50$ Hz, $U_t = 0.9457$ p.u., and $P_e = 7.335$ p.u. Figures 18–20 show the system frequency variation, node voltage variation, and load power variation.

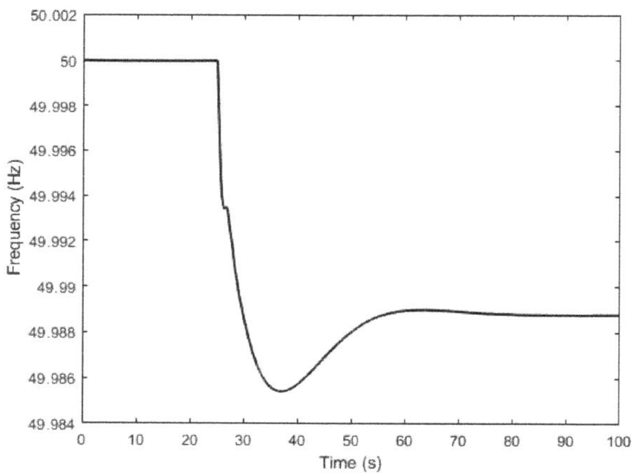

Figure 18. The curve of system frequency considering comprehensive load.

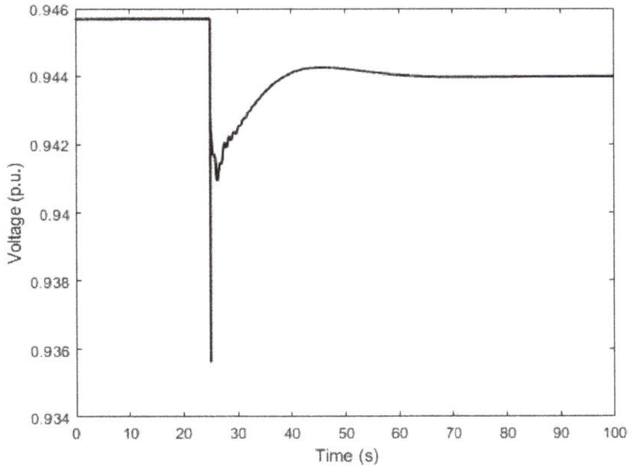

Figure 19. The node voltage of complex load.

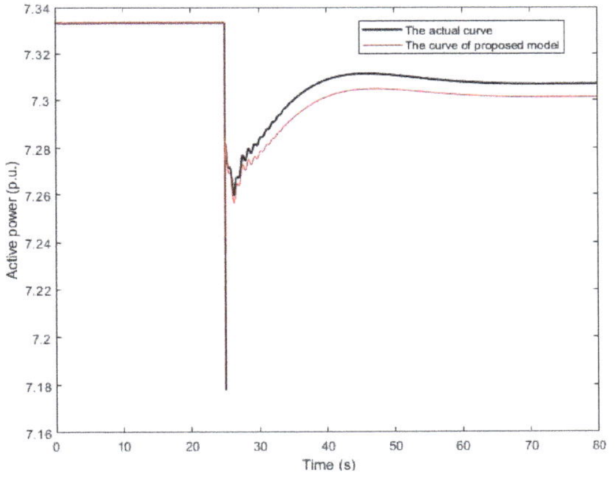

Figure 20. The curve of active power considering the actual model and the proposed model.

As depicted in Figure 18, the trend of the system frequency variation corresponds to that shown in Figures 11 and 14, which first reaches the nadir point and then recovers to the new steady state. Due to the large proportion of static load contained in the complex load, the voltage of the complex load node decreases rapidly and then recovers to the new steady state after being disturbed. Therefore, the voltage curve in Figure 19 drops rapidly at first and then rises, gradually returning to a steady state. Figure 20 shows that the active power curve of the load model proposed in this paper is consistent with the waveform of the actual curve. The final data has some deviation because the actual comprehensive load contains some unknown load parameters. The maximum error of the active power curve is 0.008 p.u., which is in line with the actual engineering error range.

5.2. Parameter Influence Analysis

To investigate the parameter impact of the power load on the system frequency, the minimum point of the system frequency varies with the system inertia value. At time $t = 0$ s, a disturbance of the same $\Delta P_d = 0.025$ p.u. is applied. The final result is shown in Figure 21.

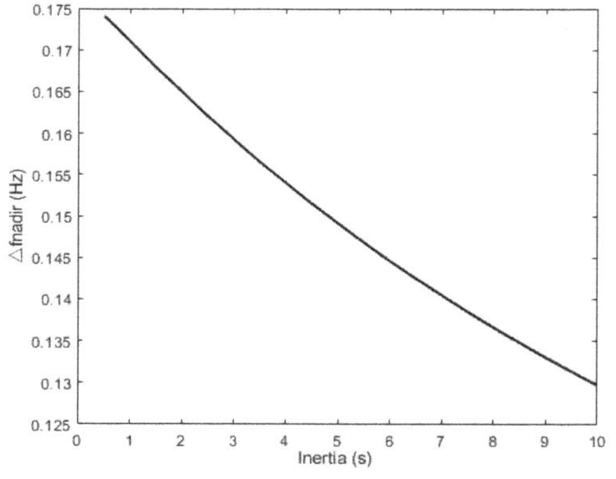

Figure 21. The relation between inertia and nadir point under PSASP.

The nadir point serves as an effective metric for assessing the system's frequency support capability. It can be seen from Figure 21 that with the increase in the inertia of the induction machine, the nadir point of the frequency of the system after the disturbance gradually decreases. It has been demonstrated that the inertia of the induction machine can effectively support the system frequency.

To validate the impact of induction machine speed on the system frequency response, the nadir point of system frequency was measured by applying a disturbance $\Delta P_d = 0.05$ p.u. By changing the different speeds of the induction machine, the results are shown in Figure 22.

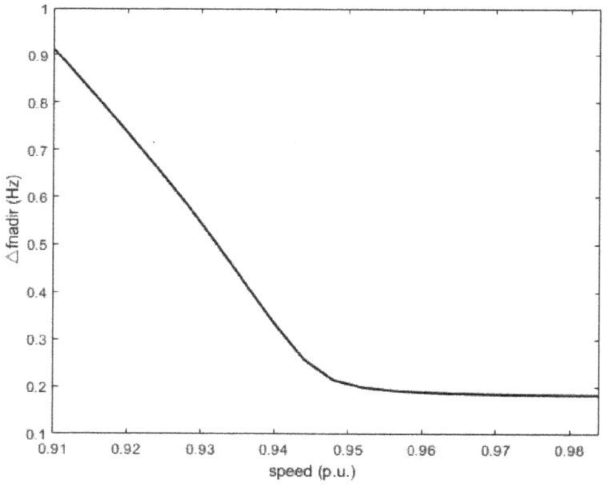

Figure 22. The relation between rotational speed and nadir point under PSASP.

Figure 22 shows the relationship between rotational speed and nadir points. With the increase in the speed, the nadir point of the system gradually decreases, which proves that the frequency support capacity of the induction machine is stronger. There are two factors contributing to this phenomenon. Firstly, the speed of the induction machine directly affects the size of the slip. Moreover, an increased rotor speed leads to a higher amount of kinetic energy stored in the rotor. When the system is disturbed, the induction motor can release more kinetic energy.

Since constant power loads have no frequency response capability, it is only necessary to adjust the proportion of constant impedance and constant current loads to analyze the frequency response capability of static loads. The initial voltages of the load node are $U_t = 1.072$ p.u. and $U_t = 0.9459$ p.u. A disturbance of $\Delta P_d = 0.05$ p.u. was applied to observe the nadir points of the system frequency under static loads of different proportions. The final results are shown in Figure 23.

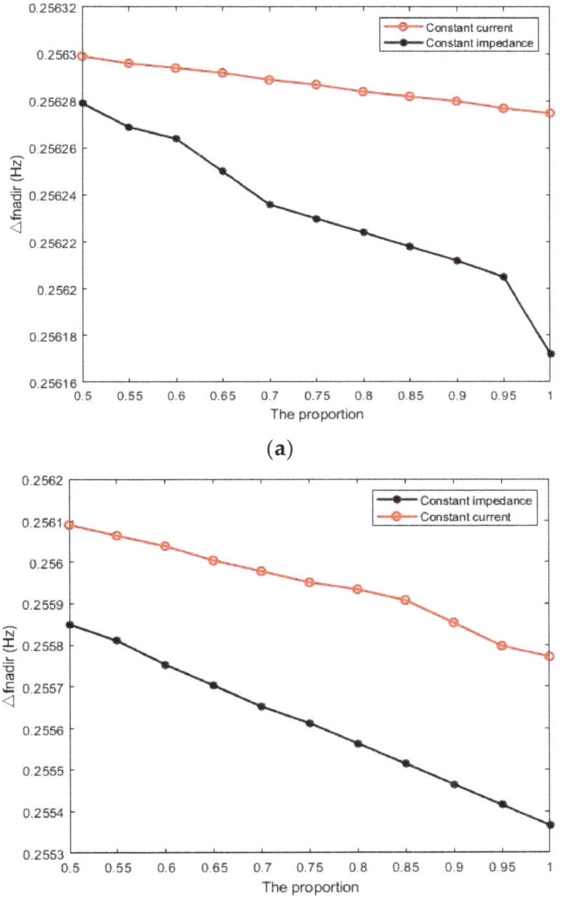

Figure 23. The relation between the proportion of different types of loads and system frequency. (**a**) $U_t = 1.072$ p.u. (**b**) $U_t = 0.9459$ p.u.

In order to exclude the mutual influence of two variables, only the constant power load is included in the static load, as the variable. As depicted in Figure 23, as the ratio of constant current and constant impedance increases, the nadir point of the system frequency gradually diminishes, leading to increased stability in the system frequency. According to the curve in the figure, no matter what the node voltage is, the nadir point of the constant impedance load is always lower than that of the constant current load. The results show that the constant impedance load has better frequency support ability than the constant current load.

5.3. Validation and Analysis in Practical System

In this section, the accuracy of the proposed model method is validated using the IEEE 39-node system on the DIgSILENT PowerFactory platform. Specifically, Node 26 is designated as static load, Node 28 is designated as dynamic load, and the parameters are detailed in Tables 2 and 3. The topology of the IEEE 39-node system is shown in Figure 24.

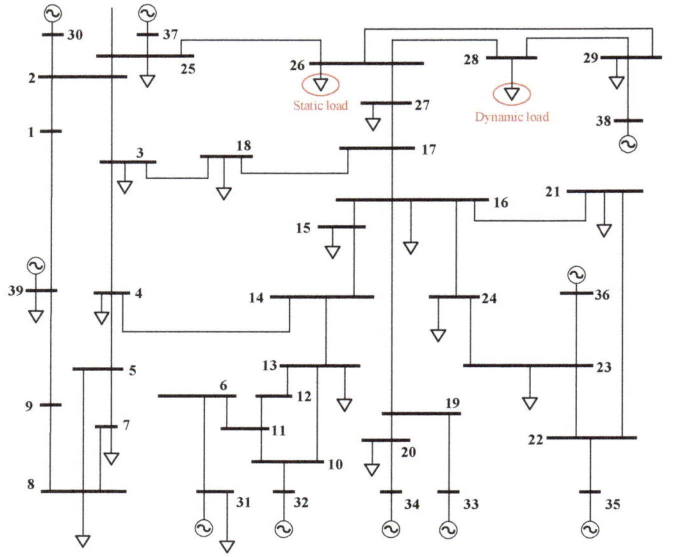

Figure 24. Topology diagram of the IEEE 39-node system.

When the power system is in a state of stability, the frequency of the system is maintained at 50 Hz. Upon disturbance by active power, Figures 25–27 depict the changes in system frequency, voltage at the load node, and comprehensive load response power curves.

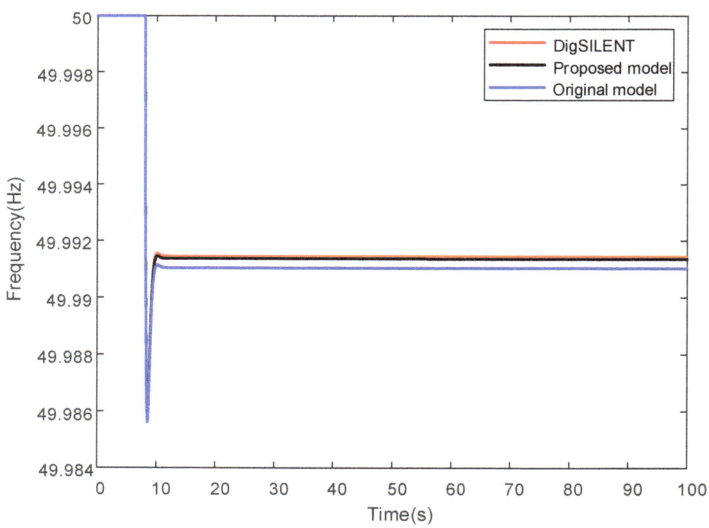

Figure 25. The system frequency curves under different models.

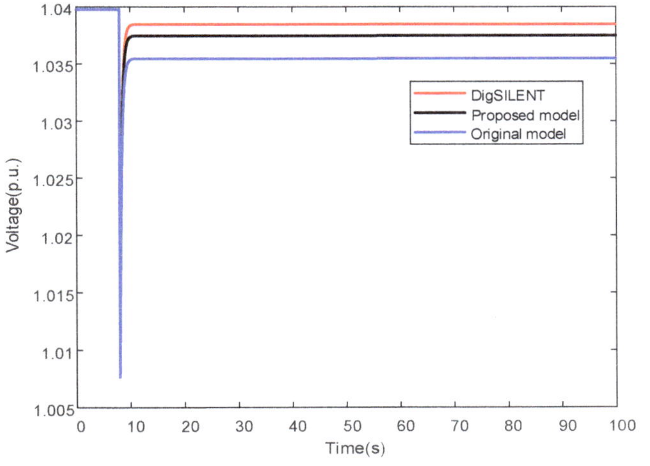

Figure 26. The load node voltage curves under different models.

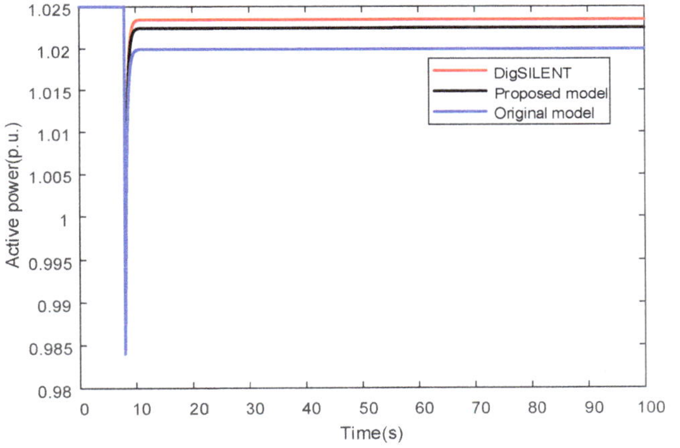

Figure 27. The active power curves under different models.

After an active power disturbance, the system frequency undergoes a change, initially dropping to its nadir point before gradually increasing through the operation of frequency modulation equipment. The voltage at the load node experiences a decrease followed by stabilization due to the combined impact of the terminal voltage and comprehensive load. The comprehensive load predominantly consists of a static load, with its total capacity significantly outweighing that of the induction machine. Furthermore, as the static load power response is heavily influenced by voltage, the overall frequency response curve closely mirrors the voltage curve for the comprehensive load. Consequently, under the dual influence of voltage and frequency, the active power curve for comprehensive load first decreases and then rises to establish a new steady state.

The inertia time constant of the comprehensive load induction machine is evaluated by using the nadir point of the system frequency as a crucial parameter index for verification. At time $t = 0$, a load disturbance of $\Delta P = 0.01$ p.u. is applied to investigate the frequency response of the induction machine under various inertia time constants, with the results depicted in Figure 28. The nadir point, as the peak of system frequency deviation, serves as an effective indicator of the frequency response and support capabilities of the induction machine. As the inertia time constant increases, the nadir point of the system frequency gradually decreases, indicating a corresponding reduction in the system's frequency deviation. With equivalent capacity, a higher partial inertia time constant for the comprehensive load induction machine enhances its ability to support frequency and further bolsters the system's frequency stability.

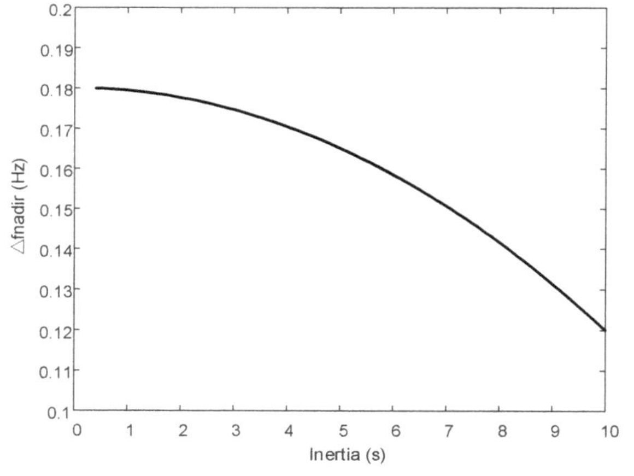

Figure 28. The relation between inertia and nadir point under DIgSILENT PowerFactory.

To investigate the impact of partial rotor speed on the frequency support capacity of a comprehensive load in an induction machine system, a uniform active power disturbance of $\Delta P = 0.02$ p.u. was applied to systems with identical capacity but different rotor speeds. The resulting changes in the nadir point of the system frequency were then observed, leading to the generation of the relationship diagram depicted in Figure 29. As the rotor speed of the induction machine increases under the same capacity, the nadir point of the system frequency gradually decreases. This observation serves as evidence that the frequency support capacity of the comprehensive load also increases gradually with the increase in the rotor speed of the induction machine.

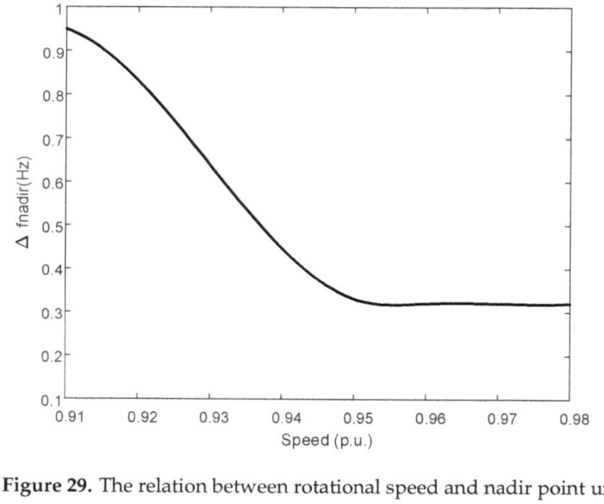

Figure 29. The relation between rotational speed and nadir point under DIgSILENT PowerFactory.

Due to the lack of frequency response capability in a constant power load, it is essential to investigate the impact of the ratio between constant impedance and constant current loads on the frequency response ability of the static load. By adjusting the ratio of various components with constant current and constant impedance, the minimum point image of the system frequency can ultimately be achieved as depicted in Figure 30. In order to mitigate the mutual influence between the two variables, all static loads, other than the variable itself, are maintained as constant power loads. As depicted in the figure, the frequency support capacity of a constant impedance load surpasses that of a constant current load. With an increasing proportion of the constant impedance load within the static load, there is also an increase in the frequency support capacity of the static load.

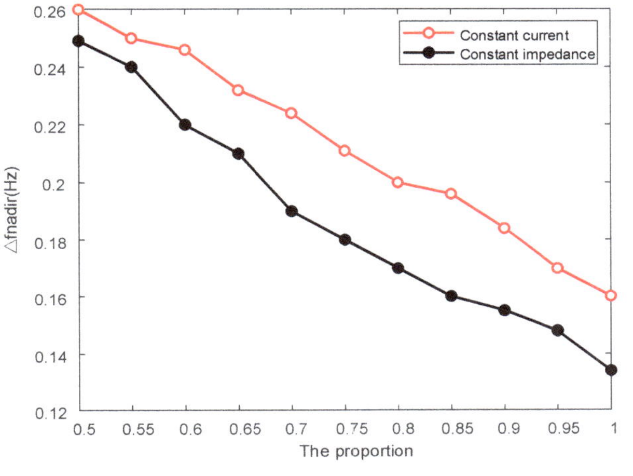

Figure 30. The relation between the proportion of different types of loads and nadir point.

6. Conclusions

This paper establishes a comprehensive load frequency response model that integrates both dynamic and static loads, accounting for fluctuations in node voltage, thereby providing a more precise depiction of the frequency response behaviors of power loads. Through the utilization of a case study, the proposed comprehensive model and its support capacity analysis results are validated, with its validity and accuracy in addressing voltage variation effects.

(1) In this paper, a frequency response model for a dynamic load is established for frequency dynamics. The simulation results demonstrate that an induction machine with frequency response capabilities can mitigate its own electromagnetic power, effectively prevent system frequency fluctuations, and support system frequency regulation during occurrences of disturbances in active power imbalances.
(2) The small-signal model of static load is established to depict its frequency dynamics, considering voltage fluctuations. Changes in the node load voltage resulting from variations in terminal voltage and system power flow subsequently affect the active power of a static load. Simulation results validate the effectiveness and precision of the proposed model with its support capability.
(3) On the basis of existing modeling and capacity analysis, a comprehensive model of complex load is established containing both dynamic and static loads. The analysis of critical parameters reveals that higher rotor speed and inertia enhance the load's frequency support capability. Additionally, the proportional coefficients of constant impedance and constant current in the static load component can enhance its frequency support capacity, thereby improving the overall system frequency stability.

In this paper, a load frequency response model for power systems is developed, taking into consideration the impact of voltage fluctuations and providing a more precise depiction of the active load response and frequency support capacity during frequency dynamics. This research can offer valuable insights and guidance for ensuring the safety and stability of power systems, with a high penetration of renewable energy sources. For future research, focus should be drawn to prioritize data-driven approaches in modeling load frequency responses, as they hold the potential to enhance the accuracy and efficiency of predicting and managing system dynamics. In actual grids, due to the random and dynamic nature of the proportion and parameters of each load type, subsequent efforts should be further dedicated using deep-learning methods to address uncertainties in node voltage and random fluctuations in load.

Author Contributions: Conceptualization, T.Z. and C.W.; methodology, Y.Z. and L.C.; software, Y.Z.; validation, Y.Z.; formal analysis, C.W.; data curation, B.L.; writing—original draft preparation, Y.Z.; writing—review and editing, T.Z. and B.L.; supervision, Z.C.; project administration, T.Z. and Z.C.; funding acquisition, T.Z. and L.C. All authors have read and agreed to the published version of the manuscript.

Funding: This project is supported by the State Key Laboratory of Power System Operation and Control (No. SKLD23KM17), the Natural Science Foundation of Jiangsu Province (No. BK20220216) and the Fundamental Research Funds for the Central Universities (No. 30922010709).

Data Availability Statement: Data are contained within the article.

Conflicts of Interest: The authors declare no conflicts of interest.

References

1. Fu, B.; Ouyang, C.; Li, C.; Wang, J.; Gul, E. An Improved Mixed Integer Linear Programming Approach Based on Symmetry Diminishing for Unit Commitment of Hybrid Power System. *Energies* **2019**, *12*, 833. [CrossRef]
2. Wu, M.; Ma, D.; Xiong, K.; Yuan, L. Deep Reinforcement Learning for Load Frequency Control in Isolated Microgrids: A Knowledge Aggregation Approach with Emphasis on Power Symmetry and Balance. *Symmetry* **2024**, *16*, 322. [CrossRef]
3. Fu, X.; Wu, X.; Zhang, C.; Fan, S.; Liu, N. Planning of Distributed Renewable Energy Systems Under Uncertainty Based on Statistical Machine Learning. *Prot. Control. Mod. Power Syst.* **2022**, *7*, 1–27. [CrossRef]
4. Angizeh, F.; Bae, J.; Chen, J.; Klebnikov, A.; Jafari, M. Impact Assessment Framework for Grid Integration of Energy Storage Systems and Renewable Energy Sources Toward Clean Energy Transition. *IEEE Access* **2023**, *11*, 134995–135005. [CrossRef]
5. Sun, Q.; Li, D.; Wang, R.; Sui, Z.; Yao, J. "Double High" Power System: A New Stability Criterion and Stability Classification. *CSEE* **2024**, *44*, 1–22.
6. Hu, J.; Liu, X.; Shahidehpour, M.; Xia, S. Optimal Operation of Energy Hubs with Large-Scale Distributed Energy Resources for Distribution Network Congestion Management. *IEEE Trans. Sustain. Energy* **2021**, *12*, 1755–1765. [CrossRef]
7. Yang, D.; Wang, X.; Chen, W.; Yan, G. Adaptive Frequency Droop Feedback Control-Based Power Tracking Operation of a DFIG for Temporary Frequency Regulation. *IEEE Trans. Power Syst.* **2024**, *39*, 2682–2692. [CrossRef]
8. Sun, X.; Bao, M.; Ding, Y.; Hui, H.; Song, Y.; Zheng, C.; Gao, X. Modeling and Evaluation of Probabilistic Carbon Emission Flow for Power Systems Considering Load and Renewable Energy Uncertainties. *Energy* **2024**, *296*, 130768. [CrossRef]
9. Gayathri, K.; Jena, K. A Practical Approach to Inertia Distribution Monitoring and Impact of Inertia Distribution on Oscillation Baselining Study for Renewable Penetrated Power Grid. *IEEE Syst.* **2023**, *17*, 3593–3601.
10. Wu, Q.; Bose, A.; Singh, C.; Chow, J. Control and Stability of Large-Scale Power System with Highly Distributed Renewable Energy Generation: Viewpoints from Six Aspects. *CSEE J. Power Energy* **2023**, *9*, 8–14.
11. Peng, Q.; Jiang, Q.; Yang, Y.; Liu, T.; Wang, H.; Blaabjerg, F. On the Stability of Power Electronics-Dominated Systems: Challenges and Potential Solutions. *IEEE Trans. Ind. Appl.* **2019**, *55*, 7657–7670. [CrossRef]
12. Yang, D.; Jun, L.; Jin, Z. Sequential Frequency Regulation Strategy for DFIG and Battery Energy Storage System Considering Artificial Deadbands. *Int. J. Electr. Power Energy Syst.* **2024**, *155*, 109503. [CrossRef]
13. Wang, C.; Ju, P.; Wu, F.; Pan, X.; Wang, Z. A Systematic Review on Power System Resilience from the Perspective of Generation, Network, and Load. *Renew. Sustain. Energy Rev.* **2022**, *167*, 112567. [CrossRef]
14. Gulzar, M. Designing of Robust Frequency Stabilization Using Optimized MPC-(1+PIDN) Controller for High Order Interconnected Renewable Energy Based Power Systems. *Prot Contr Mod Pow* **2023**, *8*, 1–14. [CrossRef]
15. Bustamante, S.; Jorge, W.; Gabriel, J.; Hugo, A. UFLS and Smart Load for Frequency Regulation in Electrical Power System: A Review. *IEEE Access* **2023**, *11*, 110967–110984. [CrossRef]
16. Gu, H.; Yan, R.; Saha, T. Review of System Strength and Inertia Requirements for the National Electricity Market of Australia. *CSEE* **2019**, *5*, 295–305.
17. Seshapalli, S. Analysis of Hybrid Power Flow Controller using Static load model under Contingency Screening. In Proceedings of the 2019 IEEE International Conference on Clean Energy and Energy Efficient Electronics Circuit for Sustainable Development (INCCES), Krishnankoil, India, 18–20 December 2019; pp. 1–6.

18. Bu, F.; Ma, Z.; Yuan, Y.; Wang, Z. Composite Load Model Parameter Identification Using Evolutionary Deep Reinforcement Learning. *IEEE Trans. Smart Grid* **2020**, *11*, 5407–5417. [CrossRef]
19. Chen, L.; Wang, X.; Min, Y. Modelling and Investigating the Impact of Asynchronous Inertia of Induction Motor on Power System Frequency Response. *Int. J. Electr. Power* **2020**, *117*, 105708.1–105708.8. [CrossRef]
20. Lu, W.; Zhang, J.; Ren, L. Influence Analysis of Different Load Models on Transient Stability of Shanghai Power Grid. *Beijing Jiaotong Univ.* **2021**, *45*, 130–142.
21. Si, D.; Qian, Y.; Sun, S. Influence of Load Model on Frequency Stability of AC/DC System in Asynchronous Network. *Yunnan Electr. Power Technol.* **2017**, *45*, 62–65.
22. Sun, H.; Wang, B.; Li, W. Research on Inertia System of Frequency Response for High-proportion Power Electronic Power Systems. *CSEE* **2020**, *40*, 5179–5192.
23. Liang, X.; Xu, W.; Chuang, C.; Freitas, W.; Xiong, K. Dynamic Load Models for Industrial Facilities. *IEEE Trans. Power Syst.* **2012**, *27*, 69–80. [CrossRef]
24. Marault, J.; Tounzi, A.; Gllon, F.; Hecquet, M. Efficient Approach Based on Equivalent Electric Circuit Model to Determine Rotor Bar Currents of Squirrel Cage Induction Machines. *IEEE Trans. Magn.* **2021**, *57*, 8101305. [CrossRef]
25. Liang, X.; He, J. Load Model for Medium Voltage Cascaded H Bridge Multi-Level Inverter Drive Systems. *IEEE Power Energy Technol. Syst.* **2016**, *3*, 13–23. [CrossRef]
26. Ge, Y.; Flueck, A.; Kim, D.; Ahn, J.; Lee, J.; Kwon, D. An Event-Oriented Method for Online Load Modeling Based on Synchrophasor Data. *IEEE Trans. Smart Grid* **2015**, *6*, 2060–2068. [CrossRef]
27. Shi, L.; Lao, W.; Wu, F.; Zheng, T.; Lee, K. Frequency Regulation Control and Parameter Optimization of Doubly-Fed Induction Machine Pumped Storage Hydro Unit. *IEEE Access* **2022**, *10*, 102586–102598. [CrossRef]
28. Batzelis, E.; Anagnostou, G.; Cole, I.; Betts, T.; Pal, B. A State-Space Dynamic Model for Photovoltaic Systems with Full Ancillary Services Support. *IEEE Trans. Sustain. Energy* **2019**, *10*, 1399–1409. [CrossRef]
29. DÖŞOĞLU, M.K.; Dursun, M. Investigation with ZIP Load Model of Voltage Stability Analysis in Wind Turbine Integrated Power System. In Proceedings of the 2018 2nd International Symposium on Multidisciplinary Studies and Innovative Technologies (ISMSIT), Ankara, Turkey, 19–21 October 2018; pp. 1–5.

symmetry

MDPI

Article

Top-Oil Temperature Prediction of Power Transformer Based on Long Short-Term Memory Neural Network with Self-Attention Mechanism Optimized by Improved Whale Optimization Algorithm

Dexu Zou [1,2], He Xu [3], Hao Quan [3,*], Jianhua Yin [3], Qingjun Peng [2], Shan Wang [2], Weiju Dai [2] and Zhihu Hong [2]

[1] School of Electrical Engineering, Chongqing University, Chongqing 400044, China; zoudexu@csg.cn
[2] Electric Power Research Institute, China Southern Power Grid Yunnan Power Grid Co., Ltd., Kunming 650217, China; penggingjun@csg.cn (Q.P.); wangshan@csg.cn (S.W.); daiweiju@csg.cn (W.D.); hongzhihu@csg.cn (Z.H.)
[3] School of Automation, Nanjing University of Science and Technology, Nanjing 210094, China; xuhe1@njust.edu.cn (H.X.); yinjhhust@163.com (J.Y.)
* Correspondence: quanhao@njust.edu.cn

Abstract: The operational stability of the power transformer is essential for maintaining the symmetry, balance, and security of power systems. Once the power transformer fails, it will lead to heightened instability within grid operations. Accurate prediction of oil temperature is crucial for efficient transformer operation. To address challenges such as the difficulty in selecting model hyperparameters and incomplete consideration of temporal information in transformer oil temperature prediction, a novel model is constructed based on the improved whale optimization algorithm (IWOA) and long short-term memory (LSTM) neural network with self-attention (SA) mechanism. To incorporate holistic and local information, the SA is integrated with the LSTM model. Furthermore, the IWOA is employed in the optimization of the hyper-parameters for the LSTM-SA model. The standard IWOA is improved by incorporating adaptive parameters, thresholds, and a Latin hypercube sampling initialization strategy. The proposed method was applied and tested using real operational data from two transformers within a practical power grid. The results of the single-step prediction experiments demonstrate that the proposed method significantly improves the accuracy of oil temperature prediction for power transformers, with enhancements ranging from 1.06% to 18.85% compared to benchmark models. Additionally, the proposed model performs effectively across various prediction steps, consistently outperforming benchmark models.

Keywords: power transformer; top-oil temperature prediction; self-attention mechanism; whale optimization algorithm; long short-term memory networks

1. Introduction

Power transformers undertake a vital role in the symmetrical operation of power systems [1]. They serve as critical infrastructure for power transmission and distribution, with extensive applications in various other fields, such as transportation [2]. Once the power transformer fails, it can severely disrupt the normality of the power system operation, potentially causing widespread power outages and significant economic losses [3]. As a vital component of the power system, the stable operation of the transformer is fundamental to maintaining the symmetry and balance of the power system [4,5].

Top oil temperature is significant for determining whether the transformer can maintain normal operation. In practice, the transformer internal faults rely on the trend of the oil temperature to make judgments [6,7]. Therefore, the good performance of oil temperature prediction helps professionals find problems promptly in the transformer's daily operation and maintenance. By reliably forecasting oil temperature, we can not only prevent unexpected failures but also optimize maintenance schedules, reduce operational risks, and extend the transformer's lifespan. Effective oil temperature prediction enhances the overall reliability and efficiency of the power system, making it an essential component in maintaining the symmetrical operation of the electrical grid.

Researchers generally study the prediction of transformer oil temperatures through mathematical and data-driven models [8–10]. Zhao et al. used the least squares method to establish a parameter identification algorithm [11], and this mathematical model can effectively predict the top oil temperature but lacks strong generalization ability. Wang et al. establish a thermal circuit model to simulate the changes in the transformer temperature over time, but it has a lengthy computation time [12].

With the development of intelligent algorithms, artificial intelligence technologies have been applied to the field of power system forecasting. Interesting studies can be found in the fields of load forecasting [13], vehicle-to-grid (V2G) scheduling prediction [14], and solar irradiance forecasting [15]. There have been some research efforts focused on predicting transformer oil temperature using these algorithms. Qing et al. developed a model based on artificial neural networks for forecasting the top oil temperature of transformers [16], and this model significantly reduces the computational time but ignores the selection of optimal hyperparameters. Tan et al. proposed a forecast model that considers path analysis and similar moments [17], but the validation dataset is small and the adaptability is difficult to confirmed. Li et al. introduced a regression model with enhanced particle swarm optimization (PSO) for transformer top oil temperature forecast [18]. However, the large sampling interval of data caused the substandard performance. Based on a similar day, Tan et al. introduced a method to predict top oil temperature. The above approach relies solely on single-day similarity for prediction and deteriorates the model prediction performance [19]. To sum up, these studies do not fully consider temporal information of different input features, thus failing to combine global and local information within transformer operational data. In addition, the optimal hyper-parameters of the model are difficult to determine.

To tackle the issues mentioned, this paper introduces a novel method: an improved whale optimization algorithm (IWOA) optimized long short-term memory (LSTM) neural network with self-attention (SA) mechanism model. The proposed method comprehensively addresses challenges related to the difficulty in selecting hyperparameters for the oil temperature prediction model and the insufficient consideration of temporal information. It integrates SA with LSTM and utilizes the IWOA to obtain the optimal hyper-parameters for the LSTM-SA model, resulting in high prediction accuracy. Finally, the proposed method is tested with actual operating data in a practical power grid. The results demonstrate that the proposed method has better forecasting performance.

The remaining sections of this paper are as below: Section 2 discusses the power transformer and top-oil temperature. Section 3 introduces the LSTM-SA model and the IWOA. Section 4 presents a case study that shows the superiority of the IWOA for optimization and the effectiveness of the proposed method for predicting top-oil temperature. Finally, conclusions and discussions are presented in Section 5.

2. Power Transformer and Top-Oil Temperature

The top oil temperature of a transformer is a crucial indicator for measuring the reliability of transformer operation, monitoring the internal insulation status. Accurately predicting the top oil temperature of the power transformer is of great significance for analyzing potential faults, carrying out transformer operation and maintenance, maintaining the symmetry and balance of the power system, and achieving early warning of transformer failures. It is a key factor in limiting the transformer's load capacity and assessing its operational lifespan.

There are two merits to considering top oil temperature as the subject of study. First, researchers can easily access real-time monitoring data for the transformer's top oil temperature, thanks to advanced sensor technologies and the widespread implementation of smart grids. This accessibility facilitates continuous monitoring and data collection, which are essential for accurate prediction and timely intervention. Second, the hot spot temperature that is difficult to obtain can be calculated from the transformer top oil temperature. Hot spot temperature is crucial, as it represents the highest temperature within the transformer and is a direct indicator of the condition of the transformer's insulation. Accurate estimation of this temperature is vital for predicting the remaining life of the insulation and planning maintenance activities.

The above advantages have made the top oil temperature highly favored by researchers, and it has now become a hot research topic [20]. The basic construction of an oil-immersed transformer is graphically represented in Figure 1. This paper focuses on improving the accuracy of oil temperature prediction, particularly in addressing the challenges posed by the nonlinearity and time-series characteristics of the data.

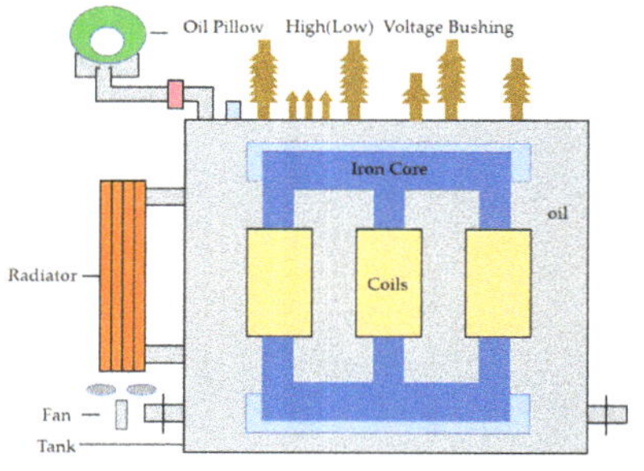

Figure 1. The basic construction of an oil-immersed transformer.

3. The Proposed IWOA-LSTM-SA Method for Top-Oil Temperature Prediction

3.1. Framework

In this study, IWOA-LSTM-SA has been developed for transformer oil temperature forecasting, in which IWOA has been employed to precisely search optimal input hyperparameters and LSTM-SA as the forecasting model to combine global and local information. The flowchart is presented in Figure 2.

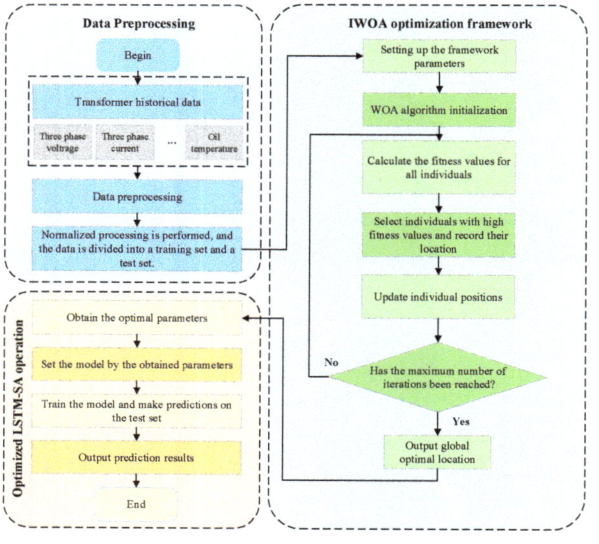

Figure 2. Flow chart of IWOA-LSTM-SA.

The main phases of the IWOA-LSTM-SA will be detailed in the following sections.

3.2. LSTM Integrated by SA

LSTM is a specialized type of recurrent neural network (RNN), specifically designed to process temporal data sequences. On the basis of traditional RNN, LSTM introduces the concept of "gating", which not only overcomes the gradient vanishing but also selects samples. Therefore, LSTM is more suitable for solving nonlinear temporal structure problems. Each memory block of an LSTM comprises one or more self-connected memory cells and three gating units: the input gate, the output gate, and the forget gate. The specific structure of the gate is shown in Figure 3. The forgetting gate is responsible for deciding which information should be discarded from the cell state, effectively determining the extent to which the previous cell state is preserved within the current cell state. The calculation equation is as below:

$$m_t = \sigma(W_m \times [r_{t-1}, x_t] + p_m) \quad (1)$$

The input gate controls which the current input is stored in the unit state. The formulas for input gates and candidate cell states is as below:

$$s_t = \sigma(W_s \times [r_{t-1}, x_t] + p_s) \quad (2)$$

The output gate regulates the current output and decides the output information. The formula for calculation is given below:

$$g_t = \sigma(W_g \cdot [r_{t-1}, x_t] + p_g) \quad (3)$$

$$r_t = o_t \cdot \tanh(C_t) \quad (4)$$

The formula for calculating the cell state is as below:

$$\widetilde{C}_t = \tanh(W_C \times [r_{t-1}, x_t] + p_C) \quad (5)$$

$$C_t = m_t \cdot C_{t-1} + s_t \cdot \widetilde{C}_t \quad (5)$$

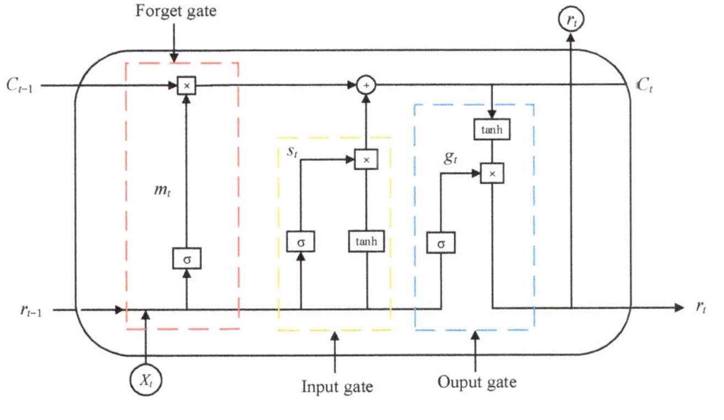

Figure 3. LSTM structure diagram.

In summary, LSTM is suitable for processing time series data, so this paper uses LSTM to establish a temperature prediction model. Furthermore, it is difficult to process long sequence data for the LSTM model that we introduce SA to solve this problem. This method considers both local and global information.

It consists of three components. Firstly, the data that come from the LSTM model is the input of the SA layer. Secondly, the matrices q, k, and v are calculated using the weight matrices W_q, W_k, and W_v. Thirdly, $a^{1,2}$ is the dot product between q_1 and k_2, and $a^{2,2}$ is the dot product between q_2 and k_2. The attention matrix M means the correlation between different time steps. The structure is shown in Figure 4.

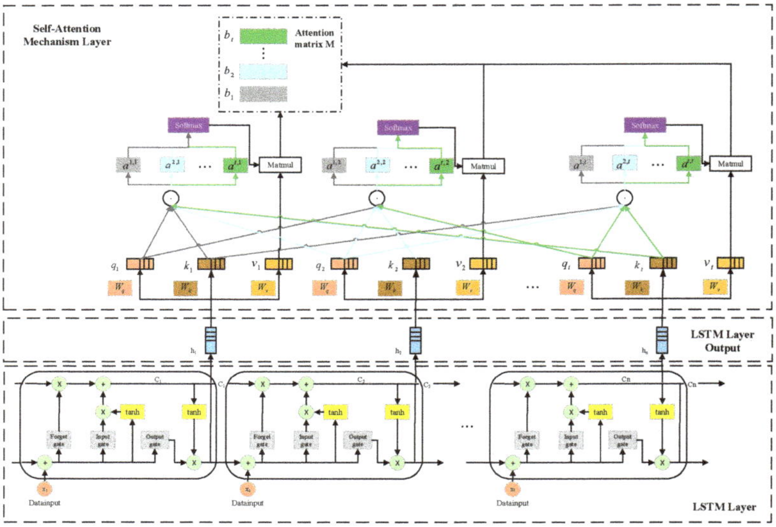

Figure 4. LSTM-SA structure.

3.3. Hyper-Parameters Optimization by IWOA

The Whale Optimization Algorithm (WOA) was introduced to deal intricate optimization problems by Mirjalili et al. [21,22]. The WOA can be formulated as the following steps: encircling prey, bubble-net attacking method and search for prey.

3.3.1. Encircling Prey

Humpback whales can identify and encircle their prey. In the population, the remaining whales will try to adjust their positions towards the direction of the best search agent as defined by the equation:

$$\vec{G}(t+1) = \vec{G}^*(t) - \vec{A}\left|\vec{C}\vec{G}^*(t) - \vec{G}(t)\right| \tag{7}$$

where t denotes the current iteration; $\vec{G}$ is a vector indicating the position; $\vec{G}^*$ is the place vector of the best solution acquired yet, $\vec{A}$ and $\vec{C}$ are calculated from the following:

$$\vec{A} = 2\vec{a}\,\vec{r}_1 - \vec{a} \tag{8}$$

$$\vec{C} = 2\vec{r_2} \tag{9}$$

where $\vec{a}$ is an adjustment vector and $\vec{a}$ is linearly decreasing from 2 to 0; the vectors $\vec{r_1}$ and $\vec{r_2}$ are random vectors that fall within the range of [0, 1].

3.3.2. Bubble-Net Attacking Method

Humpback whale predation consists of two main mechanisms: shrinkage bracketing mechanism and the spiral updating location.

(1) Shrinkage bracketing mechanism: As $\vec{a}$ decreases, $\vec{A}$ represents an any value within the range of $[-1, 1]$. The new position is determined by the distance between its

original position and the position of the currently best-so-far whale. The equation for calculation is as below:

$$\vec{a} = 2 \times \left(1 - \frac{t}{t_{\max}}\right) \tag{10}$$

(2) Spiral updating location: the WOA uses spiral updating location to launch attacks on prey, and the spiral hunting equation is as below:

$$\vec{G}(t+1) = e^{bl}\cos(2\pi l) \cdot \left|\vec{G}^*(t) - \vec{G}(t)\right| + \vec{G}^*(t) \tag{11}$$

where l is a random count within the interval $[-1, 1]$ and b represents a constant. They approach the prey using two mechanisms: a shrinking circle and a spiral-shaped path. The updated equations are as follows.

$$\vec{G}(t+1) = \begin{cases} \vec{G}^*(t) - \vec{A}\left|\vec{C} \cdot \vec{G}^*(t) - \vec{G}(t)\right|, p < 0.5 \\ e^{bl}\cos(2\pi l) \cdot \left|\vec{G}^*(t) - \vec{G}(t)\right| + \vec{G}^*(t), p \geq 0.5 \end{cases} \tag{12}$$

where p falls within the range of [0,1].

3.3.3. Search for Prey

Humpback whales search for their prey randomly, with their locations varying relative to each other. In this stage, the position of a searching whale is modified according to the position of a randomly selected whale, as opposed to being updated based on the current best whale. The calculation formula is as listed below:

$$\vec{G}(t+1) = \vec{G}_{\text{rand}}(t) - \vec{A} \cdot \left|\vec{C}\vec{G}_{\text{rand}}(t) - \vec{G}(t)\right| \tag{13}$$

where $\vec{G}_{\text{rand}}$ denotes the random location of a whale.

3.3.4. Improved Whale Optimization Algorithm

The original WOA faces certain limitations, particularly in terms of inadequate local search capabilities and insufficient population diversity. Therefore, it is necessary to further improve the strategy and adjust the algorithm [23]. For example, Naderi et al. proposed a Whale Optimization Algorithm enhanced by wavelet mutation, aimed at improving the algorithm's convergence characteristics to address the complex trade-off between generation costs and water consumption [24]. In this study, an approach takes a different direction by introducing three key improvements: Latin Hypercube Sampling for more diverse and uniform population initialization, an adaptive selection threshold to dynamically adjust the whale's movement strategy, and a nonlinear parameter adjustment to enhance local search capabilities. These modifications are designed to address different aspects of the original WOA's limitations. The specific improvements are as follows:

(1) Latin Hypercube Sampling (LHS) initialization of population: as stated in [25], population initialization plays a crucial role in swarm intelligence optimization algorithms. In WOA, population initialization follows a random approach. However, it can lead to uneven population distribution and individual overlap [26]. Therefore, it is necessary to optimize the population initialization. IWOA incorporates LHS to increase the diversity of initial population, and this method can initialize population more uniformly and efficiently.

(2) Adaptive selection threshold: in WOA, the whales choose either encircling activity or spiral movement with 50% probability. However, this method prevents the whale population from choosing the appropriate movement for the current population [27,28].

In this paper, an adaptive selection threshold is used to replace the fixed threshold. The method automatically adjusts the threshold according to the problem's characteristics throughout the search process. The calculation is given by the following formula:

$$p_a = 1 - \left[\frac{t}{(L+f)t_{\max}} \times \left(L \times \frac{e^t}{e^{t_{\max}}} + f \times \frac{t^f}{{t_{\max}}^f} \right) \right] \tag{14}$$

where t denotes the current iteration, while $t_{\max}$ denotes the maximum iteration count; L, f are control parameters, and their values are 2 and 4, respectively.

In our method, when the threshold is larger in the initial stage, the whale will preferentially choose the encircling movement strategy. With the increasing of iterations, the threshold decreases, thus the whale is more likely to choose the spiral motion strategy. Equation (12) is updated to Equation (15).

$$\vec{G}(t+1) = \begin{cases} \vec{G}^*(t) - \vec{A} \left| \vec{C} \cdot \vec{G}^*(t) - \vec{G}(t) \right|, p < p_a \\ e^{bl} \cos(2\pi l) \cdot \left| \vec{G}^*(t) - \vec{G}(t) \right| + \vec{G}^*(t), p \geq p_a \end{cases} \tag{15}$$

(3) Adaptive parameter: in traditional method, $\vec{a}$ decreases linearly from 2 to 0. In order to enhances local searching ability, this study uses a nonlinear strategy to adjust b in Equation (16), which influences the shape of the logarithmic spiral. It can significantly improve the effectiveness of local search and the speed of global search, thereby enhancing overall accuracy [29]. At the same time, we establish a relationship between b and t to achieve adaptive adjustment. Equation (10) is updated to Equation (16).

$$\begin{cases} \vec{a}(t) = 2 \times \left(1 - \tanh\left(\sqrt[k]{\frac{t}{t_{max}}}\right)\right) \\ b(t) = v - \left(\frac{v}{t_{max}}\right) \times t \end{cases} \tag{16}$$

where k, v are control parameters, and their values are 4 and 10, respectively.

The IWOA flowchart is illustrated in Figure 5.

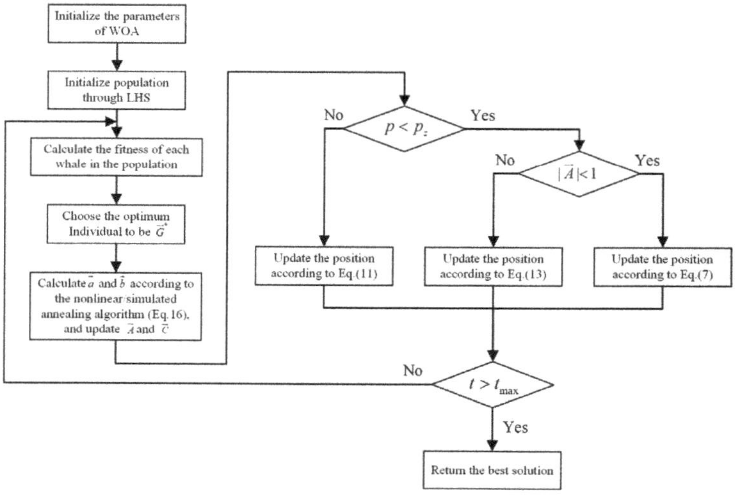

Figure 5. Flow chart of the IWOA.

4. Case Studies and Results Analysis

4.1. Data Source

This study includes two datasets. Dataset 1 consists of transformer operation data collected from a 500 kV substation from 1 April to 30 June in 2022, with a sampling period of half an hour. In total, there are 4368 samples. The characteristic parameters include high-voltage-side three-phase current (A_I, B_I, C_I), active and reactive power (P, Q), high-voltage-side three-phase voltage (A_U, B_U, C_U), and top-oil temperature (T). This paper used the Pearson correlation coefficient method to select features, and the results are shown in Table 1. Dataset 2 consists of transformer operation data collected from a 220 kV substation from 10 February 2021 to 10 February 2022, with a sampling period of half an hour. In total, there are 17,518 samples.

Table 1. Correlation matrix.

	A_I	B_I	C_I	P	Q	A_U	B_U	C_U	T
A_I	1.000	0.999	0.999	0.999	0.925	−0.862	−0.866	−0.835	0.371
B_I	0.999	1.000	0.999	0.999	0.924	−0.863	−0.866	−0.835	0.371
C_I	0.999	0.999	1.000	0.999	0.925	−0.862	−0.866	−0.835	0.371
P	0.999	0.999	0.999	1.000	0.925	−0.857	−0.859	−0.828	0.369
Q	0.925	0.924	0.925	0.925	1.000	−0.842	−0.844	−0.823	0.372
A_U	−0.862	−0.863	−0.862	−0.857	−0.842	1.000	0.979	0.964	−0.346
B_U	−0.866	−0.866	−0.866	−0.859	−0.844	0.979	1.000	0.981	−0.342
C_U	−0.835	−0.835	−0.835	−0.828	−0.823	0.964	0.981	1.000	−0.339
T	0.371	0.371	0.371	0.369	0.372	−0.346	−0.342	−0.339	1.000

As shown in Table 1, the correlation coefficient between the top-oil temperature and the high-voltage side three-phase current is 0.371, and the correlation coefficients with active power and reactive power are 0.369 and 0.372, respectively, indicating a positive correlation. The correlation coefficients between the top-oil temperature and the high-voltage side three-phase voltage are −0.346, −0.342, and −0.339, respectively, indicating a negative correlation with the top-oil temperature. This also suggests that the high-voltage side three-phase voltage, current, and active and reactive power have some influence on the transformer oil temperature. Similarly, a correlation analysis of the input features of Dataset 2 based on the Pearson correlation coefficient method is conducted. Ultimately, this paper selects high-voltage-side current, active and reactive power, voltage, and top-oil temperature as input features. The dataset is split into training and test sets, in which 80% is used for training and 20% for testing.

4.2. Comparison of Algorithm Optimization Results

This paper compared the performance of IWOA with traditional methods, which consist of GA, PSO, and the original WOA. Appendix A, Table A1 presents the ten test functions employed for evaluation, which are derived from the studies conducted in [30,31].

In Appendix A, Table A1: Each function has a dimension of 30, and the minimum value is 0. To ensure the fairness of the comparison, the iteration is set to 500. The crossover probability of GA is set to 1, and the variance probability is 0.1. Meanwhile, the learning factor c1 = c2 = 2 for PSO, and b is 10 for WOA. Each algorithm runs independently 30 times. The average and the best results are utilized for comparison, as shown in Table 2. The average convergence curve of each algorithm is shown in Figure 6.

In Table 2, the optimal value reaches 0 in the F_5, F_6 and F_8 functions, and the average values also show significant improvement. As shown in Figure 6, IWOA exhibits better convergence performance compared to traditional algorithms. These findings confirm the effectiveness of the enhancement strategies for WOA.

Table 2. Comparison of test results for each algorithm.

Function	Evaluation Index	GA	PSO	WOA	IWOA
F_1	Mean	3602.311	0.035	7.21×10^{-10}	1.46×10^{-19}
	Best	1454.955	0.001	3.32×10^{-13}	1.17×10^{-24}
F_2	Mean	21.197	32.013	5.16×10^{-9}	1.73×10^{-13}
	Best	13.936	0.081	5.12×10^{-9}	2.24×10^{-15}
F_3	Mean	3477.958	0.047	8.98×10^{-10}	4.16×10^{-20}
	Best	1771.241	0.001	1.68×10^{-12}	1.42×10^{-22}
F_4	Mean	1.432	5.176	0.015	0.00075
	Best	0.413	0.065	0.003	0.00014
F_5	Mean	28.474	51.152	0	0
	Best	5.522	0	0	0
F_6	Mean	91.831	127.257	0.462	1.78×10^{-16}
	Best	64.795	69.170	6.78×10^{-11}	0
F_7	Mean	11.337	2.028	3.936	1.49×10^{-11}
	Best	9.197	0.023	8.06×10^{-7}	1.35×10^{-12}
F_8	Mean	77.000	551.976	0.988	0
	Best	35.494	185.625	0	0
F_9	Mean	75.910	727.867	−0.898	−0.829
	Best	28.593	479.302	−0.967	−0.986
F_{10}	Mean	73.449	596.665	−0.890	−0.796
	Best	26.910	332.989	−0.980	−0.899

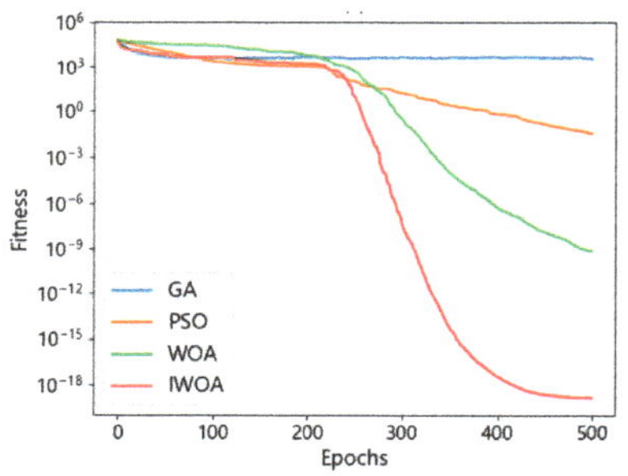

(**a**) Convergence of algorithms on function F_1

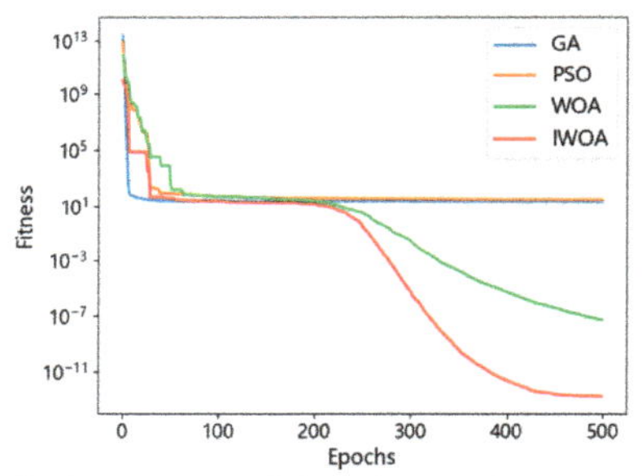

(**b**) Convergence of algorithms on function F_2

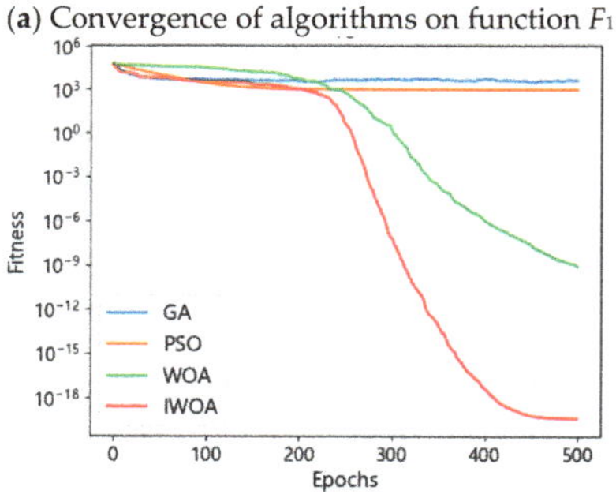

(**c**) Convergence of algorithms on function F_3

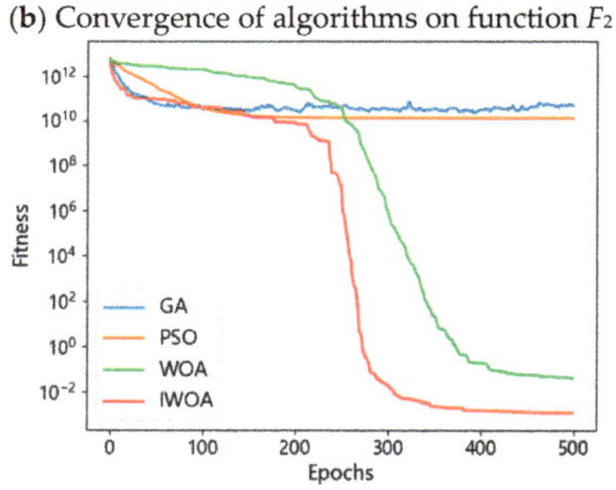

(**d**) Convergence of algorithms on function F_4

Figure 6. *Cont.*

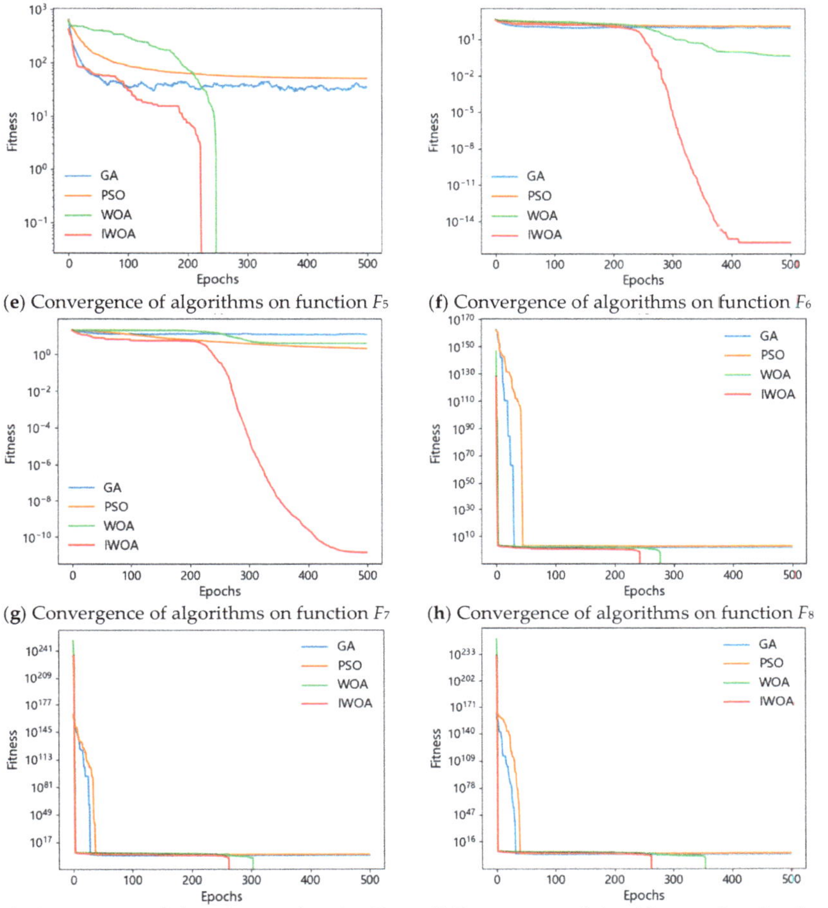

(**e**) Convergence of algorithms on function F_5 (**f**) Convergence of algorithms on function F_6

(**g**) Convergence of algorithms on function F_7 (**h**) Convergence of algorithms on function F_8

(**i**) Convergence of algorithms on function F_9 (**j**) Convergence of algorithms on function F_{10}

Figure 6. Average convergence curves for each algorithm.

4.3. One-Step Prediction

Single-step oil temperature prediction involves forecasting the transformer's top oil temperature for the next time step using historical data. In this experiment, the prediction is for 30 min into the future. To balance the training and testing errors, we introduced L2 regularization and dropout during the model training. Specifically, a dropout rate of 0.1 was applied, along with L2 regularization using a factor of 0.01. The prediction results for Dataset 1, demonstrating the effectiveness of the method, are presented in Figure 7. To further illustrate the trade-off between training and testing errors, Figure 8 provides a comparison of the training and testing errors.

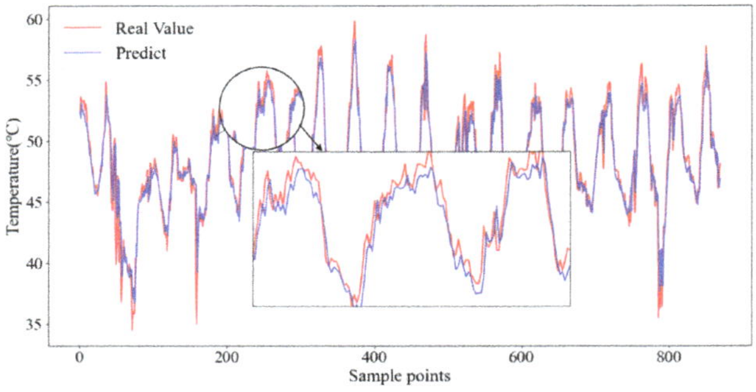

Figure 7. The prediction results of IWOA-LSTM-SA.

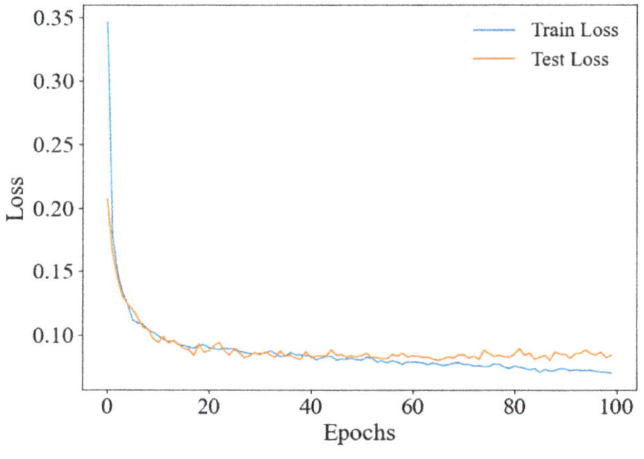

Figure 8. Training and testing errors over iterations.

Theoretically, when there is a significant gap between training and test errors, it usually indicates over-fitting, where the model performs well on the training data but struggles to generalize to unseen data. As illustrated in Figure 8, both the training and test losses decrease rapidly during the initial epochs and then converge to similar values as training progresses. This suggests that we have achieved a well-balanced trade-off between training and testing errors. This balance was successfully attained by applying regularization techniques, such as L2 regularization and dropout, which helped control model complexity, mitigate over-fitting, and enhance the model's generalization capabilities.

To assess the performance of this method, this paper compared it with benchmark methods, including BP, gate recurrent unit (GRU), convolutional neural networks (CNN), LSTM, LSTM-SA, and WOA-LSTM-SA models. In order to reduce the accidental error, this paper conducted 10 repeated experiments and averaged the results to show the forecasting performance. Figure 9 displays the prediction results for each model on Dataset 1. It is evident that the proposed model shows the best prediction result compared to all benchmark models. The reason is that the proposed approach not only combines both local and global information but also utilizes IWOA to determine the optimal hyper-parameters. Table 3 presents the comparative results.

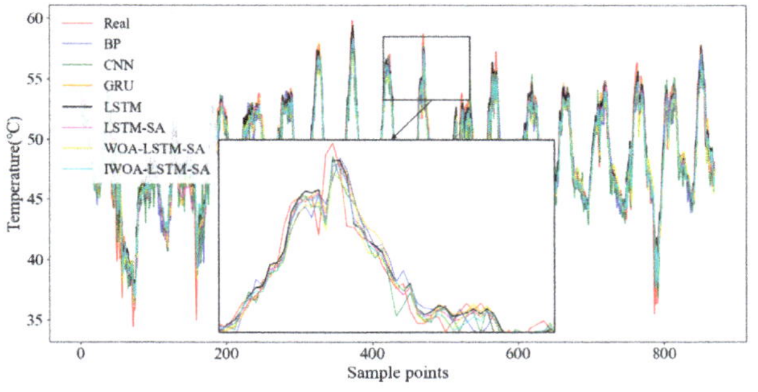

Figure 9. Performance comparison across models.

Table 3. Model prediction evaluation indexes.

	Model	RMSE	MAE	MAPE (%)	R_2	Time (s)
Dataset 1	BP	1.698	1.228	2.581	0.825	13.287
	CNN	1.646	1.170	2.462	0.836	32.317
	GRU	1.553	1.011	2.144	0.854	96.109
	LSTM	1.633	1.022	2.175	0.838	129.666
	LSTM-SA	1.537	1.031	2.253	0.861	174.497
	WOA-LSTM-SA	1.462	0.998	2.103	0.870	11,058.906
	IWOA-LSTM-SA	1.438	0.989	2.089	0.873	10,083.375
Dataset 2	BP	0.923	0.715	2.428	0.974	38.216
	CNN	0.824	0.596	1.929	0.979	80.746
	GRU	0.758	0.544	1.772	0.982	165.984
	LSTM	0.874	0.643	2.129	0.977	234.946
	LSTM-SA	0.809	0.576	1.890	0.980	383.995
	WOA-LSTM-SA	0.757	0.535	1.739	0.982	13,016.477
	IWOA-LSTM-SA	0.749	0.524	1.703	0.983	11,075.689

From Table 3, it is evident that our method does not have an advantage in terms of computation time compared to traditional machine learning models. Therefore, in scenarios where prediction accuracy is not a primary concern, traditional machine learning models can still be considered for top oil temperature prediction of transformers. The prediction model proposed in this paper, however, places a greater emphasis on improving prediction accuracy. To analyze and compare each model more comprehensively, this paper includes a residual plot. Using Dataset 1 as an example, in the residual plot (Figure 10), the true values are shown on the horizontal axis, while the vertical axis represents the residual values (percentage).

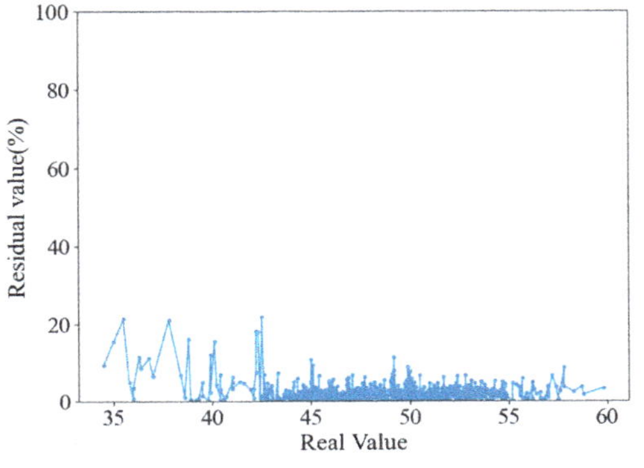

Figure 10. Model residuals.

The residual percentage is relatively higher for the data between 30 and 43 °C and 55 to 60 °C. The reason is as follows: there are about 4000 sample points within the temperature range of 43 to 55 °C, whereas the temperature ranges of 30~43 °C and 55~60 °C each contain approximately 200 sample points. This unbalanced distribution leads to low accuracy on sparse samples.

4.4. Ablation Experiment

To comprehensively validate the effectiveness of each component of the proposed method (IWOA-LSTM-SA), ablation experiments were conducted. Specifically, the experiments compared the following models: LSTM, LSTM-SA, WOA-LSTM, IWOA-LSTM, and WOA-LSTM-SA, with the LSTM model serving as the benchmark for comparison and analysis. Results are shown in Table 4.

Table 4. Ablation experiment evaluation metrics.

		LSTM	LSTM-SA	WOA-LSTM	IWOA-LSTM	WOA-LSTM-SA	IWOA-LSTM-SA
Dataset 1	RMSE	1.633	1.537	1.596	1.517	1.462	1.438
	MAPE	2.175	2.253	2.141	2.106	2.103	2.089
Dataset 2	RMSE	0.874	0.809	0.837	0.782	0.757	0.749
	MAPE	2.129	1.890	2.042	1.814	1.739	1.703

As shown in Table 4, the proposed model demonstrates higher prediction accuracy compared to the baseline model LSTM and other comparative models. Compared to LSTM, the RMSE of LSTM-SA decreased by 5.88% on Dataset 1 and by 7.44% on Dataset 2; the MAPE increased by 3.59% on Dataset 1 but decreased by 11.23% on Dataset 2. This validates the effectiveness of combining the SA algorithm with LSTM. Compared to LSTM-SA, the RMSE of WOA-LSTM-SA and IWOA-LSTM-SA decreased by 4.88% and 6.44% on Dataset 1, and by 6.43% and 7.42% on Dataset 2, respectively. The MAPE decreased by 6.66% and 7.28% on Dataset 1, and by 7.99% and 9.89% on Dataset 2, respectively. This validates the effectiveness of the optimization algorithms proposed in the models. Additionally, compared to WOA-LSTM and IWOA-LSTM, the RMSE of the proposed model decreased by 9.89% and 5.21% on Dataset 1, and by 10.51% and 4.22% on Dataset 2, respectively. The MAPE decreased by 2.43% and 0.81% on Dataset 1, and by 16.60% and 6.12% on Dataset 2, respectively.

In summary, compared to using optimization algorithms or SA individually, combining them results in a greater improvement in the performance of the prediction model.

4.5. Multi-Step Forecasting

The multi-step prediction model refers to a model that predicts a series of values rather than a single value. Multi-step prediction is more important in real-world power system operations because it provides longer-term temperature trend forecasts, which help to identify potential issues in advance. Therefore, this section conducts a multi-step prediction analysis, where the prediction steps are set to 3 steps (90 min) and 5 steps (150 min). The evaluation metrics are shown in Table 5, and the prediction results (for one week) are presented in Figure 11.

From Table 5, it can be seen that the error increases as the prediction step increases across all models. By comparing the RMSE metric, it can be concluded that the proposed model exhibits better accuracy across different prediction steps compared to the baseline model. Specifically, in Dataset 1 and Dataset 2, for the 3 step prediction, the RMSE of the proposed model is 1.537 and 1.015, respectively. This represents reductions of 12.83% and 38.65% compared to the BP model, 6.98% and 20.89% compared to the CNN model, 3.75% and 13.62% compared to the GRU model, 4.24% and 27.16% compared to the LSTM model, 1.60% and 17.93% compared to the LSTM-SA model, and 1.16% and 4.34% compared to the WOA-LSTM-SA model. For the 5 step prediction, the RMSE of the proposed model is 1.714

and 1.634, representing reductions of 12.60% and 11.11% compared to the BP model, 7.61% and 15.89% compared to the CNN model, 6.49% and 17.30% compared to the GRU model, 5.19% and 14.14% compared to the LSTM model, 4.56% and 12.82% compared to the LSTM-SA model, and 3.06% and 1.80% compared to the WOA-LSTM-SA model. By analyzing the multi-step prediction metrics, we conclude that the proposed model demonstrates good performance across different prediction steps compared to traditional models.

Table 5. Multi-step prediction evaluation metrics.

	Step	Model	RMSE	MAE	MAPE (%)	Time (s)
Dataset 1	1 (30 min)	BP	1.698	1.228	2.581	13.287
		CNN	1.646	1.170	2.462	32.317
		GRU	1.553	1.011	2.144	96.109
		LSTM	1.633	1.022	2.175	129.666
		LSTM-SA	1.537	1.031	2.253	174.497
		WOA-LSTM-SA	1.462	0.998	2.103	11,058.906
		IWOA-LSTM-SA	1.438	0.989	2.089	10,083.375
	3 (90 min)	BP	1.763	1.382	2.873	14.082
		CNN	1.652	1.221	2.557	22.572
		GRU	1.597	1.133	2.409	95.775
		LSTM	1.605	1.164	2.453	179.898
		LSTM-SA	1.562	1.162	2.448	229.012
		WOA-LSTM-SA	1.555	1.102	2.311	11,746.135
		IWOA-LSTM-SA	1.537	1.088	2.308	10,149.217
	5 (150 min)	BP	1.961	1.611	3.351	13.617
		CNN	1.855	1.411	2.973	21.579
		GRU	1.833	1.387	2.943	98.763
		LSTM	1.808	1.367	2.878	197.507
		LSTM-SA	1.796	1.345	2.832	240.519
		WOA-LSTM-SA	1.768	1.352	2.859	12,212.086
		IWOA-LSTM-SA	1.714	1.294	2.702	10,778.976
Dataset 2	1 (30 min)	BP	0.923	0.715	2.428	38.216
		CNN	0.824	0.596	1.929	80.746
		GRU	0.758	0.544	1.772	165.984
		LSTM	0.874	0.643	2.129	234.946
		LSTM-SA	0.809	0.576	1.890	383.995
		WOA-LSTM-SA	0.757	0.535	1.739	13,016.477
		IWOA-LSTM-SA	0.749	0.524	1.703	11,075.689
	3 (90 min)	BP	1.654	1.124	4.225	37.313
		CNN	1.283	1.012	3.166	79.190
		GRU	1.175	0.831	2.821	229.788
		LSTM	1.394	1.080	3.674	320.336
		LSTM-SA	1.237	0.923	3.111	433.645
		WOA-LSTM-SA	1.061	0.833	2.746	13,623.563
		IWOA-LSTM-SA	1.015	0.750	2.537	11,284.158
	5(150 min)	BP	1.838	1.568	4.854	37.081
		CNN	1.943	1.403	4.933	77.883
		GRU	1.976	1.387	4.801	264.860
		LSTM	1.903	1.414	4.765	171.239
		LSTM-SA	1.874	1.365	4.810	414.213
		WOA-LSTM-SA	1.664	1.249	4.298	12,823.645
		IWOA-LSTM-SA	1.634	1.229	4.162	10,984.776

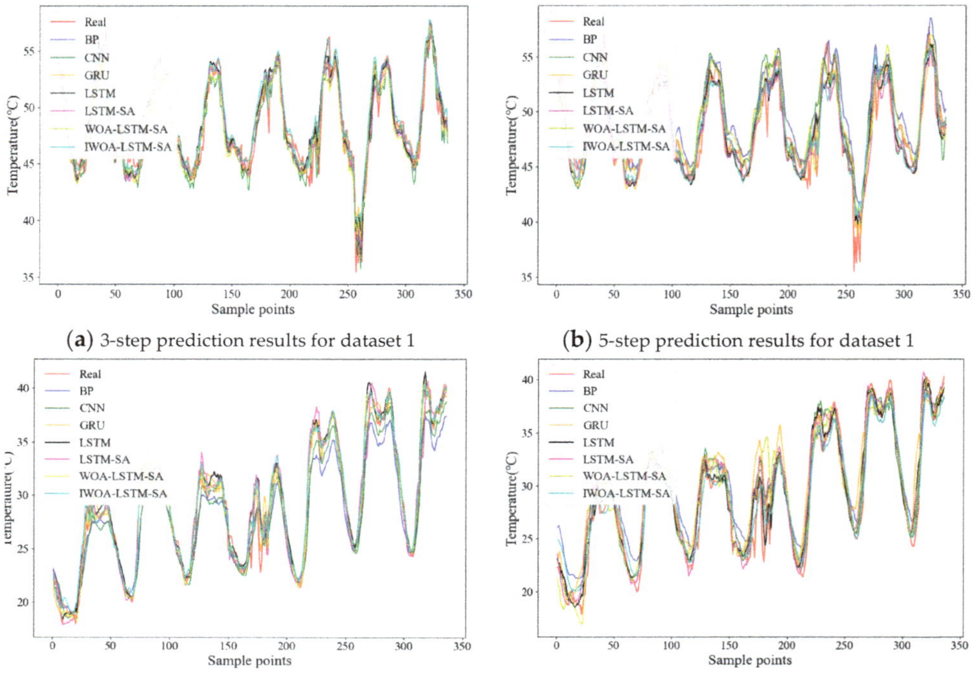

(**a**) 3-step prediction results for dataset 1 (**b**) 5-step prediction results for dataset 1 (**c**) 3-step prediction results for dataset 2 (**d**) 5-step prediction results for dataset 2

Figure 11. Multi-step prediction performance comparison across models (one week).

5. Conclusions

Oil temperature prediction can effectively prevent symmetrical and asymmetrical faults in transformers. This paper adopts a novel approach to improve the performance of top-oil temperature prediction during transformer operations. The proposed model has been tested using actual data, and some conclusions can be obtained as follows:

(1) To verify the efficacy of the IWOA, this paper conducts tests with eight test functions. The findings demonstrate that the IWOA outperforms GA, PSO, and WOA in terms of convergence speed and accuracy.

(2) To verify the effectiveness of the proposed model, extensive experiments were conducted using actual operating data. The experimental results indicate that the proposed approach outperforms current state-of-the-art methods. On Dataset 1, the model achieved reductions in RMSE of 15.31%, 12.64%, 7.41%, 11.94%, 6.44%, and 1.98% compared to the BP, CNN, GRU, LSTM, LSTM-SA, and WOA-LSTM-SA methods, respectively. Similarly, on Dataset 2, the model demonstrated significant improvements, with RMSE reductions of 18.85%, 9.09%, 1.19%, 14.29%, 7.42%, and 1.06% compared to the same benchmark methods.

(3) The proposed model performs effectively across various prediction steps compared to benchmark models. Specifically, for the 3-step prediction, the RMSE of the proposed model is 1.537 and 1.015 for Dataset 1 and Dataset 2, respectively, reflecting reductions of 12.83% and 38.65% compared to the BP model, 6.98% and 20.89% compared to the CNN model, 3.75% and 13.62% compared to the GRU model, 4.24% and 27.16% compared to the LSTM model, 1.60% and 17.93% compared to the LSTM-SA model, and 1.16% and 4.34% compared to the WOA-LSTM-SA model. For the 5-step prediction, the RMSE of the proposed model is 1.714 and 1.634, representing reductions of 12.60% and 11.11% compared to the BP model, 7.61% and 15.89% compared to

the CNN model, 6.49% and 17.30% compared to the GRU model, 5.19% and 14.14% compared to the LSTM model, 4.56% and 12.82% compared to the LSTM-SA model, and 3.06% and 1.80% compared to the WOA-LSTM-SA model.

Author Contributions: D.Z. led the conceptualization, methodology, software development, and original draft preparation. Validation was carried out by D.Z., H.X. and H.Q., while H.X. and D.Z. handled formal analysis. H.Q. managed the investigation, and Z.H. and W.D. provided resources. S.W. was responsible for data curation. Writing—review and editing involved D.Z., H.X., H.Q., Q.P. and J.Y., with visualization by D.Z., H.X. and J.Y. Supervision was provided by D.Z. and H.Q., project administration by Q.P. and S.W., and funding acquisition by D.Z. All authors have read and agreed to the published version of the manuscript.

Funding: This work was supported by the Electric Power Research Institute of Yunnan Power Grid Co., Ltd., Kunming, Yunnan, China (No. YNKJXM20220009).

Data Availability Statement: Data are contained in the article.

Conflicts of Interest: Authors Dexu Zou, Qingjun Peng, Shan Wang, Weiju Dai, and Zhihu Hong were employed by the company China Southern Power Grid Yunnan Power Grid Co., Ltd. The remaining authors declare that the research was conducted in the absence of any commercial or financial relationships that could be construed as a potential conflict of interest.

Appendix A

Table A1 displays the ten test functions used in this study.

Table A1. Test functions.

Function	Range
$F_1(x) = \sum_{n=1}^{k} x_n^2$	[−100, 100]
$F_2(x) = \sum_{n=1}^{k} \lvert x_n \rvert + \prod_{n=1}^{k} \lvert x_n \rvert$	[−10, 10]
$F_3(x) = \sum_{n=1}^{k} \left(\sum_{i-1}^{n} x_i \right)^2$	[−100, 100]
$F_4(x) = \sum_{n=1}^{k} n x_n^4 + random[0, 1)$	[−1.28, 1.28
$F_5(x) = 1 + \frac{1}{4000} * \sum (x_n^2) - \prod \left(\cos \left(\frac{x_n}{\sqrt{n}} \right) \right)$	[−600, 600]
$F_6(x) = [x_n^2 - 10\cos(2\pi x_n) + 10]$	[−5.12, 5.12
$F_7(x) = 20 - 20\exp\left(-0.2\sqrt{\frac{1}{k}\sum_{n=1}^{k} x_n^2} \right) - \exp\left[\frac{1}{k}\sum_{n=1}^{k} \cos(2\pi x_n) \right] + e$	[−32, 32]
$F_8(x) = \frac{\pi}{k}\left\{ 10\sin(\pi y_1) + \sum_{n=1}^{k-1} (y_n - 1)^2 [1 + 10\sin^2(\pi y_{n+1})] + (y_n - 1)^2 \right\} + \sum_{n=1}^{k} \mu(x_n, 10, 100, 4)$	[−50, 50]
$F_9(x) = \sum_{i=d}^{d} \left(-x_i \times \sin\left(\sqrt{\lvert x_i \rvert} \right) \right) + 418.98288727243369 \times d$	[−500, 500]
$F_{10}(x) = \sum_{i=1}^{d} \left((\ln(x_i - 2))^2 + (\ln(10 - x_i))^2 \right) - \left(\prod_{i=1}^{10} x_i \right)^{0.2}$	[2, 10]

References

1. Xu, X.; He, Y.; Li, X.; Peng, F.; Xu, Y. Overload Capacity for Distribution Transformers with Natural-Ester Immersed High-Temperature Resistant Insulating Paper. *Power Sys. Technol.* **2018**, *42*, 1001–1006.
2. Wang, S.; Gao, M.; Zhuo, R. Research on high efficient order reduction algorithm for temperature coupling simulation model of transformer. *High Volt. Appar.* **2023**, *59*, 115–126.
3. Liu, X.; Xie, J.; Luo, Y. A novel power transformer fault diagnosis method based on data augmentation for KPCA and deep residual network. *Energy Rep.* **2023**, *9*, 620–627. [CrossRef]
4. Chen, T.; Chen, Y.; Li, X. Prediction for dissolved gas concentration in power transformer oil based on CEEMDAN-SG-BiLSTM. *High Volt. Appar.* **2023**, *59*, 168–175.
5. Zang, C.; Zeng, J.; Li, P. Intelligent diagnosis model of mechanical fault for power transformer based on SVM algorithm. *High Volt. Appar.* **2023**, *59*, 216–222.
6. Ji, H.; Wu, X.; Wang, H. A New Prediction Method of Transformer Oil Temperature Based on C-Prophet. *Adv. Power Syst. Hyd. Eng.* **2023**, *39*, 48–55.

7. Tan, F.; Xu, G.; Zhang, P. Research on Top Oil Temperature Prediction Method of Similar Day Transformer Based on Topsis and Entropy Method. *Elect. Power Sci. Eng.* **2021**, *37*, 62–69.
8. Amoda, O.A.; Tylavsky, D.J.; McCulla, G.A.; Knuth, W.A. Acceptability of three transformer hottest-spot temperature models. *IEEE Trans. Power Deliv.* **2011**, *27*, 13–22. [CrossRef]
9. Zhou, L.; Wang, J.; Wang, L.; Yuan, S.; Huang, L.; Wand, D.; Guo, L. A Method for Hot-Spot Temperature Prediction and Thermal Capacity Estimation for Traction Transformers in High-Speed Railway Based on Genetic Programming. *IEEE Trans. Transp. Electrif.* **2019**, *5*, 1319–1328. [CrossRef]
10. Deng, Y.; Ruan, J.; Quan, Y.; Gong, R.; Huang, D.; Duan, C.; Xie, Y. A Method for Hot Spot Temperature Prediction of a 10 kV Oil-Immersed Transformer. *IEEE Access* **2019**, *7*, 107380. [CrossRef]
11. Zhao, B.; Zhang, X. Parameter Identification of Transformer Top Oil Temperature Model and Prediction of Top Oil Tempeature. *High. Volt. Eng.* **2004**, *30*, 9–10.
12. Wang, H.; Su, P.; Wang, X. Prediction of Surface Temperatures of Large Oil-Immersed Power Transformers. *J. Tsinghua Univ. Sci. Technol.* **2005**, *45*, 569–572.
13. Tan, M.; Hu, C.; Chen, J.; Wang, L.; Li, Z. Multi-node load forecasting based on multi-task learning with modal feature extraction. *Eng. Appl. Artif. Intell.* **2022**, *112*, 104856. [CrossRef]
14. Shang, Y.; Li, S. FedPT-V2G: Security enhanced federated transformer learning for real-time V2G dispatch with non-IID data. *Appl. Energy* **2024**, *358*, 122626. [CrossRef]
15. Bai, M.; Yao, P.; Dong, H.; Fang, Z.; Jin, W.; Yang, X.; Liu, J.; Yu, D. Spatial-temporal characteristics analysis of solar irradiance forecast errors in Europe and North America. *Energy* **2024**, *297*, 131187. [CrossRef]
16. Qing, H.; Jennie, S.; Daniel, J. Prediction of top-oil temperature for transformers using neural network. *IEEE Trans. Power Deliv.* **2000**, *15*, 1205–1211.
17. Tan, F.; Chen, H.; He, J. Top oil temperature forecasting of UHV transformer based on path analysis and similar time. *Elect. Power Autom. Equip.* **2021**, *41*, 217–224.
18. Li, S.; Xue, J.; Wu, M.; Xie, R.; Jin, B.; Zhang, H.; Li, Q. Prediction of Transformer Top-oil Temperature with the Improved Weighted Support Vector Regression Based on Particle Swarm Optimization. *High Volt. Appar.* **2021**, *57*, 103–109.
19. Tan, F.L.; Xu, G.; Li, Y.F.; Chen, H.; He, J.H. A method of transformer top oil temperature forecasting based on similar day and similar hour. *Elect. Power Eng. Tech.* **2022**, *41*, 193–200.
20. Yi, Y. Research on Prediction Method of Transformer Top-Oil Temperature Based on Assisting Dispatchers in Decision-Making. Master's Thesis, Southwest Jiaotong University, Chengdu, China, 2017.
21. Gharehchopogh, F.S.; Gholizadeh, H. A comprehensive survey: Whale Optimization Algorithm and its applications. *Swarm Evol. Comput.* **2019**, *48*, 1–24. [CrossRef]
22. Brodzicki, A.; Piekarski, M.; Jaworek-Korjakowska, J. The whale optimization algorithm approach for deep neural networks. *Sensors* **2021**, *21*, 8003. [CrossRef] [PubMed]
23. Mostafa Bozorgi, S.; Yazdani, S. IWOA: An improved whale optimization algorithm for optimization problems. *J. Comput. Des. Eng.* **2019**, *6*, 243–259. [CrossRef]
24. Naderi, E.; Azizivahed, A.; Asrari, A. A step toward cleaner energy production: A water saving-based optimization approach for economic dispatch in modern power systems. *Electr. Power Syst. Res.* **2022**, *204*, 107689. [CrossRef]
25. Gao, W.; Liu, S.; Huang, L. Inspired artificial bee colony algorithm for global optimization problems. *Acta Electron. Sin.* **2012**, *40*, 2396.
26. Shi, X.; Li, M.; Wei, Q. Application of Quadratic Interpolation Whale Optimization Algorithm in Cylindricity Error evaluation. *Metrol. Meas. Tech.* **2019**, *46*, 58–60.
27. He, Q.; Wei, K.; Xu, Q. Mixed strategy based improved whale optimization algorithm. *Appl. Res. Comput.* **2019**, *36*, 3647–3651.
28. Qiu, X.; Wang, R.; Zhang, W.; Zhang, Z.; Zhang, Q. Improved Whale Optimizer Algorithm Based on Hybrid Strategy. *Comput. Eng. Appl.* **2022**, *58*, 70–78.
29. Chen, Y.; Han, B.; Xu, G.; Kan, Y.; Zhao, Z. Spatial Straightness Error Evaluation with Improved Whale Optimization Algorithm. *Mech. Sci. Technol. Aero. Eng.* **2022**, *41*, 1102–1111.
30. Xu, J.; Yan, F. The Application of Improved Whale Optimization Algorithm in Power Load Dispatching. *Oper. Res. Manag. Sci.* **2020**, *29*, 149–159.
31. Naderi, E.; Mirzaei, L.; Pourakbari-Kasmaei, M.; Cerna, F.V.; Lehtonen, M. Optimization of active power dispatch considering unified power flow controller: Application of evolutionary algorithms in a fuzzy framework. *Evol. Intell.* **2024**, *17*, 1357–1387. [CrossRef]

symmetry

MDPI

Article

Inertia Support Capability Evaluation for Wind Turbine Generators Based on Symmetrical Operation

Zaiyu Chen [1,*], Yang Li [1] and Qian Zhou [2]

[1] School of Automation, Nanjing University of Science and Technology, Nanjing 210094, China; liyang970926@foxmail.com
[2] State Grid Jiangsu Electric Power Co., Ltd. Research Institute, Nanjing 211103, China; xjtu@163.com
* Correspondence: zaiyu.chen@njust.edu.cn

Abstract: With the increasing integration of new energy into the grid, the level of system inertia has been significantly reduced, posing a severe challenge to frequency stability. Consequently, there is an urgent need for wind turbine generators (WTGs) to actively provide inertia support through virtual inertia control. Assessing the inertia support capability of WTGs reasonably and setting appropriate controller parameters based on this assessment is a topic worthy of discussion. As WTGs' characteristics are mostly ignored in the evaluation of inertia support capability for WTGs, an evaluation method based on symmetrical operation is proposed. The proposed method considers the impact of real inertia and aerodynamic characteristics, thereby helping to determine reasonable virtual inertia coefficients and de-loading reserve capacity for WTGs. With the proposed method, it can be determined that large WTGs can provide inertia support capabilities close to those of synchronous generators to the grid without exceeding a 0.1% reduction in reserve capacity during de-loading operation.

Keywords: wind turbine generators; virtual inertia control; inertia support capability; symmetrical operation; equivalent inertia; aerodynamic efficiency

1. Introduction

In recent years, with the continuous maturation of new energy generation technologies, the proportion of new energy in the grid has been increasing. New energy sources operate dynamically decoupled from the grid frequency and no longer exhibit the inertia characteristics that hinder the change of rotor speed (i.e., grid frequency) of traditional synchronous generators (SGs), which are predominantly interfaced to the grid through power electronic converters [1,2]. Consequently, the grid with a high penetration of new energy increasingly exhibits "low-inertia" characteristics, with a diminishing capacity to withstand disturbances, thereby threatening system frequency stability [3,4].

A feasible solution to mitigate the issue of reduced inertia in the grid is to employ virtual inertia control, effectively simulating the inertia characteristics of synchronous generators [5,6]. Virtual inertia control integrates the rate of change of the grid frequency into the power control of new energy sources, thereby endowing new energy sources with a supporting effect that hinders the frequency drop of the grid (i.e., inertia support). When frequency oscillations occur in the grid due to an imbalance between generation and load, these sources adjust their output power to suppress the frequency variations [7].

The provision of inertia support to the grid by new energy sources is primarily determined by the gain coefficient (referred to as the virtual inertia coefficient hereinafter) of the virtual inertia control [8–10]. Concurrently, new energy sources must address the energy source for inertia support through measures such as de-loading [11] or the integration of energy storage systems [12]. In this context, while a higher virtual inertia coefficient can enhance the equivalent inertia provided by renewable energy sources to the grid, it also necessitates the expenditure of greater energy for inertia support.

Distinct from photovoltaic generation, wind power generation retains a physical rotor that can mitigate the imbalanced power between the prime mover (i.e., wind rotor) and the generator through the absorption and release of rotor kinetic energy [11]. With the increasing capacity of single wind turbine generators (WTGs), the dimensions of the blades and the inertia of the wind rotor are also progressively expanding [13], allowing the significant kinetic energy contained in the wind rotor to be utilized as an energy source for virtual inertia control in WTGs.

On this basis, research results have suggested using wind rotor inertia to offer grid inertia support without de-loading [14], but improper virtual inertia coefficient settings can excessively decrease rotor speed, leading to instability or power drop in WTGs [15,16]. To address this, References [17–19] set the virtual inertia coefficient based on the impact of control parameters on frequency stability, guiding practical parameter tuning. References [20,21] adjust the virtual inertia coefficient in real time according to kinetic energy release, improving rotor stability for WTGs. Reference [15] enhanced traditional virtual inertia control with a nonlinear controller, dynamically regulating support power based on wind and rotor speed variations, stabilizing rotor stability for WTGs, and preventing secondary frequency drop in the grid due to power drop. In summary, the above research has focused on tuning the virtual inertia coefficient, variable coefficient schemes, and enhanced virtual inertia control to maintain stability for WTGs.

Therefore, the inertia support provided by WTGs is primarily determined by the virtual inertia coefficient, as their rotor speed is decoupled from grid frequency, rather than being constrained by actual rotor inertia. However, the magnitude of the real rotor inertia influences the rate of speed change under virtual inertia control. Additionally, the unique aerodynamic characteristics of the wind rotor couple the rotor speed with the mechanical power output, indirectly affecting the inertia support capability. Current research on virtual inertia control mainly focuses on the rational setting of the virtual inertia coefficient to maximize inertia support while ensuring stable operation for WTGs. However, most research evaluates the appropriateness of the virtual inertia coefficient based on practical outcomes, with limited attention given to the effects of real rotor inertia and aerodynamic properties on virtual inertia control. Therefore, it is difficult to evaluate the ability of WTGs to support inertia based on their own inertia.

To address the aforementioned issues, this paper investigates the impact of prime mover characteristics on the traditional virtual inertia control of WTGs, aiming to explore whether the kinetic energy contained in the rotor of the WTG has the potential to serve as an energy source for providing inertia support to the grid. The contributions of this paper are as follows:

(1) Considering the coupling effects among power, inertia, rotor speed, and aerodynamics, this paper analyzes the energy transfer process during the active inertia support of WTGs under virtual inertia control, identifying two key factors that influence the inertia support capability, which are real inertia and aerodynamic characteristics.
(2) Furthermore, a symmetrical operation mode for WTGs is developed to assess the inertia support capability while accounting for prime mover characteristics, which is defined as an operation mode with a fixed ratio of WTGs' rotor speed to SGs' rotor speed.

(3) Analysis results based on a standard IEEE 10-machine 39-bus system and an NREL 5MW WTG simulation model indicate that the inertia support provided by large WTGs has a minimal impact on their aerodynamic efficiency under non-fault conditions, allowing them to deliver inertia support equivalent to that of SGs with a power reduction of no more than 0.1%.

2. Mathematical Model and Virtual Inertia Control of WTGs

This section mainly focuses on the mathematical model and virtual inertia control of WTGs, which constitutes the basis for the analysis in the later sections.

2.1. Mathematical Model of the WTG

A block diagram of the WTG structure based on a permanent magnet synchronous generator [22] is shown in Figure 1. The WTG utilizes the kinetic energy inherent in the moving air, which is converted into rotor kinetic energy by the lift force exerted on the blades. Thereafter, the rotor, via a gearbox, drives the rotation of the generator, and the electromagnetic power output from the generator is fed into the power grid through the power electronic converters.

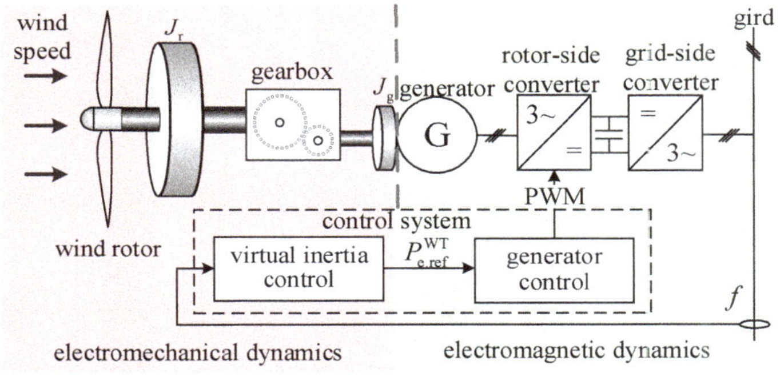

Figure 1. Schematic diagram of the WTG.

The aerodynamic power P_m^{WT} can be expressed as

$$P_m^{WT} = 0.5\rho\pi R^2 v^3 C_P(\lambda,\ \beta) \tag{1}$$

where ρ is the air density, R is the wind rotor radius, v is the wind speed, and $C_P(\lambda,\ \beta)$ is the wind energy utilization coefficient, which is a function of the tip speed ratio λ and the pitch angle β. The term λ is defined as the ratio of the linear velocity of the wind turbine blade tip to the wind speed

$$\lambda = \frac{\omega_r^{WT} R}{v} \tag{2}$$

where ω_r^{WT} is the rotor speed of the WTG. When the wind speed is lower than the rated wind speed, the pitch angle is generally fixed at 0° and C_P can be regarded as a function of λ. At this time, the $C_P(\lambda)$ has a single-peak characteristic, that is, there is an optimal tip speed ratio λ_{opt} so that the wind energy utilization coefficient obtains the maximum C_{Pmax}, and the WTG can capture the maximum wind energy.

The drive chain consists of a low-speed shaft on the wind rotor side and a high-speed shaft on the generator side which are connected via a gearbox. If the drive chain is approximated as a rigid shaft, the drive chain model based on a simplified single mass block [23] is obtained as follows:

$$\begin{cases} J_{\mathrm{h}}^{\mathrm{WT}} \omega_{\mathrm{g}}^{\mathrm{WT}} \frac{d\omega_{\mathrm{g}}^{\mathrm{WT}}}{dt} = P_{\mathrm{m}}^{\mathrm{WT}} - P_{\mathrm{e}}^{\mathrm{WT}} \\ J_{\mathrm{h}}^{\mathrm{WT}} = \frac{J_{\mathrm{r}}}{N_{\mathrm{g}}^2} + J_{\mathrm{g}} \end{cases} \tag{3}$$

where J_{r} and J_{g} are the rotor inertia of the wind rotor and the generator, respectively, N_{g} is the gear ratio, and $J_{\mathrm{h}}^{\mathrm{WT}}$ is the overall rotor inertia of the wind turbine converted to the high-speed side. The above parameters can be obtained according to the hardware parameters of the WTG when it leaves the factory. $P_{\mathrm{e}}^{\mathrm{WT}}$ is the generator electromagnetic power, and $\omega_{\mathrm{g}}^{\mathrm{WT}}$ is generator speed, which has

$$\omega_{\mathrm{g}}^{\mathrm{WT}} = \omega_{\mathrm{r}}^{\mathrm{WT}} \cdot N_{\mathrm{g}} \tag{4}$$

The comprehensive dynamic response process of the WTG in Figure 1 includes both electromechanical dynamics and electromagnetic dynamics. The former concentrates on the energy conversion process from wind to electrical energy within the wind turbine system, while the latter addresses the variations in the states of the electrical components during power generation. Given that the regulation process and response time of the electromagnetic dynamics are on the millisecond timescale, which is significantly lower than the second-scale timescale associated with virtual inertia control for WTGs, it is feasible within the scope of this study to decouple the fast and slow subsystems, thereby neglecting the electromagnetic dynamics and assuming that the generator can instantaneously output the reference power command $P_{\mathrm{e.ref}}^{\mathrm{WT}}$ [14], which has $P_{\mathrm{e}}^{\mathrm{WT}} = P_{\mathrm{e.ref}}^{\mathrm{WT}}$. Therefore, for the rotor-side converter in Figure 1, the virtual inertia control power command of the WTG through the generator control response is also applicable to the grid-side converter, which can be referred to [24,25].

2.2. Virtual Inertia Control

Virtual inertia control of the WTG is used to adjust the active power output according to the grid frequency change, thus emulating the inertia response characteristics of the SG and providing inertia support to the grid. As shown in Figure 2, the reference power $P_{\mathrm{e.ref}}^{\mathrm{WT}}$ of the WTG under virtual inertia control can be expressed as

$$\begin{cases} P_{\mathrm{e.ref}}^{\mathrm{WT}} = P_0^{\mathrm{WT}} + \Delta P_{\mathrm{VIC}}(t) \\ \Delta P_{\mathrm{VIC}}(t) = -K_{\mathrm{pf}} \cdot \frac{df(t)}{dt} \cdot \frac{S_{\mathrm{N}}^{\mathrm{WT}}}{f_{\mathrm{N}}} \end{cases} \tag{5}$$

where P_0^{WT} is the output power in the initial equilibrium state, which is related to the operating scenario of the WTG. In the mode of maximal wind energy capture, the WTG is operated under maximum power point tracking (MPPT), where P_0^{WT} is the output power when the WTG is operated at the MPPT (defined as the MPPT operation hereafter). In the mode of de-loading control, the speed regulation-based active power control is adopted for the WTG, where P_0^{WT} is the output power when the WTG is operated at the balance point of de-loading (defined as the de-loading operation hereafter). ΔP_{VIC} is the inertia support power calculated by Equation (5), K_{pf} is the virtual inertia coefficient to be set, $S_{\mathrm{N}}^{\mathrm{WT}}$ is the rated power of the WTG determined when the WTG leaves the factory, f is the real-time grid frequency obtained by measurement, and f_{N} is the nominal frequency of the grid determined by the power grid dispatching. In either scenario, the method relies on rapid regulation of the output power of the WTG, changing its speed to release/absorb rotor kinetic energy and buffering the unbalanced power between aerodynamic and electromagnetic power, so as to provide inertial support to the grid.

It is pointed out that according to the principle of frequency regulation in the grid, the frequency change is negatively correlated with the regulation of the power supply, that is, when the power shortage of power grid leads to a decrease in the grid frequency, an increase in the wind power output hinders the decrease in frequency. Therefore, there is a negative sign for calculating ΔP_{VIC} in Equation (5) and Figure 2 [11].

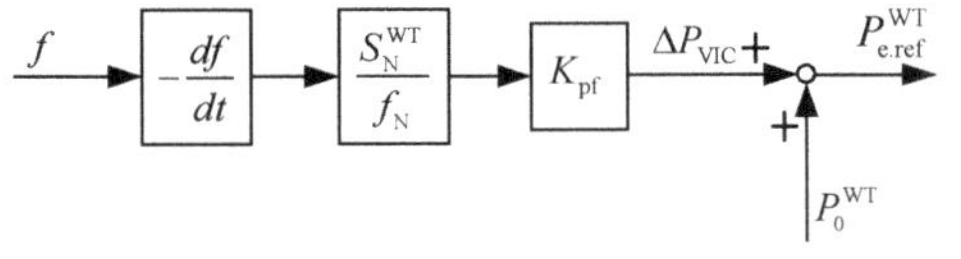

Figure 2. Virtual inertia control for the WTG.

3. Analysis of Influencing Factors for the Inertia Support Capacity of WTGs

The WTGs actively adjust their output power in response to grid frequency changes through virtual inertia control, and thereby provide inertia support. This section will further analyze the energy transfer in the active support process, and discuss the difference in energy utilization requirements between inertia support based on virtual inertia control and primary frequency regulation based on droop control. The impact of the real inertia and the inertia support capability of WTGs considering aerodynamic characteristics are also analyzed.

3.1. Energy Transfer in Inertia Support Processes

As can be seen from Section 2.1, the WTG converts the captured wind energy into rotor kinetic energy. Furthermore, the kinetic energy is converted into electrical energy using the generator and transmitted to the electrical grid. However, during the process of providing inertial support to the grid, the aerodynamic power of the WTG changes with the wind speed and the rotor speed. The electromagnetic power has to respond to the grid frequency changes, and thus the aerodynamic power and electromagnetic power cannot be guaranteed to be equal in real time, which makes the WTG accelerate/decelerate to store/release the rotor kinetic energy to achieve power balance.

For the mathematical model introduced in Section 2.1, the rotor kinetic energy E_{k} of the WTG at time t is

$$E_{\mathrm{k}}(t) = \frac{1}{2}J_{\mathrm{h}}^{\mathrm{WT}}\left(\omega_{\mathrm{g}}^{\mathrm{WT}}\right)^{2}(t) \tag{6}$$

Neglecting energy losses, the rate of change of the kinetic energy $\dot{E}_{\mathrm{k}}$ can be expressed as

$$\dot{E}_{\mathrm{k}}(t) = P_{\mathrm{m}}^{\mathrm{WT}}(t) - P_{\mathrm{e}}^{\mathrm{WT}}(t) \tag{7}$$

When the electromagnetic power is greater than the aerodynamic power, the kinetic energy of the WTG decreases, and vice versa. Combining Equations (1) and (5), the following equation can be obtained by substituting the aerodynamic power expression of the WTG and the virtual inertia control expression into Equation (7).

$$\dot{E}_{\mathrm{k}}(t) = \frac{1}{2}\rho\pi R^{5}\omega_{\mathrm{r}}^{3}(t)\frac{C_{\mathrm{P}}(\lambda)}{\lambda^{3}} - \left(P_{0}^{\mathrm{WT}} + \Delta P_{\mathrm{VIC}}(t)\right) \tag{8}$$

A schematic diagram of a typical inertia support dynamic process for WTGs that deals with frequency dip in the grid under the MPPT operating scenario is shown in Figure 3. For virtual inertia control, the WTGs operate at a stable equilibrium state at the beginning of the frequency disturbance. At this time, $\dot{E}_k$ is mainly determined by ΔP_{VIC}. As the rotor kinetic energy is released, the rotor speed of WTGs will reach a minimum point, followed by a gradual increase in aerodynamic and electromagnetic power. When the grid frequency reaches the steady state again, the rate of change of frequency will become 0. At this time, the ΔP_{VIC} becomes 0 and the WTGs return to the equilibrium state before the disturbance occurs. Therefore, the WTGs not only provide a certain degree of inertial support in the early stage but also reduce their power output for speed recovery in the later stage. Compared with the case where no inertial support is provided, the process of recovering the rotor speed of WTGs is equivalent to "absorbing" some energy from the grid.

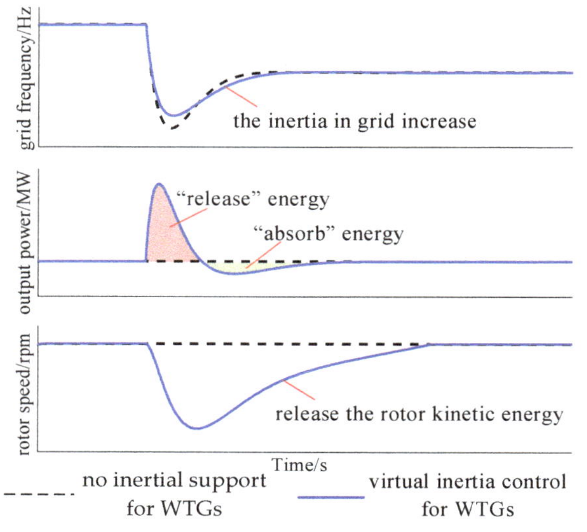

Figure 3. The schematic diagram of inertia support for WTGs during grid frequency dips.

3.2. Difference in Energy Utilization Requirements Between Virtual Inertia Control and Primary Frequency Regulation

Section 3.1 analyzes the energy transfer in the inertial support process for WTGs. This section will analyze the difference in kinetic energy release requirements between inertia support and primary frequency regulation, to reveal the inertia support characteristics under inertia support control and to provide the basis for a summary of the factors influencing inertia support capacity in Section 3.3.

In analogy to the primary frequency regulation of SGs, the WTGs adjust their output power by introducing deviations from the grid frequency, that is, droop control [11]. As shown in Figure 4, the output power of the WTG under this control can be expressed as

$$\begin{cases} P_{\text{e.ref}}^{\text{WT}} = P_0^{\text{WT}} + \Delta P_{\text{DC}}(t) \\ \Delta P_{\text{DC}}(t) = -K_{\text{df}} \cdot \Delta f(t) \cdot \frac{S_{\text{N}}^{\text{WT}}}{f_{\text{N}}} \end{cases} \tag{9}$$

where ΔP_{DC} is the support power for primary frequency regulation, K_{df} is the gain coefficient of droop control, and Δf is the deviation of f from f_N, that is $\Delta f(t) = f(t) - f_N$.

Comparing (5) with (9), it can be found that both controls adjust the output power according to the grid frequency, and therefore, both are often regarded as the same type of control, called integrated inertia control [26]. However, in terms of control objectives, the former aims to reduce the frequency variation caused by power imbalance under load disturbance, while the latter seeks to compensate for the power deficit after the frequency exceeds the dead zone.

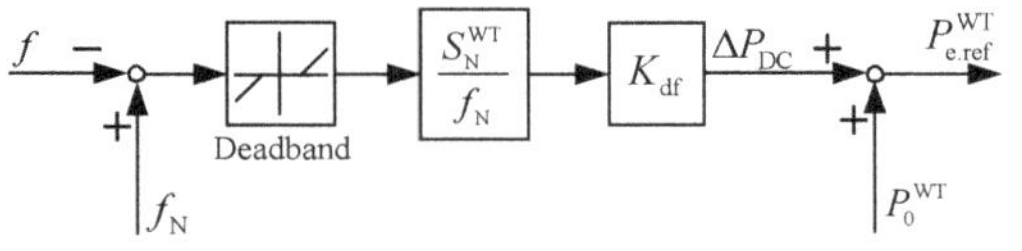

Figure 4. The droop control of the WTG.

The difference in energy utilization requirements between the two types of control described above is analyzed below. It is assumed that the grid initially operates steadily at f_N, that a frequency event occurs at time 0, and that $\Delta f(t)$ is bounded stable. For virtual inertia control,

$$\Delta E_{VIC} = \int_0^{\infty} \Delta P_{VIC}(t)\, dt = -\int_0^{\infty} K_{pf} \cdot \frac{df(t)}{dt} \cdot \frac{S_N^{WT}}{f_N}\, dt = -\frac{K_{pf} S_N^{WT}}{f_N} \cdot \lim_{t\to\infty} \Delta f(t) \tag{10}$$

At this point, the ΔE_{VIC} is bounded. Assuming that the grid frequency is out of the dead zone at t_0, for droop control, $\exists t_1 \in (t_0, \infty)$ that satisfies $\left|\Delta f(t) - \lim_{t\to\infty} \Delta f(t)\right| \leq \varepsilon$ when $t \geq t_1$ (ε is a very small positive number), so $\Delta f(t) \leq \varepsilon + \lim_{t\to\infty} \Delta f(t)$. Thus,

$$\begin{aligned}
\Delta E_{DC} &= \textstyle\int_{t_0}^{\infty} \Delta P_{DC}(t)\, dt = -\int_{t_0}^{\infty} K_{df} \cdot \Delta f(t) \cdot \dfrac{S_N^{WT}}{f_N} dt \\
&= -\frac{K_{df} S_N^{WT}}{f_N} \cdot \textstyle\int_{t_0}^{t_1} \Delta f(t)\, dt - \dfrac{K_{df} S_N^{WT}}{f_N} \cdot \int_{t_1}^{\infty} \Delta f(t)\, dt \\
&\geq \alpha - \frac{K_{df} S_N^{WT}}{f_N} \cdot \textstyle\int_{t_1}^{\infty} \left(\varepsilon + \lim_{t\to\infty} \Delta f(t)\right) dt \\
&= \alpha - \frac{K_{df} S_N^{WT}}{f_N} \cdot \left(\varepsilon + \lim_{t\to\infty} \Delta f(t)\right) \cdot t \Big|_{t_1}^{\infty}
\end{aligned} \tag{11}$$

where α is constant and $\alpha = -K_{df} S_N^{WT} / f_N \cdot \int_{t_0}^{t_1} \Delta f(t)\, dt$. At this point, ΔE_{DC} has no lower boundary. This shows that the output energy of WTGs under inertial support is a finite value, but the output energy under primary frequency regulation will be monotonically increased or decreased, and a continuous energy supply is needed.

Furthermore, a typical grid frequency dip event is used as an example to analyze the difference in energy utilization requirements between the two types of control under the MPPT operation scenario. For the inertia support dynamics of WTGs shown in Figure 3, the sum of the energy "released" and "absorbed" by WTGs is limited, and ΔE_{VIC} is bounded. For the schematic diagram of primary frequency regulation dynamics of WTGs shown in Figure 5, the energy released continues to increase, but the aerodynamic power under the MPPT operation scenario gradually decreases with the decreasing rotor speed, which results in severe kinetic energy overdraft and causes rotor speed instability. Therefore, it is necessary to exit the frequency regulation process and reduce the output power in time, but this also causes a secondary drop in the grid frequency.

The true inertia of WTGs is therefore more suitable for providing inertial support to the grid. To avoid a secondary drop in the frequency or instability of WTGs, a safe way for the WTGs to participate in primary frequency regulation is to reserve a certain capacity through de-loading control so that the output energy can be supplemented in time by the increase in aerodynamic power.

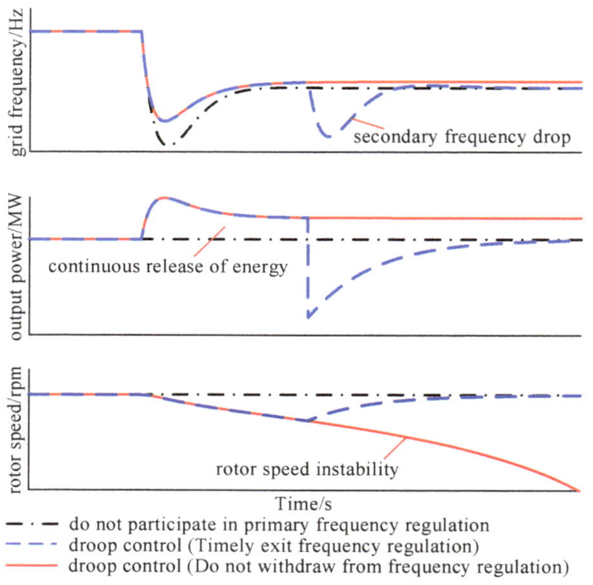

Figure 5. The schematic diagram of primary frequency regulation for WTGs during grid frequency dips.

3.3. Factors That Influence Inertia Support Capacity

Based on the analysis of energy transfer in virtual inertia control and its differences from primary frequency regulation in Sections 3.1 and 3.2, this section will further discuss two important factors affecting the inertia support capability for WTGs—real inertia and aerodynamic characteristics.

3.3.1. Real Inertia

Equation (5) shows that the WTG with larger rotor inertia can store more kinetic energy, while the rotor kinetic energy change will slow down with the increase in rotor inertia. Therefore, the magnitude of rotor inertia is one of the key factors that determine the inertia support capability of WTGs, and larger rotor inertia allows the WTG to provide more inertia support within the same speed variation range. With an increase in the capacity, the size and real inertia of the WTG gradually increase. The following example calculates and compares the real inertia of ten SGs in a standard IEEE 10-machine 39-bus system, an NREL 5 MW WTG [27], and an NREL 1.5 MW WTG [28]. The details can be seen in Appendix A.

Assuming the standard IEEE 10-machine 39-bus system f_N is 50 Hz (that is, the rated speed of all SGs ω_{fN}^{SG} is 3000 rpm), checking reference [29], the rated capacity of the G2 S_N^{SG2} is 700 MVA, the time constant of inertia H^{SG2} is 4.329 s and the number of pole pairs p is 1, and the inertia of this SG can be calculated as

$$J^{SG2} = \frac{2H^{SG2}S_N^{SG2}p^2}{\left(\omega_{fN}^{SG}\right)^2} = 4.27 \times 10^4 \text{ kgm}^2 \tag{12}$$

It is known that J_r of NREL 5 MW WTG is 3.5444067×10^7 kgm^2, J_g is 534.116 kgm^2, N_g is 97 and ω_{gN}^{WT} is 1173.7 rpm. The following calculation of the true inertia of the WTG is based on the rated speed of the SG because the rated speed of the WTG is different from that of the SG. Therefore, from (3), the true inertia of a 5 MW WTG converted to the high-speed side is calculated as

$$J_{ac}^{WT} = J_h^{WT} \cdot \left(\frac{\omega_{gN}^{WT}}{\omega_{fN}^{SG}} \right)^2 = 658.35 \text{ kgm}^2 \tag{13}$$

As the rated capacity of the SG is 140 times larger than that of the WTG, the true inertia of the WTG with the same capacity size as the SG is 6.4×10^4 kgm^2. A comparison of the data of the true inertia of the WTGs and all SGs in the system is shown in Table 1. Except for G5, the true inertia of the WTGs with the same capacity is greater than the true inertia of the SGs.

Table 1. Comparison of real inertia of WTGs and SGs.

SGs in the Standard IEEE 39-Bus System	Real Inertia of Each SG/kgm^2	Real Inertia of the NREL 5 MW WTG Under the Same Capacity/kgm^2	Real Inertia of the NREL 1.5 MW WTG Under the Same Capacity/kgm^2
G1	7.04×10^5	1.32×10^6	1.05×10^6
G2	4.27×10^4	9.22×10^4	7.32×10^4
G3	5.04×10^4	1.05×10^5	8.36×10^4
G4	4.03×10^4	1.05×10^5	8.36×10^4
G5	3.66×10^4	3.95×10^4	3.14×10^4
G6	4.90×10^4	1.05×10^5	8.36×10^4
G7	3.72×10^4	9.22×10^4	7.32×10^4
G8	3.42×10^4	9.22×10^4	7.32×10^4
G9	4.86×10^4	1.32×10^5	1.05×10^5
G10	5.91×10^4	1.32×10^5	1.05×10^5

3.3.2. Aerodynamic Characteristics

For the mathematical model in Section 2.1, the aerodynamic characteristics of different WTGs can be expressed with different functions of $C_P(\lambda)$. The function of $C_P(\lambda)$ has a single-peaked characteristic and $\lambda = \omega_r^{WT} R/v$, so P_m^{WT} is a function of ω_r^{WT} and $P_m^{WT}(\omega_r^{WT})$ still has a single-peaked characteristic. This shows that, unlike the mechanical characteristics of the SG, the aerodynamic power of the WTG is influenced by its rotor speed and the aerodynamic power will further influence its rotor speed. This coupling relationship between aerodynamic power and rotor speed makes the operation dynamics of the WTG more complex. In addition, different WTGs have specific $C_P(\lambda)$, so the variations in their aerodynamic power influenced by fluctuations in rotor speed are also related to their own $C_P(\lambda)$. The common $C_P(\lambda)$ curves [27,28,30] are shown in Figure 6.

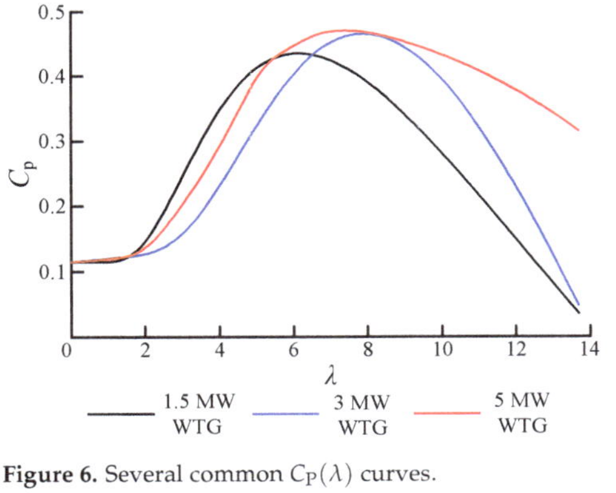

Figure 6. Several common $C_P(\lambda)$ curves.

In summary, the rotor speed dynamics of WTGs under virtual inertia control are influenced by the two factors mentioned above, specifically by the real inertia preventing large fluctuations in the rotor speed, and by the aerodynamic characteristics coupling the aerodynamic power with the rotor speed, which indirectly influences the inertia support capacity of WTGs to the grid.

4. Inertia Support Capability Evaluation for WTGs Based on Symmetrical Operation

To evaluate the inertia support capabilities of WTGs, this section constructs a symmetric operation mode that mirrors the rotor speed dynamic changes of conventional SGs, based on the principle of inertia support for SGs. In this mode, the inertia provided by WTGs without passing through a converter can be directly calculated. For WTGs that are more commonly grid-connected through converters, the symmetric operation serves as an approximation of the dynamic behavior for WTGs under virtual inertia control. However, this mode bridges the real and virtual inertia of WTGs, facilitating the analysis of the impact on aerodynamic efficiency when providing varying levels of inertia support.

4.1. Symmetrical Operation Mode of WTGs

In the absence of converters, the frequency of electricity generated by WTGs matches the grid frequency. Consequently, the speed variations of WTGs are in a fixed ratio to those of the SGs, governed by the generator's pole pairs and gearbox ratio. This paper denotes this operational mode as the symmetric operation mode, that is,

$$\frac{\omega_{\mathrm{g}}^{\mathrm{WT}}-\omega_{\mathrm{g0}}^{\mathrm{WT}}}{\omega_{\mathrm{g0}}^{\mathrm{WT}}}=k\cdot\frac{\omega_{\mathrm{f}}^{\mathrm{SG}}-\omega_{\mathrm{fN}}^{\mathrm{SG}}}{\omega_{\mathrm{fN}}^{\mathrm{SG}}} \tag{14}$$

where k is the proportional coefficient of rotor speed change, $k>0$, $\omega_{\mathrm{g0}}^{\mathrm{WT}}$ is the reference generator speed of WTGs, and $\omega_{\mathrm{f}}^{\mathrm{SG}}$ is the actual speed of SGs.

Under symmetric operation, the speed variation of WTGs is consistent with that of SGs, and WTGs interfaced through converters under virtual inertia control exhibit similar dynamic characteristics. Thus, the concept of symmetric operation can also be applied to converter-interfaced WTGs, where control is employed to align the speed of WTGs with that of SGs as described in Equation (14), termed as symmetric operation for WTGs. Due to the presence of converters, the rotor speed of WTGs is decoupled from grid frequency. Hence, the symmetric operation is considered an approximation of the dynamic behavior under virtual inertia control. The primary distinction lies in that the aerodynamic power variations are reflected in the output power for the former, while for the latter, they are manifested in rotor speed changes.

4.2. Equivalent Inertia Calculation of WTGs Under Symmetrical Operation

It can be seen from Equation (14) that the rotor speed for WTGs under symmetrical operation can be expressed as

$$\omega_{\mathrm{g}}^{\mathrm{WT}}=k\cdot\omega_{\mathrm{f}}^{\mathrm{SG}}\frac{\omega_{\mathrm{g0}}^{\mathrm{WT}}}{\omega_{\mathrm{fN}}^{\mathrm{SG}}}-(k-1)\omega_{\mathrm{g0}}^{\mathrm{WT}} \tag{15}$$

If the SG and the WTG are initially operated in a stable equilibrium state, the dynamic expression of the SG is

$$\Delta P_{\mathrm{m}}^{\mathrm{SG}}-\Delta P_{\mathrm{e}}^{\mathrm{SG}}=J^{\mathrm{SG}}\omega_{\mathrm{f}}^{\mathrm{SG}}\frac{d\omega_{\mathrm{f}}^{\mathrm{SG}}}{dt} \tag{16}$$

where $\Delta P_{\mathrm{m}}^{\mathrm{SG}}$ is the mechanical power deviation of SGs and $\Delta P_{\mathrm{e}}^{\mathrm{SG}}$ is the electromagnetic power deviation of SGs. The deviation of the speed fluctuation relative to $\omega_{\mathrm{fN}}^{\mathrm{SG}}$ can be ignored during the normal operation of the SGs [31], so it can be regarded as $\omega_{f}^{SG} = \omega_{fN}^{SG}$. Combined with Equations (3) and (15), the dynamic expression of the WTG is

$$\begin{aligned}\Delta P_{m}^{WT} - \Delta P_{e}^{WT} &= J_{h}^{WT}\omega_{g}^{WT}\frac{d\omega_{g}^{WT}}{dt} \\ &= J_{h}^{WT}\cdot\left(k\cdot\omega_{f}^{SG}\frac{\omega_{g0}^{WT}}{\omega_{fN}^{SG}} - (k-1)\omega_{g0}^{WT}\right)\cdot\left(k\cdot\frac{\omega_{g0}^{WT}}{\omega_{fN}^{SG}}\cdot\frac{d\omega_{f}^{SG}}{dt}\right) \\ &= k\cdot J_{h}^{WT}\cdot\left(\frac{\omega_{g0}^{WT}}{\omega_{fN}^{SG}}\right)^{2}\cdot\omega_{fN}^{SG}\frac{d\omega_{f}^{SG}}{dt}\end{aligned} \tag{17}$$

where ΔP_{m}^{WT} is the aerodynamic power deviation of WTGs and ΔP_{e}^{WT} is the electromagnetic power deviation of WTGs. The expression of equivalent inertia for WTGs is obtained by comparing Equations (16) and (17); it can be obtained that

$$J_{eq}^{WT} = k\cdot J_{h}^{WT}\left(\frac{\omega_{g0}^{WT}}{\omega_{fN}^{SG}}\right)^{2} \tag{18}$$

It should be noted that the symmetric operation can be achieved through the closed-loop speed control of WTGs, with the specific implementation process detailed in reference [32], and thus will not be reiterated here. The simulation results are presented in Section 4.3. Consequently, Equation (18) calculates the equivalent inertia under ideal conditions where there is no deviation in closed-loop speed control.

4.3. *Equivalent Inertia of WTGs Under Symmetrical Operation and Its Influence on Aerodynamic Efficiency*

As indicated in Section 4.2, the equivalent inertia of WTGs under symmetric operation is related to their initial operating conditions. Therefore, two constant wind speed scenarios of 10 m/s and 7 m/s are set, with the grid frequency fluctuating within the range of 49.5 Hz to 50.5 Hz and $k = 1$. This case evaluates and analyzes the equivalent inertia of WTGs operating under MPPT operation and its impact on aerodynamic efficiency. The deviation of ±0.5 Hz is the frequency deviation limit of power system under normal operating conditions specified in the Chinese standard [33], which can be used on demand.

4.3.1. Equivalent Inertia Evaluation

Building upon the equivalent inertia calculation method for the symmetric operation detailed in Section 4.2, this section compares the equivalent inertia of WTGs with the real inertia of SGs within the standard IEEE 10-machine 39-bus system, as mentioned in Section 3.3 (refer to Appendix A). The real inertia of the SGs has been calculated in Section 3.3, and thus will not be reiterated here.

By combining Equations (2) and (3) and substituting the relevant data, the calculation result for the equivalent inertia on the high-speed side of a single 5 MW WTG at a constant wind speed of 10 m/s is obtained as

$$J_{eq1}^{WT} = J_{h}^{WT}\left(\frac{\omega_{g0}^{WT}}{\omega_{fN}^{SG}}\right)^{2} = 596.81\ \mathrm{kgm^{2}} \tag{19}$$

Given that the rated capacity of G2 is 140 times that of the WTG, the equivalent inertia of the WTG with the same capacity as G2 should be 8.36×10^{4} kgm^2. Furthermore, the equivalent inertia of both WTGs at the same capacity is calculated and compared with the real inertia of all SGs in the system, as shown in Table 2. Consistent with the results from Section 3.3 regarding real inertia, the equivalent inertia of WTGs at the same capacity exceeds that of the real inertia of SGs, with the exception of G5.

Table 2. Comparison of real inertia of WTGs and SGs under 10 m/s wind speed scenario.

SGs in the Standard IEEE 39-Bus System	Real Inertia of Each SG/kgm^2	Real Inertia of the NREL 5 MW WTG Under the Same Capacity/kgm^2	Real Inertia of the NREL 1.5 MW WTG Under the Same Capacity/kgm^2
G1	7.04×10^5	1.19×10^6	7.42×10^5
G2	4.27×10^4	8.36×10^4	5.20×10^4
G3	5.04×10^4	9.55×10^4	5.94×10^4
G4	4.03×10^4	9.55×10^4	5.94×10^4
G5	3.66×10^4	3.58×10^4	2.23×10^4
G6	4.90×10^4	9.55×10^4	5.94×10^4
G7	3.72×10^4	8.36×10^4	5.20×10^4
G8	3.42×10^4	8.36×10^4	5.20×10^4
G9	4.86×10^4	1.19×10^5	7.42×10^4
G10	5.91×10^4	1.19×10^5	7.42×10^4

Similarly, the equivalent inertia provided by both WTGs under a constant wind speed scenario of 7 m/s is calculated and compared with the real inertia of all SGs in the system, with the results presented in Table 3. Although the equivalent inertia of the NREL 5 MW WTG is less than the real inertia of the SGs, it remains within the same order of magnitude. The equivalent inertia of the NREL 1.5 MW WTG is close to that of the SGs. In summary, under different scenarios, WTGs operating under the symmetric operation mode can provide inertia support to the grid that is comparable to, or even greater than, that of SGs.

Table 3. Comparison of real inertia of WTGs and SGs under 7 m/s wind speed scenario.

SGs in the Standard IEEE 39-Bus System	Real Inertia of Each SG/kgm^2	Real Inertia of the NREL 5 MW WTG Under the Same Capacity/kgm^2	Real Inertia of the NREL 1.5 MW WTG Under the Same Capacity/kgm^2
G1	7.04×10^5	5.85×10^5	3.64×10^5
G2	4.27×10^4	4.09×10^4	2.55×10^4
G3	5.04×10^4	4.68×10^4	2.91×10^4
G4	4.03×10^4	4.68×10^4	2.91×10^4
G5	3.66×10^4	1.75×10^4	1.09×10^4
G6	4.90×10^4	4.68×10^4	2.91×10^4
G7	3.72×10^4	4.09×10^4	2.55×10^4
G8	3.42×10^4	4.09×10^4	2.55×10^4
G9	4.86×10^4	5.85×10^4	3.64×10^4
G10	5.91×10^4	5.85×10^4	3.64×10^4

4.3.2. The Influence of Inertia Support on Aerodynamic Efficiency

Furthermore, under 10 m/s and 7 m/s wind speed scenarios, the WTGs achieve symmetric operation with the grid frequency through closed-loop speed control, with simulation results depicted in Figures 7 and 8. As shown in the figures, the rotor speed of the WTG varies in near consistency with the grid frequency (that is, the rotor speed of SGs) under both wind speed scenarios. Concurrently, the fluctuation in the C_p of the WTGs is minimal, with the maximum aerodynamic power loss being only 0.022% at a wind speed of 10 m/s and 0.023% at a wind speed of 7 m/s. Therefore, the impact of inertia support on the aerodynamic efficiency of WTGs under symmetric operation is negligible, implying that changes in the operating state for WTGs during the inertia support process have a minimal effect on aerodynamic power.

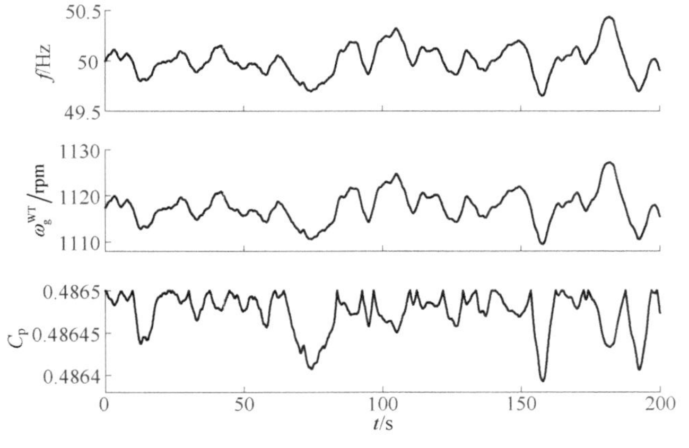

Figure 7. Grid frequency, rotor speed, and Cp curve for WTGs under 10 m/s wind speed scenario.

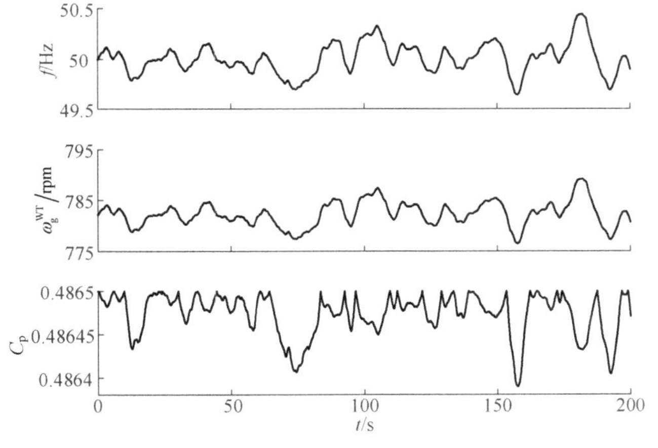

Figure 8. Grid frequency, rotor speed, and Cp curve for WTGs under 7 m/s wind speed scenario.

4.4. The Influence of WTG Speed Variation Range on Equivalent Inertia and Aerodynamic Efficiency

Furthermore, the relationship between equivalent inertia under symmetric operation mode and virtual inertia control is derived. As indicated by Equation (5), the electromagnetic power deviation of WTGs under virtual inertia control is equal to the inertia support power, that is,

$$\Delta P_{\mathrm{e}}^{\mathrm{WT}} = \Delta P_{\mathrm{VIC}}(t) = -K_{\mathrm{pf}} \frac{S_{\mathrm{N}}^{\mathrm{WT}}}{\left(\omega_{\mathrm{fN}}^{\mathrm{SG}}\right)^2} \omega_{\mathrm{fN}}^{\mathrm{SG}} \frac{d\omega_{\mathrm{f}}^{\mathrm{SG}}}{dt} \tag{20}$$

Based on the analysis in Section 4.3, by neglecting the impact of inertia support in the symmetric operation on $\Delta P_{\mathrm{m}}^{\mathrm{WT}}$, and combining Equations (17), (18), and (20), we can obtain

$$K_{\mathrm{pf}} = \frac{J_{\mathrm{eq}}^{\mathrm{WT}} \left(\omega_{\mathrm{fN}}^{\mathrm{SG}}\right)^2}{S_{\mathrm{N}}^{\mathrm{WT}}} \tag{21}$$

It is evident that under the symmetric operation, the equivalent inertia of WTGs corresponds directly to the virtual inertia coefficient. Furthermore, by combining Equation (18) with Equation (21), we can derive

$$k = \frac{K_{\text{pf}} S_{\text{N}}^{\text{WT}}}{J_{\text{h}}^{\text{WT}} \left(\omega_{\text{g0}}^{\text{WT}} \right)^2} \tag{22}$$

In other words, the virtual inertia coefficient determines the k. When the grid frequency (i.e., the rotor speed of SGs) is within the normal operating range, the virtual inertia coefficient can be used to establish the range of speed variation for WTGs. Concurrently, increasing the virtual inertia coefficient not only amplifies the equivalent inertia of the WTGs but also broadens the range of their speed variation, which in turn increases the impact on aerodynamic efficiency. Therefore, it is necessary to set aside a reserve capacity that exceeds the loss in aerodynamic efficiency to enable WTGs to provide inertia support under de-loading operation, ensuring the stability of the WTGs.

4.5. Verification of Virtual Inertia Control for WTGs Under De-Loading Operation

Section 4.4 establishes the relationship between the virtual inertia control coefficient of WTGs, the equivalent inertia under symmetric operation, and the range of speed variation. In this section, based on the simulation model provided in Appendix B, two case studies will be set up for WTGs in de-loading operation. The relationships derived in Section 4.4 will be utilized to calculate the virtual inertia coefficient, thereby validating that the virtual inertia control of WTGs can provide adequate inertia support to the grid.

4.5.1. Case 1: Operating in the Same Rotor Speed Range

In this case, the grid frequency is set to operate normally between 49.5 Hz and 50.5 Hz, with the speed variation range of WTGs being the same as that of the grid frequency (i.e., $k = 1$). If the WTGs initially operate at a constant wind speed of 10 m/s under MPPT, Section 4.3 indicates a loss of 0.022% in aerodynamic power, thus allowing for the determination to reserve a 0.1% reserve capacity. Furthermore, Section 4.3 calculates the equivalent inertia provided by the WTGs as $J_{\text{eq}}^{\text{WT}} = 596.81$ kgm^2, and $K_{\text{pf}} = 11.78$ can be derived from Equation (21). Simulations based on these calculations are conducted, with the system operation curves under virtual inertia control shown in Figure 9. Consequently, the WTGs effectively reduce the frequency fluctuations during normal grid operation while maintaining stable operation, demonstrating the effectiveness of the virtual inertia coefficient setting under the same rotor speed variation range.

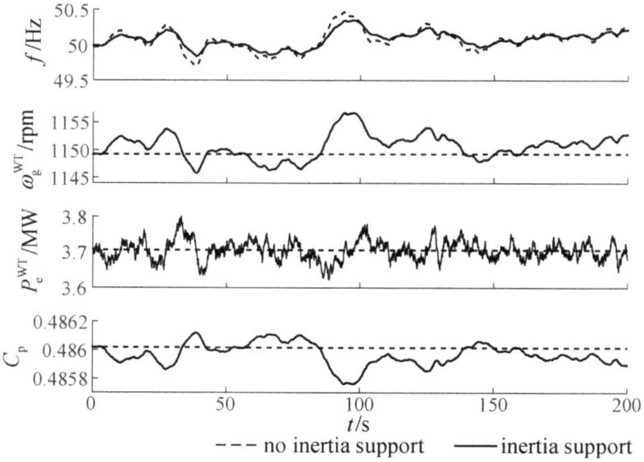

Figure 9. System operation curve for Case 1.

4.5.2. Case 2: Provide the Same Equivalent Inertia as G1

In this case, the grid frequency is set to operate normally between 49.5 Hz and 50.5 Hz, and the WTGs are required to provide an equivalent inertia comparable to that of G1, hence calculating the equivalent inertia provided by the WTGs as $J_{eq}^{WT} = 352$ kgm^2. If the WTGs initially operate at a constant wind speed of 7 m/s under MPPT control, then according to Equation (18), $k = 1.2$. The time-domain curves of the grid frequency, rotor speed, and C_p curve for WTGs under symmetrical operation are shown in Figure 10. The aerodynamic power loss is statistically 0.03%, thus allowing for the determination to reserve a 0.1% reserve capacity.

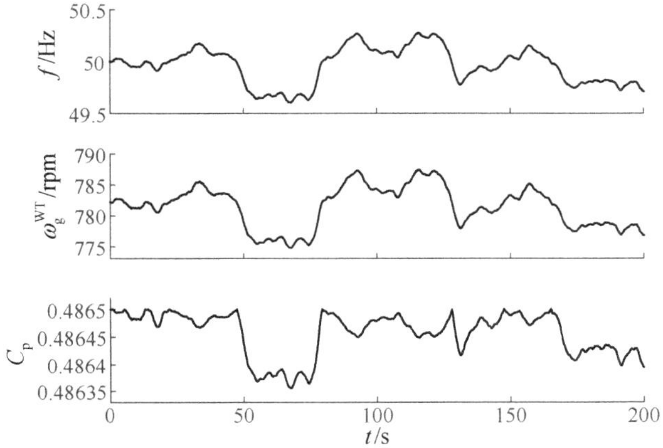

Figure 10. Grid frequency, rotor speed, and C_p curve for WTGs under symmetrical operation.

Furthermore, by substituting J_{eq}^{WT} into Equation (21), $K_{pf} = 6.94$ is calculated. Based on the aforementioned computational results, simulations are conducted, and the system operation curves under virtual inertia control are depicted in Figure 11. Similarly, the virtual inertia control of WTGs mitigates frequency fluctuations in the grid during normal operation while ensuring stable running, indicating that the virtual inertia coefficient is effective when tuned to provide an equivalent inertia comparable to that of G1.

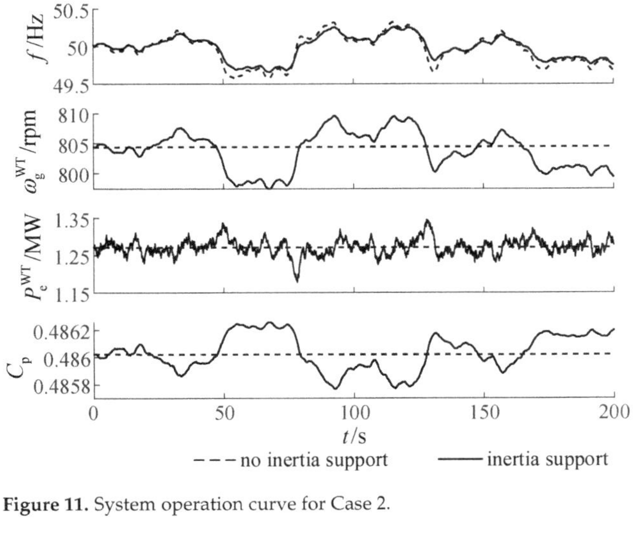

Figure 11. System operation curve for Case 2.

Finally, it can be seen from the above simulation that in the large-scale power systems dominated by SGs, the fluctuation range of output power and rotor speed for virtual inertia control of the WTGs under symmetrical operation mode is acceptable. Since the virtual inertia control provides inertia support for the grid, the WTGs will no longer be able to achieve maximum power tracking with MPPT control, which will affect the tip speed ratio and aerodynamic power extraction. However, the inertia support provided by large WTGs has a minimal impact on their aerodynamic efficiency under non-fault conditions, allowing them to deliver inertia support equivalent to that of SGs with power reduction of no more than 0.1%. When facing the actual turbulent wind speed, the power change and speed fluctuation of the WTG only under MPPT control are much larger than those in the above simulation [34,35].

5. Conclusions

In this paper, following an analysis of the energy utilization process and influencing factors of inertia support in WTGs, a symmetric operation mode is constructed to calculate and analyze the equivalent inertia under various scenarios and control parameters, as well as its impact on aerodynamic efficiency. The results indicate that, unlike primary frequency regulation which relies on continuous kinetic energy release, large WTGs can provide inertia support to the grid approaching that of SGs under de-loading operation with a reserve capacity not exceeding 0.1%. Currently, the design dimensions of WTGs are continually increasing with rotor diameters now exceeding 250 m [36], particularly offshore WTGs, resulting in a more substantial equivalent inertia and thus a stronger inertia support capability. Finally, considering the distinct energy requirements of primary frequency regulation and inertia support, it is recommended that WTGs utilize the reduced reserve capacity reserved for overspeed or pitch control to participate in the grid's primary frequency regulation while reserving their inherent inertia solely for virtual inertia support.

Author Contributions: Problem extraction, Z.C. and Q.Z.; conceptualization, Z.C. and Q.Z.; methodology, Z.C. and Y.L.; software, Z.C. and Y.L.; validation, Z.C. and Y.L.; formal analysis, Z.C. and Y.L.; data curation, Z.C. and Y.L.; writing—original draft preparation, Z.C. and Y.L.; writing—review and editing, Z.C. and Y.L.; supervision, Q.Z.; project administration, Q.Z.; funding acquisition, Q.Z. All authors have read and agreed to the published version of the manuscript.

Funding: This project is supported by the Science &Technology Program of SGCC (Research on Dynamic Low Frequency Offshore Wind Power System Coordinated with Onshore Converters, No. 4000-202218073A-1-1-ZN).

Data Availability Statement: Data are contained within the article.

Conflicts of Interest: Author Mr. Qian Zhou was employed by the company "State Grid Jiangsu Electric Power Co., Ltd. Research Institute". The authors declare that this study received funding from the Science &Technology Program of SGCC (Research on Dynamic Low Frequency Offshore Wind Power System Coordinated with Onshore Converters, No. 4000-202218073A-1-1-ZN). The funder had the following involvement with the study: the project administration, supervision, funding acquisition, problem extraction and conceptualization.

Appendix A

Appendix A.1. Parameters of Each SG in IEEE 10-Machine 39-Bus System

The SGs are derived from the standard IEEE 10-machine 39-bus New England system, with the main parameters of each SG presented in Table A1 [29]. Specifically, the inertia constant of the SGs is defined as the ratio of the kinetic energy to the rated capacity under rated operating conditions.

Table A1. Main parameters of each SG in the standard IEEE 10-machine 39-bus system.

SGs in the Standard IEEE 39-Bus System	Rated Capacity /MVA	Inertia Time Constant /s
G1	10,000	5
G2	700	4.329
G3	800	4.475
G4	800	3.575
G5	300	4.333
G6	800	4.35
G7	700	3.771
G8	700	3.471
G9	1000	3.45
G10	1000	4.2

Appendix A.2. Parameters of 1.5 MW WTG and 5 MW WTG

The 1.5 MW WTG and 5 MW WTG referenced are sourced from the National Renewable Energy Laboratory (NREL) [27,28], with their main parameters detailed in Table A2.

Table A2. Main parameters of 1.5 MW WTG and 5 MW WTG.

Parameter	Value of 1.5 MW WTG	Value of 5 MW WTG
wind rotor radius	35 m	63 m
wind rotor inertia	2.96×10^6 kgm^2	3.54×10^7 kgm^2
generator inertia	53 kgm^2	534.116 kgm^2
rated capacity	1.5 MW	5 MW
variable speed ratio	87.965	97
optimal tip speed ratio	6.32	7.6
maximum wind energy utilization coefficient	0.4382	0.4865

Appendix B

This paper constructs an IEEE 10-machine 39-bus system with an aggregated wind farm in DIgSILENT/PowerFactory based on the standard model outlined in reference [37], with the model structure depicted in Figure A1. The simulation model comprises ten SGs and an aggregated wind farm, with G1 equivalently aggregated from the U.S. and Canadian regions. Except for G1, all other SGs are equipped with governors, the structures and specific parameters of which are shown in Figure A2 and Table A3, respectively. The aggregated wind farms are connected to bus 9 within the system, consisting of 2000 NREL 5 MW WTGs. Since the operational dynamics of identical WTGs are consistent under the same wind speed scenario, this paper aggregates them into a single WTG with the same capacity as G1, resulting in a wind power penetration of approximately 37.3%.

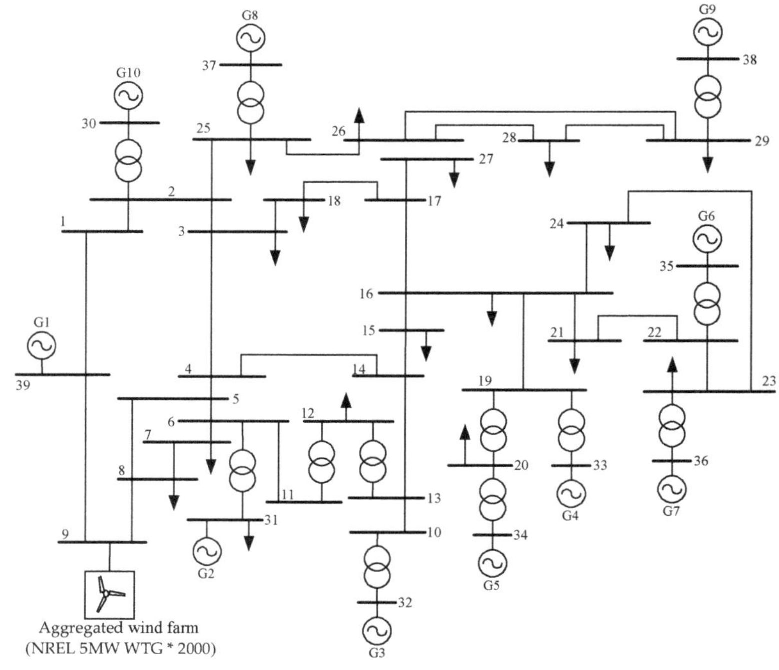

Figure A1. IEEE 10-machine 39-bus system with aggregated wind farm.

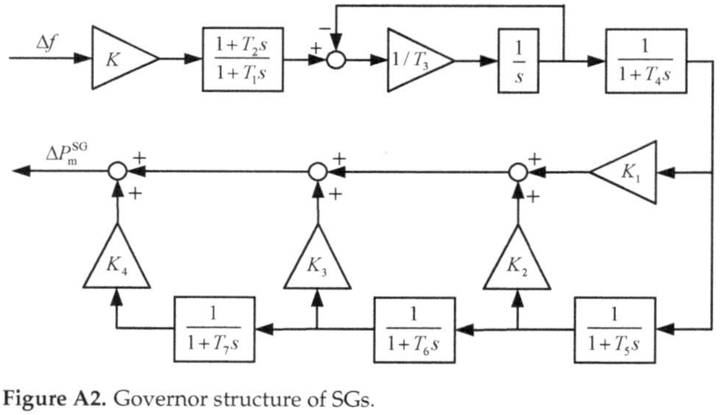

Figure A2. Governor structure of SGs.

Table A3. Governor parameters of SGs.

Parameter	Value
controller gain K	5
governor time constant T_1	0.2
governor derivative time constant T_2	1
servo time constant T_3	0.6
high-pressure turbine time constant T_4	0.6
intermediate-pressure turbine time constant T_5	0.5
medium-pressure turbine time constant T_6	0.8
low-pressure turbine time constant T_7	1
high-pressure turbine factor K_1	0.3
intermediate-pressure turbine factor K_2	0.25
medium-pressure turbine factor K_3	0.3
low-pressure turbine factor K_4	0.15

References

1. Shobug, M.A.; Chowdhury, N.A.; Hossain, M.A.; Sanjari, M.J.; Lu, J.; Yang, F. Virtual inertia control for power electronics-integrated power systems: Challenges and prospects. *Energies* **2024**, *17*, 2737. [CrossRef]
2. Yap, K.Y.; Sarimuthu, C.R.; Lim, J.M.Y. Virtual inertia-based inverters for mitigating frequency instability in grid-connected renewable energy system: A review. *Appl. Sci.* **2019**, *9*, 5300. [CrossRef]
3. Wang, H.; He, H.; Yang, Z.; Li, G.; Chen, Z.; Yang, J.; Zhang, G.; Dong, H. Frequency response methods for grid-connected wind power generations: A review. *Electr. Power Syst. Res.* **2023**, *221*, 109396. [CrossRef]
4. Jafari, M.; Gharehpetian, G.B.; Anvari-Moghaddam, A. On the role of virtual inertia units in modern power systems: A review of control strategies, applications and recent developments. *Int. J. Electr. Power Energy Syst.* **2024**, *159*, 110067. [CrossRef]
5. Fernández-Guillamón, A.; Gómez-Lázaro, E.; Muljadi, E.; Molina-Garcia, Á. A review of virtual inertia techniques for renewable energy-based generators. In *Renewable Energy–Technologies and Applications*; IntechOpen: Rijeka, Croatia, 2021.
6. Hu, P.; Li, Y.; Yu, Y.; Blaabjerg, F. Inertia estimation of renewable-energy-dominated power system. *Renew. Sustain. Energy Rev.* **2023**, *183*, 113481. [CrossRef]
7. Ekanayake, J.; Jenkins, N. Comparison of the response of doubly fed and fixed-speed induction generator wind turbines to changes in network frequency. *IEEE Trans. Energy Convers.* **2004**, *19*, 800–802. [CrossRef]
8. Sun, D.; Liu, H.; Gao, S.; Wu, L.; Song, P.; Wang, X. Comparison of different virtual inertia control methods for inverter-based generators. *J. Mod. Power Syst. Clean Energy* **2020**, *8*, 768–777. [CrossRef]
9. Tamrakar, U.; Shrestha, D.; Maharjan, M.; Bhattarai, B.P.; Hansen, T.M.; Tonkoski, R. Virtual inertia: Current trends and future directions. *Appl. Sci.* **2017**, *7*, 654. [CrossRef]
10. Hosseini, S.A. Frequency control using electric vehicles with adaptive latency compensation and variable speed wind turbines using modified virtual inertia controller. *Int. J. Electr. Power Energy Syst.* **2024**, *155*, 109535. [CrossRef]
11. Cheng, Y.; Azizipanah-Abarghooee, R.; Azizi, S.; Ding, L.; Terzija, V. Smart frequency control in low inertia energy systems based on frequency response techniques: A review. *Appl. Energy* **2020**, *279*, 115798. [CrossRef]
12. Wang, X.; Li, K.; Liu, Y.; Cheng, H. Virtual inertia control of energy storage system in wind turbine based on extended state observer. *Tran. China Electrotechnol. Soc.* **2018**, *33*, 1257–1264.
13. Bošnjaković, M.; Katinić, M.; Santa, R.; Marić, D. Wind turbine technology trends. *Appl. Sci.* **2022**, *12*, 8653. [CrossRef]
14. Morren, J.; Pierik, J.; De Haan, S.W. Inertial response of variable speed wind turbines. *Electr. Power Syst. Res.* **2006**, *76*, 980–987. [CrossRef]
15. Liu, B.; Zhao, J.; Huang, Q.; Milano, F.; Zhang, Y.; Hu, W. Nonlinear virtual inertia control of WTGs for enhancing primary frequency response and suppressing drivetrain torsional oscillations. *IEEE Trans. Power Syst.* **2021**, *36*, 4102–4113. [CrossRef]
16. Harisha, K.; Jayasankar, V.N. Frequency stability of a wind-based energy system by virtual inertia controller of an inverter connected to grid. *IETE J. Res.* **2024**, *70*, 3134–3149. [CrossRef]
17. Qiao, Y.; Guo, X.; Lu, Z.; Sun, R. Parameter setting of auxiliary frequency regulation of wind turbines considering secondary frequency drop. *Power Syst. Technol.* **2020**, *44*, 807–815.
18. Jiang, C.; Cai, G.; Yang, D.; Liu, X.; Hao, S.; Li, B. Multi-objective configuration and evaluation of dynamic virtual inertia from DFIG based wind farm for frequency regulation. *Int. J. Electr. Power Energy Syst.* **2024**, *158*, 109956. [CrossRef]
19. Tingyun, G.; Houyi, Z.; Bowen, L.; Junyi, M.; Qiang, F.; Yutao, X.; Qihui, F. Virtual inertia control to active support of the variable-speed wind turbine in variable frequency limit time. *Front. Energy Res.* **2024**, *12*, 1352385. [CrossRef]
20. Lee, J.; Muljadi, E.; Srensen, P.; Kang, Y.C. Releasable kinetic energy-based inertial control of a DFIG wind power plant. *IEEE Trans. Sustain. Energy* **2016**, *7*, 279–288. [CrossRef]
21. Qi, X.; Lei, L.; Yu, C.; Ma, Z.; Qu, T.; Du, M.; Gu, M. Adaptive distributed MPC based load frequency control with dynamic virtual inertia of offshore wind farms. *IET Control. Theory Appl.* **2024**, *18*, 2228–2238. [CrossRef]
22. Khan, A.; Aragon, D.A.; Seyedmahmoudian, M.; Mekhilef, S.; Stojcevski, A. Inertia emulation control of PMSG-based wind turbines for enhanced grid stability in low inertia power systems. *Int. J. Electr. Power Energy Syst.* **2024**, *156*, 109740. [CrossRef]
23. Mérida, J.; Aguilar, L.T.; Dávila, J. Analysis and synthesis of sliding mode control for large scale variable speed wind turbine for power optimization. *Renew. Energy.* **2014**, *71*, 715–728. [CrossRef]
24. Wang, Y.; Meng, J.; Zhang, X.; Xu, L. Control of PMSG-based wind turbines for system inertial response and power oscillation damping. *IEEE Trans. Sustain. Energy* **2015**, *6*, 565–574. [CrossRef]
25. Liu, H.; Li, M.; Liu, L.; Shi, J. Frequency trajectory planning-based transient frequency regulation strategy for wind turbine systems. *IEEE J. Emerg. Sel. Top. Power Electron.* **2021**, *10*, 3987–4000. [CrossRef]
26. Zhang, X.; Chen, Y.; Yue, S.; Zha, X.; Zhang, D.; Xue, L. Retrospect and prospect of research on frequency regulation technology of power system by wind power. *Power Syst. Technol.* **2018**, *42*, 1793–1803.
27. Jonkman, J. *Definition of a 5-MW Reference wind Turbine for Offshore System Development*; National Renewable Energy Laboratory: Golden, CO, USA, 2009.

28. Ren, Y.; Li, L.; Brindley, J.; Jiang, L. Nonlinear PI control for variable pitch wind turbine. *Control Eng. Pract.* **2016**, *50*, 84–94. [CrossRef]
29. Pai, M.A. *Energy Function Analysis for Power System Stability*; Kluwer Academic Publishers: Boston, MA, USA, 1989.
30. Huang, C.; Li, F.; Jin, Z. Maximum power point tracking strategy for large-scale wind generation systems considering wind turbine dynamics. *IEEE Trans. Ind. Electron.* **2015**, *62*, 2530–2539. [CrossRef]
31. Qin, X.; Su, L.; Chi, Y.; Guo, Q.; Xu, X. Functional orientation discrimination of inertia support and primary frequency regulation of virtual synchronous generator in large power grid. *Autom. Electr. Power Syst.* **2018**, *42*, 36–43.
32. Boukhezzar, B.; Siguerdidjane, H.; Hand, M.M. Nonlinear control of variable-speed wind turbines for generator torque limiting and power optimization. *J. Sol. Energy Eng.* **2006**, *128*, 516–530. [CrossRef]
33. *GB/T 15945-2008*; Power Quality—Frequency Deviation for Power System. Standardization Administration of China: Beijing, China, 2008.
34. Abolvafaei, M.; Ganjefar, S. Maximum power extraction from a wind turbine using second-order fast terminal sliding mode control. *Renew. Energy* **2019**, *139*, 1437–1446. [CrossRef]
35. Chen, Z.; Yin, M.; Zhou, L.; Xia, Y.; Liu, J.; Zou, Y. Variable parameter nonlinear control for maximum power point tracking considering mitigation of drive-train load. *IEEE/CAA J. Autom. Sin.* **2017**, *4*, 252–259. [CrossRef]
36. Yao, S.; Chetan, M.; Griffith, D.T.; Escalera Mendoza, A.S.; Selig, M.S.; Martin, D.; Kianbakht, S.; Johnson, K.; Loth, E. Aero-structural design and optimization of 50 MW wind turbine with over 250-m blades. *Wind Eng.* **2022**, *46*, 273–295. [CrossRef]
37. Guo, Y.; Bao, W.; Ding, L.; Liu, Z.; Kheshti, M.; Wu, Q.; Terzija, V. Analytically derived fixed termination time for stepwise inertial control of wind turbines—Part II: Application strategy. *Int. J. Electr. Power Energy Syst.* **2020**, *121*, 106106. [CrossRef]

symmetry

MDPI

Article

Bi-Level Optimization Scheduling Strategy for PIES Considering Uncertainties of Price-Based Demand Response

Xiaoyuan Chen, Jiazhi Lei * and Xiangliang Zhang

School of Intelligent Manufacturing, Huzhou College, Huzhou 313000, China; chenxiaoyuan@zjhzu.edu.cn (X.C.); zhangxiangliang@zjhzu.edu.cn (X.Z.)
* Correspondence: leijiazhi@126.com; Tel.: +86-15623559820

Abstract: The asymmetry-induced uncertainty in both sources and loads is a crucial and continuously spotlighted issue within modern power systems. Applying optimization scheduling method to deal with this asymmetry is a feasible solution. Accordingly, this paper proposes a bi-level park-level integrated energy system (PIES) optimization strategy considering uncertainties of price-based load demand response (PLDR). Firstly, a model for characterizing the uncertainties of the PLDR is developed based on fuzzy theory. Secondly, a bi-level two-stage PIES optimization model that includes multiple device models is established. In the first stage, the dynamic pricing optimization is carried out with the aim of maximizing user satisfaction. In the second stage, the PIES scheduling strategy optimization is performed with the aim of minimizing the operation costs of PIES. Finally, multiple scenarios are set to conduct comparative validation, which demonstrates that the proposed method not only improves the renewable energy integration capacity of the system, optimizes the load profiles, and enhances the economic and low-carbon performance, but also increases user satisfaction, thus providing a reference for the dispatch and operation of the park-level integrated energy system.

Keywords: uncertainty of asymmetry; demand response; price decision; uncertainty; fuzzy theory

1. Introduction

To achieve the "dual carbon" goals, alleviate energy supply–demand asymmetry and improve energy utilization, integration of multiple complementary resources by applying the optimization method has become a significant direction for future energy revolution [1]. Park-level integrated energy systems, which are a modern power system characterized by coupling multiple energy resources and achieving cascaded energy utilization, play a crucial role in this process and are also one of the key trends in the field of energy development. Within PIES, on one hand, there are multiple flexible energy conversion devices, and on the other hand, there are many potential sources of flexibility. Demand response on the consumer side is one particularly important form of flexibility resource. Regarding demand response, both domestic and foreign governments have introduced several incentive-based support policies, thus laying a solid foundation for the participation of demand response mechanisms in ancillary service markets [2–4]. This approach, which incentivizes users to change their consumption behaviors by providing them with economic compensation, can partially alleviate the supply–demand imbalance during peak load periods. It provides an available approach for demand-side resources in PIES strategy optimization.

 https://doi.org/10.3390/sym17010043

Recently, extensive research on strategy optimization problems of PIES has been conducted. The concept of an Energy Hub (EH) was introduced in [5] and a standardized modeling of PIES was presented for planning and scheduling studies. Literature [6] proposed an IES scheduling method considering energy-carbon comprehensive pricing model. It establishes an energy scheduling model using a Stackelberg game approach to maximize profits. This method can enhance economic efficiency and reduce carbon emissions. Literature [7] proposes a two-stage coordinated optimization method with long-term and short-term scheduling strategy for PIES. The results show that under long-term operating conditions, two-stage scheduling strategy of PIES can acquire more energy revenue. This two-stage scheduling strategy determines the optimum capacity for long-term scheduling while ensuring effective integration with power grid, thereby cutting down decision-making complexity. Literature [8] proposes a comprehensive planning strategy considering the coupling relationship of electricity, natural gas, and thermal load. In the aforementioned literature, researchers have studied system optimization problems from various perspectives, including optimization scheduling methods, equipment model modeling, and solution algorithms. However, none of these studies considered the influences of demand response on optimization process. Indeed, in practical operation of integrated energy systems, system operators can guide users to actively adjust their energy consumption and participate in demand response programs, which can improve system's economic and low-carbon performance. Given this, ref. [9] analyzes the implementation of integrated energy demand response based on cooling, heating, and electricity loads. It examines how demand response can enhance the system economy and flexibility. Literature [10] explores an implementation of demand response in real-time electricity markets. It demonstrates that demand response can maximize the profit return of energy loads and alleviate the supply–demand imbalance of IES. Literature [11] establishes an intra-day optimization scheduling strategy for regional IES considering LDR. The results demonstrate that this scheduling strategy mitigates the intra-day supply–demand imbalances of renewable resources and loads, enhances system stability, and further reduces operational costs.

Indeed, it is evident that the optimal planning problem for integrated energy systems has gained significant attention from the academic community both domestically and internationally. Among these research efforts, there is a particular focus on fully exploiting the potential of demand-side response to improve economic and stability. This area of research is becoming a key hotspot in PIES. Although existing research has made preliminary explorations in this area, there are still some issues that need to be addressed. Firstly, most research efforts predominantly focus on deterministic demand response programs without considering the potential impact of various uncertainties that may exist in practical applications. However, for non-direct control demand response programs (such as demand response based on real-time price), users need to adjust their energy consumption plans based on real-time market signals. The inherent differences in users' lifestyle habits, rationality, and behavioral preferences create a high degree of uncertainty in their demand response capabilities. Therefore, disregarding the uncertainty impact of demand response would result in planning solutions that are not truly globally optimal in nature. In addition, in existing research, stochastic optimization or robust optimization methods are commonly used to model uncertainty factors. Stochastic optimization involves extracting statistical characteristics of uncertain factors from a large amount of reliable historical data and generating expected scenarios based on the obtained probability distribution functions. However, for this study, the acquisition of demand-side classified time-of-use data is limited owing to simultaneous involvement of energy consumption timing and type of energy consumption [12] in demand response under integrated energy systems. As a result, acquiring and analyzing demand response data becomes challenging, making it difficult to obtain reliable probability distributions for user response parameters. Furthermore, user response behavior often exhibits randomness and fuzziness [13,14], which is difficult to effectively describe using fixed probability distribution functions. Therefore, these reasons present significant obstacles for the application of stochastic optimization methods in this study. On the other hand, traditional robust optimization approaches typically focus on the most unfavorable uncertain scenarios for achieving system objectives, leading to conservative solution outcomes that may not satisfy the planning requirements for system economics effectively.

To address the asymmetry of sources and loads in modern power systems, this paper proposes a bi-level PIES optimization scheduling method considering the uncertainties of PLDR. This paper focuses on demand response programs based on dynamic pricing as an example. By taking into account both the configuration of PIES and demand-side management strategies in a comprehensive way, the proposed model aims to optimize the configuration of PIES. Additionally, the impact of uncertainty in demand response is described by using fuzzy theory. Finally, the validity of the proposed method is verified by means of setting multiple scenarios.

2. Park-Level Integrated Energy System

The PIES is a multi-input and multi-output system integrating multiple energy conversion facilities, energy resources, and loads [15]. Different types of energy flows are conducted in PIES and multiple load demands are satisfied. In the PIES designed in this paper, wind power, photovoltaics, electric boilers (EB), gas boilers (GB), battery energy storage devices, and combined heat and power (CHP) are included. Additionally, energy storage devices, under different operating conditions, can store excess energy, thereby enhancing the system's capacity to accommodate renewable energy and forming beneficial interactions with the grid.

3. Uncertainties of PLDR

In PLDR, users voluntarily adjust their energy consumption based on dynamic prices provided by the system operator. Due to subjective factors, demand response exhibits a certain level of volatility. This paper takes into account the uncertainty of demand response by using fuzzy theory. From the perspective of demand characteristics, the loads in PIES can be approximately classified into four types: essential loads, reducible loads, transferable loads, and substitutable loads [16,17]. The following section discusses the modeling of these types of loads.

3.1. Essential Loads (Els)

ELs refer to loads that could not be interrupted under a specific time period. They possess fixed and uninterrupted characteristics, meaning they do not respond to dynamic electricity or gas prices. With the presence of ELs in PIES, their operating characteristics of ELs can be represented by Equations (1) and (2).

$$P_{el,t} = \gamma_{el,t} P_t \tag{1}$$

$$P_{hl,t} = \gamma_{hl,t} P_t \tag{2}$$

where P_t represents the total load value, $P_{el,t}$ represents the value of essential electric load, $P_{hl,t}$ represents the value of essential thermal load, $\gamma_{el,t}$ represents the ratio of essential electric load to total load value, and $\gamma_{hl,t}$ represents the ratio of essential thermal load to total load value at time t.

3.2. Reducible Loads and Transferable Loads

Reducible loads are defined as the loads that users can voluntarily reduce according to their own interests and needs without sacrificing the fulfillment of their own demand. These loads have a certain degree of flexibility and can respond to dynamic electricity and natural gas prices. Common reducible loads include some air conditioning equipment, household appliances, and certain types of lighting fixtures used in residential buildings. Transferable loads are the loads whose supply periods can be changed according to a schedule. Within a certain time frame, the total amount of power supplied remains constant, but the specific hours of supply can be adjusted without significantly impacting the users. Transferable loads exhibit strong controllability. Common examples of transferable loads include washing machines and water heaters. Considering the existence of these reducible and transferable loads in PIES, this paper aims to establish a mapping relationship between the reducible load quantity, transferable load quantity, and dynamic electricity and natural gas prices (ENGPs). This mapping relationship is described as follows:

1. Dynamic ENGPs are positively correlated with the reducible load quantity. This means that as the ENGPs increase, the reduction rate of load at a given time also increases. Similarly, during periods with higher ENGPs, the load tends to be shifted to periods with lower prices. In other words, as the ENGPs increase, the transfer rate of load at a given time also increases;

2. The reduction rate and transfer rate of the load range between zero and one. As the electricity and natural gas prices increase, these rates tend to approach one, while they tend to approach zero when the prices are lower;

3. In an ideal scenario, assuming that user behavior follows a normal distribution pattern, the reduction rate and transfer rate of the load are symmetric about a center point of electricity and natural gas prices.

The sigmoid function has the properties of monotonicity, a constrained range, and symmetry, which are in line with the aforementioned mapping relationship. Given the complexity of this mapping relationship, a non-standard type of sigmoid function is used for modeling [18]. Additionally, considering the uncertainty in user behavior, most users are sensitive to benefits, but the sensitivity varies among individuals. Moreover, non-benefit factors such as psychological factors and unexpected events can also influence user behavior. As a result, there is significant fluctuation in the reduction rate and transfer rate of the load. In general, when prices are lower, the benefits are smaller, and subjective factors play a dominant role. This leads to higher uncertainty in user behavior, resulting in larger fluctuations in the reduction rate and transfer rate of the load. Conversely, when prices are higher, the benefits are greater, and the influence of benefits becomes dominant. This leads to lower uncertainty in user behavior, resulting in smaller fluctuations in the reduction rate and transfer rate of the load. This paper utilizes triangular fuzzy numbers to characterize this uncertainty. Therefore, the modeling of the reduction rate, transfer rate, and their uncertainties is as follows:

$$\lambda_{xc,t} = \frac{1}{1 + e^{-\alpha_{x1}(x_{xc,t} - \beta_{x1})}} + \varepsilon_{xc,t} \tag{3}$$

$$\lambda_{xs,tt'} = \frac{1}{1 + e^{-\alpha_{x2}(x_{xs,tt'} - \beta_{x2})}} + \varepsilon_{xs,tt'} \tag{4}$$

$$\begin{cases} \varepsilon_{xc,t} = (-d_{xc,t}, 0, d_{xc,t}) \\ d_{xc,t} = \begin{cases} k_{x1}, x_{xc,t} \leq 0 \\ k_{x1} e^{-k_{x2} x_{xc,t}}, x_{xc,t} \geq 0 \end{cases} \end{cases} \tag{5}$$

$$\begin{cases} \varepsilon_{xs,tt'} = (-d_{xs,tt'}, 0, d_{xs,tt'}) \\ d_{xs,tt'} = \begin{cases} k_{x3}, x_{xs,tt'} \le 0 \\ k_{x3}e^{-k_{x4}x_{xc,tt'}}, x_{xs,tt'} \ge 0 \end{cases} \end{cases} \tag{6}$$

where α_{x1}, α_{x2}, β_{x1}, β_{x2} are the parameters of the sigmoid functions, $\lambda_{xc,t}$ represents the reduction rate of the xth type of load (electrical or thermal loads) at time t, $\lambda_{xs,tt'}$ represents the transfer rate of the xth type of load from t to t', $x_{xc,t}$ represents the price difference between electricity or natural gas prices and the baseline electricity or natural gas price, $x_{xs,tt'}$ represents the price difference between electricity or natural gas prices and electricity or natural gas prices, $\varepsilon_{xc,t}$ represents the uncertainty error in the reducible load of the xth type, $\varepsilon_{xs,tt'}$ represents the uncertainty error in transfer rate of the xth type of load from time t to t', k_{x1}, k_{x2}, k_{x3}, k_{x4} are proportional coefficients, $d_{xc,t}$ represents the maximum error in the reduction rate of the xth type of load at time t, $d_{xs,tt'}$ represents the maximum error in the transfer rate of the xth type of load from time t to t';

Taking all the above formulas into consideration, we can obtain the values of electricity/thermal load after reduction and transfer for each period, as shown below.

$$P_{xc,t} = (1 - \lambda_{xc,t})\gamma_{xc,t}P_t \tag{7}$$

$$P_{xs,t} = (1 - \sum_{t' \in T, t' \neq t} \lambda_{xs,tt'})\gamma_{xs,t}P_t + \sum_{t' \in T, t' \neq t} \lambda_{xs,t't}\gamma_{xs,t'}P_{t'} \tag{8}$$

where $P_{xc,t}$ represents the value of the xth type of load after reduction at time t, $P_{xs,t}$ represents value of the xth type of load after transfer, $\gamma_{xc,t}$ represents the ratio of the value of the xth type of load that can be reduced to the total load value, $\gamma_{xs,t}$ represents the ratio of the value of the xth type load that can be transferred to the total load value.

3.3. Substitutable Loads

A substitutable load refers to a load where users have the flexibility to choose the form of energy based on their specific needs. Common examples of substitutable loads are heating equipment, kitchen appliances, and certain air conditioning systems. In PIES, these loads exist, and users can adaptively choose the form of energy according to different electricity and natural gas prices during different time periods. Therefore, it is essential to establish a model for the substitution rate of the load with respect to price. In an ideal scenario, if the electricity price is high, users choose natural gas as the energy source, while if the electricity price is low, they choose electricity as the energy source. However, apart from considering the benefits, users are also influenced by subjective factors and can voluntarily choose the energy supply method. Taking all these considerations into account, the mapping relationship between the substitution rate of the electricity load (denoted as λ_{et}) and the difference between ENGPs (denoted as x_{eh}) can be described as follows:

1. The substitution rate of the electricity load is positively correlated with the differences between ENGPs. The larger the price difference, the closer the substitution rate of the electricity load is to one, and the smaller the price difference, the closer it is to xero.

2. In the substitutable load category, assuming that electricity and natural gas are interchangeable and users have no preference in usage, user behavior follows a normal distribution pattern. When x_{eh} equals 0, λ_{et} equals 0.5, and the mapping relationship λ_{et} is symmetric about the center (0, 0.5), with its rate of change being symmetric about the y-axis. In other words, the derivative of the substitution rate of the electricity load is an even function.

3. When x_{eh} approaches zero, the user's benefit is small, indicating strong subjectivity and high uncertainty. When x_{eh} approaches infinity, the user's benefit is large, indicating

weak subjectivity and low uncertainty. Therefore, the uncertainty model is an even function with a monotonically increasing and then decreasing trend.

Based on the above relationships, this paper adopts the arctangent function to describe the relationship between λ_{et} and x_{eh}. The uncertainty is modeled using a Gaussian function. The modeling is as follows:

$$\lambda_{et,t} = \frac{1}{\pi}\left(\arctan x_{eh,t} + \frac{\pi}{2}\right) + \varepsilon_{eh,t} \tag{9}$$

$$\begin{cases} \varepsilon_{eh,t} = (-d_{eh,t}, 0, d_{eh,t}) \\ d_{eh,t} = ae^{-\frac{x_{eh,t}^2}{2c^2}} \end{cases} \tag{10}$$

where a and c are parameters of Gaussian function, $\varepsilon_{eh,t}$ represents the uncertainty error in the substitution rate of electricity load, $d_{eh,t}$ represents maximum error in the substitution rate of electricity load.

Taking all the above formulas into consideration, we can obtain the values of electricity/thermal load after substitution for each period, as shown below.

$$P_{et,t} = (1 - \lambda_{et,t})(\gamma_{et,t} + \gamma_{ht,t})P_t \tag{11}$$

$$P_{ht,t} = \lambda_{et,t}(\gamma_{et,t} + \gamma_{ht,t})P_t \tag{12}$$

where $P_{et,t}$, $P_{ht,t}$ represent the values of electricity/thermal load after substitution, $\gamma_{et,t}$, $\gamma_{ht,t}$ represent the ratio of substitutable electricity/thermal load to total load value.

3.4. The System's Total Load Demand Considering Dynamic Pricing

Considering the dynamic pricing and user uncertainty based on Equations (1)–(12), the values of electricity/thermal load at time t, denoted as $P_{e,t}$ and $P_{h,t}$, are as follows:

$$P_{e,t} = P_{el,t} + P_{ec,t} + P_{es,t} + P_{et,t} \tag{13}$$

$$P_{h,t} = P_{hl,t} + P_{hc,t} + P_{hs,t} + P_{ht,t} \tag{14}$$

4. Bi-Level Optimization Scheduling Model for PIES

In this section, a bi-level optimization scheduling strategy for PIES is put forward, taking into account the diverse interests and demands of different user groups. The model considers both supply and user side while addressing the uncertainties associated with price-based load demand response.

4.1. The Upper-Level Price Decision Model

On the user side, it is desirable to adjust the response quantity of electricity and thermal load based on dynamic pricing to achieve economic benefits while ensuring a certain level of energy comfort. The target is constructed as

$$\max C_1 = C_m + C_{pr} \tag{15}$$

$$C_m = \frac{\sum\limits_{t \in T} |P_{pr,t}|}{\sum\limits_{t \in T} P_t} \tag{16}$$

$$C_{pr} = 1 - \frac{\sum\limits_{t \in T} (P_{e,t}x_{e,t} + P_{h,t}x_{h,t})}{\sum\limits_{t \in T} \left(P'_{e,t}x_{e,0} + P'_{h,t}x_{h,0}\right)} \tag{17}$$

where C_m represents the user's satisfaction with electricity usage, C_{pr} represents the user's satisfaction with the economic benefits, $P_{pr,t}$ is load response quantity, $P'_{e,t}$, $P'_{h,t}$ represent electricity/thermal load values before the response at time t, $x_{e,0}/x_{h,0}$ is baseline electricity/natural gas prices, $x_{e,t}/x_{h,t}$ is dynamic electricity/natural gas prices.

4.2. The Lower-Level Park Energy Supply Decision Model

The objective of the park energy supply side is to meet the user's energy demand at the lowest cost:

$$\min C_2 = C_{ma} + C_b + C_{wpv} + C_{dr} \tag{18}$$

$$\begin{cases} C_{ma} = \sum\limits_{i \in I} R_i W_i \\ C_b = \sum\limits_{t \in T} (\alpha_{e,t} P_{eb,t} + \alpha_{h,t} P_{hb,t}) \\ C_{wpv} = \sum\limits_{t \in T} (\beta_w P_{qw,t} + \beta_{pv} P_{qpv,t}) \\ C_{dr} = \sum\limits_{t \in T} \begin{pmatrix} (P_{e,t} x_{e,t} + P_{h,t} x_{h,t}) - \\ \left(P'_{e,t} x_{e,0} + P'_{h,t} x_{h,0}\right) \end{pmatrix} \end{cases} \tag{19}$$

where C_{ma} is PIES annual maintenance cost, C_b is PIES annual energy procurement cost, C_{wpv} is annual penalty cost for curtailed wind and solar power, C_{dr} is PIES annual demand response cost, i represents equipment category, I represents the set of equipment types, R_i represents the capacity of equipment i, W_i is annual maintenance cost per unit capacity for equipment i, $\alpha_{e,t}/\alpha_{h,t}$ represents the price of electricity/natural gas, $P_{eb,t}/P_{hb,t}$ represents the amount of purchased electricity/natural gas, β_w/β_{pv} represents the penalty cost of curtailed wind/solar power per unit at time t, $P_{qw,t}/P_{qpv,t}$ represents curtailed wind/solar power at time t.

4.3. Constraints

(1) Pricing constraints

Due to the influence of objective factors, the electricity price and natural gas price should satisfy certain upper and lower limits.

$$x_{e,t}^{\min} \le x_{e,t} \le x_{e,t}^{\max} \tag{20}$$

$$x_{h,t}^{\min} \le x_{h,t} \le x_{h,t}^{\max} \tag{21}$$

(2) User constraints

Considering the user's own interests, the satisfaction with the electricity usage and the economic benefits should meet certain threshold values.

$$\begin{cases} C_m \le C_m^{\max} \\ C_{pr} \ge C_{pr}^{\min} \end{cases} \tag{22}$$

(3) Load transfer rate constraint

The sum of the transfer rates of transferable loads from a specific time to other times should be less than one.

$$\sum_{t' \in T, t' \ne t} \lambda_{xs,tt'}{}' \le 1 \tag{23}$$

(4) Constraints on energy supply

The constraints on the energy supply side of the PIES mainly consist of wind and photovoltaic power output constraints, gas boiler constraints, electric boiler constraints, and energy storage constraints.

4.4. Evaluation Indicators

Based on the above bi-level optimization scheduling model, this paper can effectively evaluate the economic performance and user satisfaction of the system from the perspectives of the total cost and user satisfaction with the electricity usage. The PIES as an important direction of the future energy revolution is evaluated for its low-carbon performance and peak shaving ability with the help of two indicators, namely carbon emissions and peak-valley difference rate. This evaluation is performed considering the achievement of dual carbon goals and stable operation of the power grid.

(1) Total cost: calculated by (18).

$$C = C_2 \tag{24}$$

(2) User satisfaction: user satisfaction is a comprehensive evaluation formed by users based on their actual experience after using the electric heating and gas services.

$$C_{yh} = 1 - \frac{\sum\limits_{t \in T} |P_{pr,t}|}{\sum\limits_{t \in T} P_t} \tag{25}$$

(3) Carbon emission

Assuming that external electricity purchases are all from coal-fired power units, the main carbon emissions sources include coal-fired power units, CHP, GB. The formula for calculating carbon emissions is as follows:

$$\begin{cases} E = E_{buy} + E_{cg} \\ E_{buy} = \sum\limits_{t=1}^{T} \left(a + bP_{eb,t} + cP_{eb,t}{}^2 \right) \\ E_{cg} = \sum\limits_{t=1}^{T} d(Q_{CHP} + Q_{GB}) \end{cases} \tag{26}$$

where E, E_{buy}, E_{cg} represent, respectively, total carbon emissions, carbon emissions from external electricity purchases, and carbon emissions from CHP and GB; a, b, c represent the carbon emission coefficients of external electricity purchasing; d is carbon emission coefficient for natural gas-consuming devices; Q_{CHP} and Q_{GB} are, respectively, the gas consumption of CHP and GB.

(4) Peak-valley difference rate

$$h_x = \frac{P_x^{\max} - P_x^{\min}}{P_x^{\max}} \tag{27}$$

where h_x represents the peak-valley difference rate of the xth load, $P_x^{\max}/P_x^{\min}$ represents the maximum/minimum value of the xth load within the scheduling period.

5. Case Study

To validate the effectiveness of the proposed method, a simulation analysis is conducted based on the PIES, as described in [18,19]. The scheduling period is set to a 1 h time step with a total of 24 h per day. Appendix A provides the predicted values of wind, photovoltaic power output, and electricity and thermal load demand within PIES. Specific information regarding energy conversion devices and storage equipment parameters inside the PIES can be found in Appendix B.

5.1. Comparative Analysis Across Multiple Cases

To validate the validity of demand response incentive mechanism in the optimized scheduling model for the PIES, this paper conducts the scheduling strategy analysis in three cases. The case settings are as follows:

Case 1: Without considering demand response optimization.

Case 2: Considering price-based demand response optimization, but without considering uncertainty.

Case 3: Considering price-based demand response optimization while also considering uncertainty.

From Table 1 and Figure 1, it can be observed that in Case 2 compared to Case 1, the total cost decreases by 4090 yuan, carbon emissions reduce by 7785 kg, wind and solar power curtailment penalties decrease by 3988 yuan, electricity load peak-to-valley ratio decreases by 2.21%, thermal load peak-to-valley ratio decreases by 14.73%, and the overall load peak-to-valley ratio decreases by 7.92%. This validates the finding that considering price-based demand response mechanisms can reduce the overall cost of the integrated energy system, decrease carbon emissions, lower the load peak-to-valley difference, and smooth out the load curve. In Case 3 compared to Case 2, although there is an increase of 583 yuan in operating costs, the demand response compensation cost decreases by 2110 yuan and wind and solar power curtailment penalties decrease further by 220 yuan, resulting in a decrease of 1747 yuan in total cost. Carbon emissions decrease by 13,803 kg. In terms of user satisfaction, Case 2 has a satisfaction rate of 83.61% while Case 3 has a satisfaction rate of 86.94%, an increase of 3.33%. The electricity load peak-to-valley ratio decreases by 27.47%, the thermal load peak-to-valley ratio decreases by 4.41%, and the overall load peak-to-valley ratio decreases by 14.73%. This validates the finding that the proposed price-based demand response mechanism considering uncertainty can improve user satisfaction while also ensuring the economic efficiency, low-carbon operation, and peak shaving/filling performance of the integrated energy system.

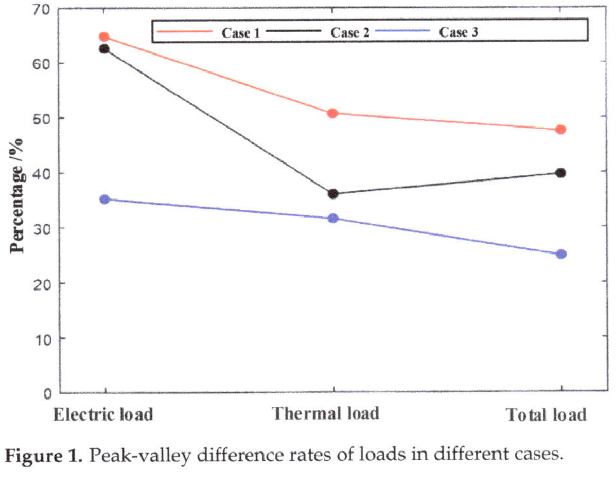

Figure 1. Peak-valley difference rates of loads in different cases.

Table 1. Indicators in Cases 1–3.

---	Case 1	Case 2	Case 3
Operating cost (yuan)	26,500	22,246	22,829
Demand response compensation cost (yuan)	0	4152	2046
Wind and solar power curtailment penalties (yuan)	4217	229	9
Total cost (yuan)	30,717	26,627	24,880
Carbon emission (kg)	39,979	32,194	18,391
User satisfaction (%)	-	83.61	86.94

In this paper, we use CPLEX to solve the proposed model. The computing times of Case 1, Case 2 and Case 3 are, respectively, 307.3 s, 422.7 s and 435.1 s. As the uncertainty increases, the computation time becomes longer. The convergence proofs of the proposed method in Case 1, Case 2, and Case 3 are shown in Figure 2. The numbers of iterations in different cases are less than 100.

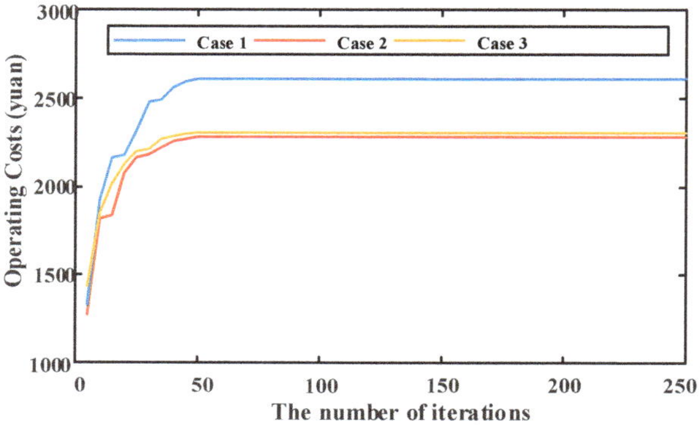

Figure 2. Numbers of iterations in different cases.

5.2. Upper-Level Price Decision Analysis

According to the simulation results, it can be observed that significant changes occur in electricity prices during different time periods after implementing demand response, as shown in Figure 3. The high peak electricity prices mainly occur from 9:00 to 21:00, while the low valley electricity prices are concentrated in the range from 22:00 to 8:00. This indicates that during periods of high electricity demand, dynamic prices are often higher than the baseline prices, while during periods of low electricity demand, dynamic prices are often lower than the baseline prices. This economically encourages users to shift their load peaks. Similarly, as shown in Figure 4, during periods of high thermal load demand, dynamic gas prices are higher than the baseline prices, and during periods of low thermal load demand, dynamic gas prices are lower than the baseline prices. This reveals a consistent positive correlation between the overall daily load curve and the price curve under demand response. After the price decision analysis, the results in Table 2 show that the average electricity price increased by 0.02 yuan, the average gas price decreased by 0.0441 yuan, and the user energy purchase cost decreased by 5.71%. This encourages users to cut down load under demand peaking periods and shift loads to low-demand periods, thereby improving system's ability for peak shaving or filling. Therefore, price-based demand response mechanism allows dynamic prices to be flexibly adjusted based on energy consumption status on demand side, stimulating user response potential.

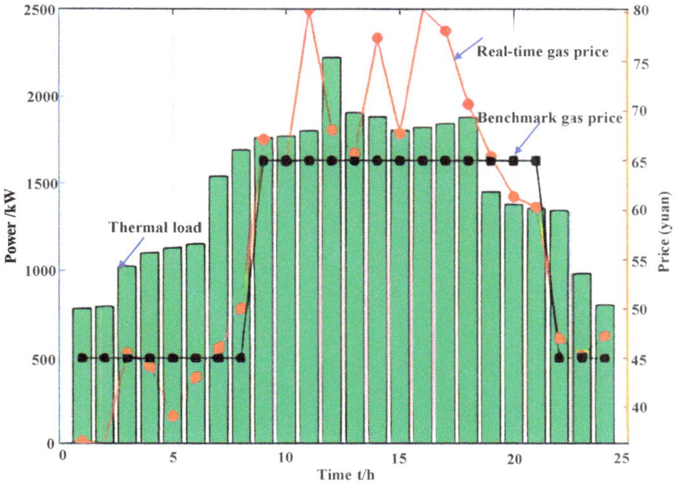

Figure 3. Comparison of electricity prices and initial electric loads.

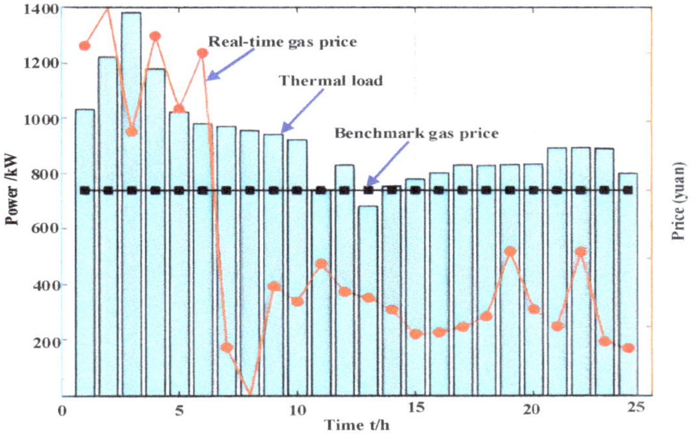

Figure 4. Comparison of natural gas prices and initial thermal loads.

Table 2. Indicators before and after price decision.

	Before Price Decision	After Price Decision
average electricity price (yuan)	0.5583	0.5783
average natural gas price (yuan)	0.7	0.6359
user energy purchase cost (yuan)	35,787	33,745
average electricity price (yuan)	0.5583	0.5783

By comparing the electricity and gas prices in Figure 5, it can be observed that during the high peak electricity price period from 9:00 to 18:00, the gas price is relatively low. At this time, the electricity load reaches a high peak while the thermal load is at a low valley. However, during the low valley electricity price period from 22:00 to 7:00, the gas price is higher, and the electricity load is at a low valley while the thermal load reaches a high peak. This is because substitutable loads are taken into account when implementing demand response. Therefore, during high peak electricity price periods, part of the electricity load can be replaced by the thermal load, which reduces the peak electricity load and increases the load during the low valley of the thermal load. Similarly, during periods of high peak thermal load, part of the thermal load can be replaced by the electricity load, reducing the peak thermal load and increasing the load during the low valley of the electricity load, further enhancing the system's ability for peak shaving and filling.

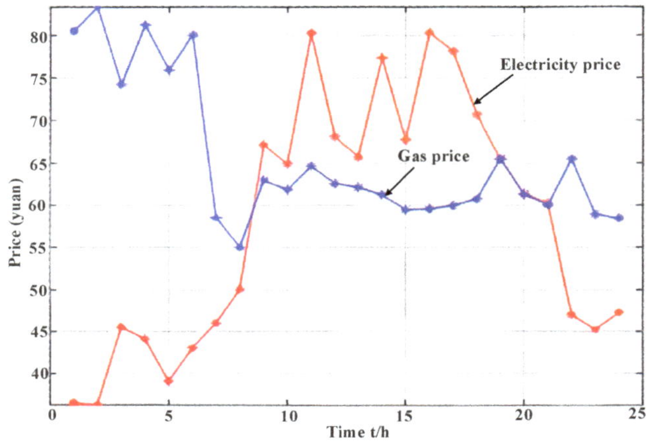

Figure 5. Comparison of electricity price and natural gas price.

5.3. Load Demand Response and Energy Supply Decision Analysis

Figure 6 compares the three cases in terms of electricity consumption, gas consumption, and carbon emissions. It can be observed that Case 2 significantly reduces electricity consumption by 8708 kWh compared to Case 1, increases gas consumption by 677 cubic meters, and decreases carbon emissions by 7785 kg. In Case 3 compared to Case 2, electricity consumption continues to decrease significantly by 2984 kWh, while gas consumption slightly increases by 155 cubic meters, and carbon emissions decrease significantly by 13,803 kg. This is because price-based demand response reduces the total electricity and thermal load. Meanwhile, the output of gas-consuming devices increases, which leads to a reduction in purchased electricity and lower carbon emission factors for gas-consuming devices, thus decreasing carbon emissions. During high electricity price periods, the load is shifted to low-price periods and periods with high renewable energy output. Additionally, a portion of the load is reduced, resulting in a decrease in operating costs. The reduction in electricity consumption and increase in gas consumption for IES improve the overall energy utilization efficiency as it relies more on natural gas to meet the load demand. Further analysis is conducted on Case 3.

Figure 7 reflects the results of electricity load participation in demand response under Case 3. From the graph, it can be observed that the peak electricity load periods are mainly concentrated from 8:00 to 18:00. After implementing the price-based demand response incentive mechanism, some electricity load is reduced and shifted during the period from 8:00 to 21:00, while the electricity load significantly increases during the time periods of 1:00 to 7:00 and 22:00 to 24:00. Some electricity load during the period from 9:00 to 21:00 is transferred to the low-demand periods of 1:00 to 8:00 and 22:00 to 24:00, resulting in a transfer of 3313 kW of electricity load. After implementing demand response, the high peak electricity price periods mainly occur from 9:00 to 21:00. During this time period, a significant amount of electricity load is reduced, while the low valley electricity prices are mainly concentrated from 22:00 to 8:00, with lesser reduction in electricity load. The total reduction in electricity load amounts to 1560 kW. Additionally, as shown in Figure 4, during the period from 9:00 to 18:00, the electricity price is significantly higher than the natural gas price. Therefore, during this time period, a substantial amount of electricity load is replaced by thermal load. Conversely, during the period from 19:00 to 8:00, the natural gas price is higher than the electricity price, leading to a significant replacement of thermal load with electricity load. The total load substitution amounts to 4343 kW. The electricity load is shifted from daytime high-price periods to nighttime low-price periods, reducing the load during high-price periods. At the same time, the high-price periods are substituted with thermal load, effectively smoothing the electricity load curve and reducing the overall operating costs of the integrated energy system.

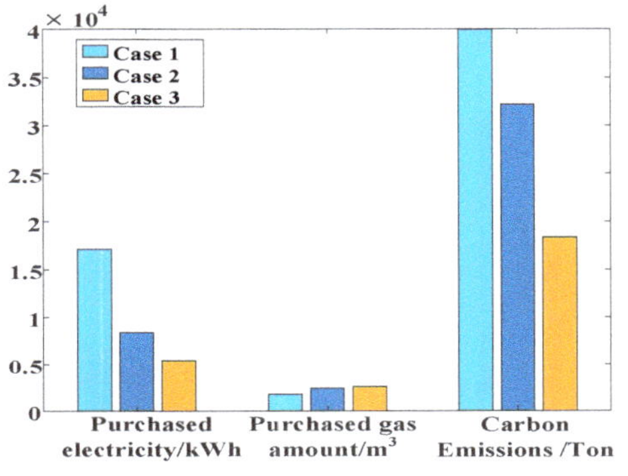

Figure 6. Indicators in Cases 1–3.

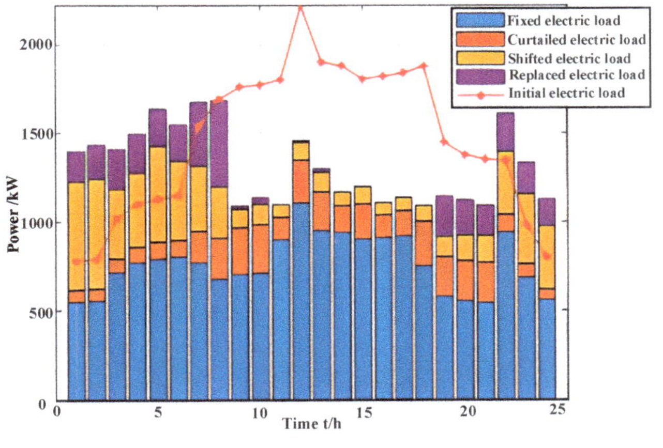

Figure 7. The results after demand response of electric loads.

Figure 8 reflects the power balance of IES. From the graph, it is observed that there are three time periods with relatively small electricity purchases. The first period is during the high electricity price period, specifically from 9:00 to 18:00. This period is the peak load period before demand response. By transferring and reducing the electricity load, most of electricity load demand is met through CHP, wind power, and photovoltaics, resulting in a decrease in electricity purchases. The second period is during the period of abundant wind resources, specifically from 24:00 to 4:00. This period is a low-demand period before demand response. By transferring and substituting, the electricity load demand increases, and most of it can be met by wind power. If demand response is not implemented, a significant amount of wind resources is wasted. Therefore, this improves the integration capacity of renewable energy consumption. The third period is during the time when electricity prices are similar to natural gas prices, specifically from 19:00 to 21:00. During this time period, natural gas prices are comparable to electricity prices. While meeting the power balance, it also provides a considerable amount of power for thermal balance. This effectively reduces PIES operation costs. Therefore, electricity load demand can be met by relying on renewable energy and CHP. After demand response, electricity load is reduced during these three time periods, smoothing electricity load curve, and improving the economic efficiency and renewable resource integration capacity of the system.

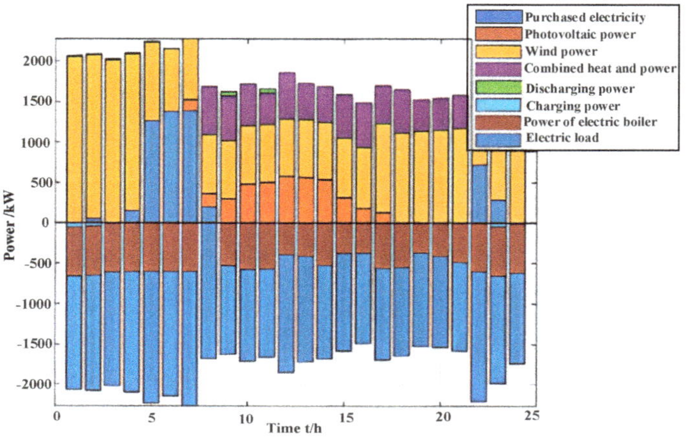

Figure 8. The power balance.

Figure 9 reflects the results of thermal load participation in demand response under Case 3. From the graph, it can be seen that peak thermal load periods are mainly during the nighttime from 1:00 to 5:00. After implementing the PLDR mechanism, thermal load during the peak periods is reduced, and a portion of the thermal load is shifted to the low-demand periods during the daytime. Additionally, during high natural gas price periods, the thermal load is substituted with electricity load. The reduction in thermal load amounts to 399 kW, the transferred thermal load amounts to 636 kW, and substituted thermal load amounts to 4343 kW. This achieves the effect of peak shaving and valley filling, effectively reducing carbon emissions.

Figure 10 reflects the thermal power balance of IES. During the period from 9:00 to 21:00, the thermal load increases after valley filling. During this time period, the electricity price is higher than or close to the natural gas price. Therefore, the cost of electricity generation from CHP is lower than purchasing electricity. Increasing the output of CHP in the energy supply has advantages. From Figure 8, it can also be seen that during this time period, most of the purchased natural gas is supplied to the CHP. From the graph, it can be observed that during the period from 22:00 to 7:00, there are two factors. Firstly, the electricity price is low and wind resources are abundant during this time period. Secondly, the thermal efficiency of gas boilers and electric boilers is higher than that of CHP. As a result, the CHP have minimal output during this time period, while the gas boilers contribute more to the heating supply. Heating is mainly provided by gas boilers and electric boilers during this time period.

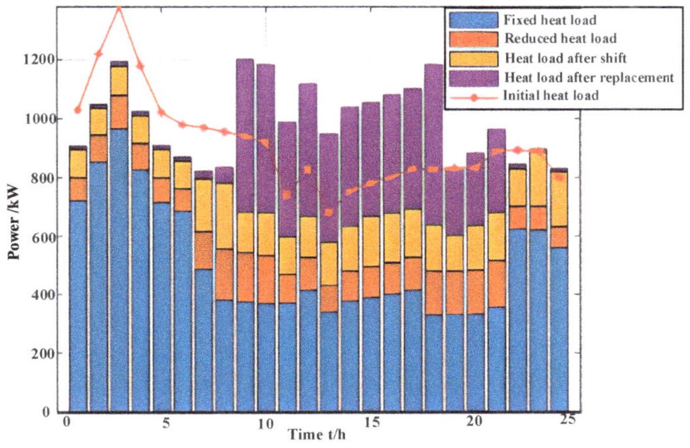

Figure 9. The results after demand response of thermal loads.

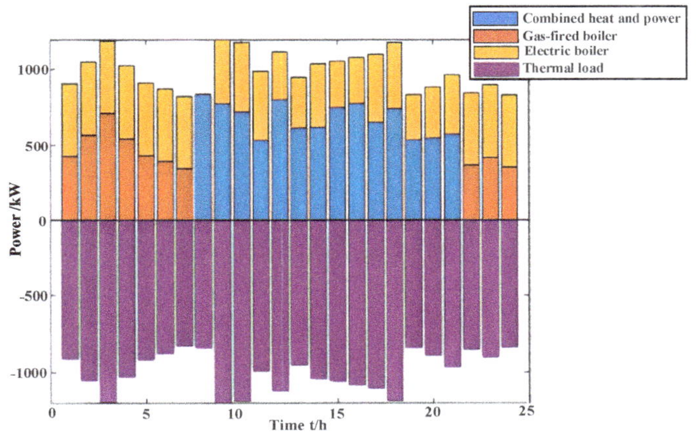

Figure 10. The thermal power balance.

In the above study, the proportions of different types of loads in end-demand were assumed to be fixed. However, in reality, the benefits vary due to natural differences in user load composition and usage habits. To reveal the impact of these variations, three additional cases are introduced with different compositions of demand response:

Case 4: Only consider transferable loads.

Case 5: Only consider substitutable loads.

Case 6: Consider both transferable and substitutable loads.

The results obtained from these cases are compared with Case 3, as shown in the table below.

From Table 3, it can be observed that the benefits, from highest to lowest, are as follows: Case 3, Case 6, Case 4, and Case 5. In comparison to Case 5, Case 6 reduces the total cost by 2508 yuan and increases user satisfaction by 9.24%. In comparison to Case 6, Case 3 reduces the total cost by 2071 yuan and increases user satisfaction by 5.16%. Therefore, considering a greater variety of load types not only reduces the overall system cost but also improves user satisfaction. The main reason for these results is that in cases where only transferable loads or substitutable loads are considered, the PIES can only transfer or substitute a portion of the load when affected by dynamic price changes. A single type of load is unable to fully utilize the flexible demand response capability. However, when both transferable and substitutable loads are considered, the complementary characteristics of different forms of demand response resources can enable the system to have control over both the timing and form of load energy consumption. This allows for more flexible adjustment of load demand and improved economic efficiency while meeting operational constraints. Case 3 also considers reducible loads, allowing for the reduction in some loads during high-price periods, further enhancing peak shaving capability and economic performance of the system. Therefore, considering demand response with various types of loads can lead to greater benefits and improvements.

Table 3. Indicators in Cases 3–6.

	Case 4	Case 5	Case 6	Case 3
Total cost (yuan)	27,949	29,459	26,951	24,880
Improvement percentage (%)	9.01	4.10	12.26	19.00
User satisfaction (%)	74.16	72.54	81.78	86.94

6. Conclusions

This paper proposes a bi-level PIES optimization strategy considering the uncertainties of PLDR. The goal is to deal with the asymmetry between sources and loads in modern power systems by applying the optimization scheduling method. The paper focuses on establishing an uncertainty model for PLDR using fuzzy theory, and through the coordinated optimization of park equipment configuration and dynamic pricing on the demand side, it aims to achieve the maximum economic performance of the PIES. Conclusions are drawn as follows:

1. Compared to not taking price decisions into account, the implementation of price decisions can decrease the overall system cost by 13.3%, decrease carbon emissions by 7785 kg, and decrease the peak-to-valley variances in electricity and heat by 2.21% and 14.73%, respectively.

2. Compared to the situation without uncertainties of PLDR, when considering the uncertainties of PLDR, the overall cost is reduced by 6.56%, carbon emissions are decreased by 13,803 kg, and the peak-valley differences in electricity and heat are reduced by 27.41% and 4.41%, respectively. In addition, user satisfaction is increased by 3.33%.

3. The more diverse the types of responsive loads, the more significant the effect of implementing demand response tends to be. Compared to considering only transferable and substitutable loads, the overall system cost is reduced by 7.68% and user satisfaction is improved by 5.16%.

Therefore, the planning solution obtained through the proposed method is effective in dealing with the asymmetry of sources and loads in modern power systems. This research can optimize the scheduling and coordinated control of multiple energy sources within the park to promote the park's integrated energy system to cope with the uncertainty of demand response.

Author Contributions: Formal analysis, X.Z.; Data curation, X.Z.; Writing—original draft, X.C.; Writing—review & editing, J.L. All authors have read and agreed to the published version of the manuscript.

Funding: This work is supported by Laboratory Work Research Project of Universities in Zhejiang Province (YB202310) and Research Project of Education and Teaching Reform of Huzhou College (hyjg202301).

Data Availability Statement: Data are contained within this article.

Conflicts of Interest: The authors declare no conflicts of interest.

Appendix A

Equipment parameters:

Equipment	Capacity/kW	Energy Conversion Efficiency/%	Maintenance Price/yuan/kW
Combined heat and power	1500	35 (Gas-to-Electricity) 50 (Gas-to-Heat)	0.05
Gas-fired boiler	1000	75	0.03
Electrical boiler	600	80	0.03
Storage battery	300	90	0.02

Appendix B

Prediction Curves of Wind Turbine, Photovoltaic and Load:

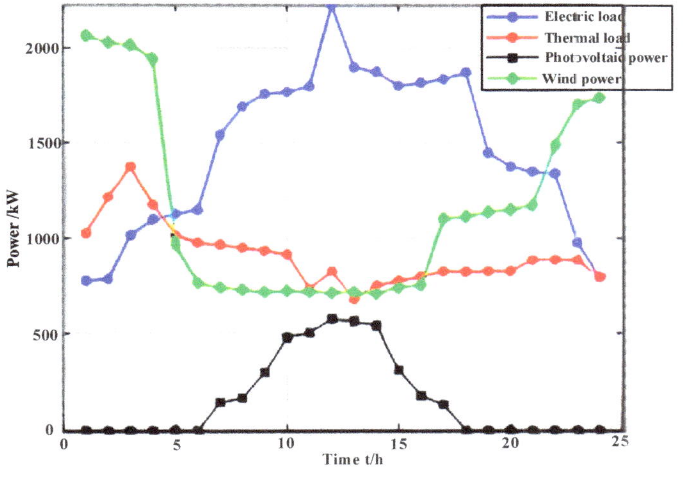

References

1. Wang, J.; Sun, Y.; Mahfoud, R.J.; Alhelou, H.H.; Siano, P. Integrated modeling of regional and park-level multi-heterogeneous energy systems. *Energy Rep.* **2022**, *8*, 3141–3155. [CrossRef]
2. Wang, L.; Cheng, J.; Luo, X. Optimal scheduling model using the IGDT method for park integrated energy systems considering P2G-CCS and cloud energy storage. *Sci. Rep.* **2024**, *14*, 17580. [CrossRef] [PubMed]
3. Xu, H.; Jin, X.; Kong, F.; Deng, Q. Two level colocation demand response with renewable energy. *IEEE Trans. Sustain. Comput.* **2020**, *5*, 147–159. [CrossRef]
4. Ji, Z.; Tian, J.; Liu, S.; Yang, L.; Dai, Y.; Banerjee, A. Optimal scheduling of park-level integrated energy system considering multiple uncertainties: A comprehensive risk strategy-information gap decision theory method. *Appl. Energy* **2025**, *377*, 124700. [CrossRef]
5. Geidl, M.; Koeppel, G.; Favre-Perrod, P.; Klockl, B.; Andersson, G.; Frohlich, K. Energy hubs for the future. *IEEE Power Energy Mag.* **2007**, *5*, 24–30. [CrossRef]
6. Wang, Y.; Hu, J.; Liu, N. Energy Management in Integrated Energy System Using Energy–Carbon Integrated Pricing Method. *IEEE Trans. Sustain. Energy* **2023**, *14*, 1992–2005. [CrossRef]
7. Zheng, L.; Wang, J.; Chen, J.; Ye, C.; Gong, Y. Two-Stage Co-Optimization of a Park-Level Integrated Energy System Considering Grid Interaction. *IEEE Access* **2023**, *11*, 66400–66414. [CrossRef]
8. Yang, W.; Liu, W.; Chung, C.Y.; Wen, F. Coordinated Planning Strategy for Integrated Energy Systems in a District Energy Sector. *IEEE Trans. Sustain. Energy* **2020**, *11*, 1807–1819. [CrossRef]

9. Qiu, Z.; Wang, B.; Huang, J.; Xie, Z. Optimal configuration and sizing of regional energy service company's energy hub with integrated demand response. *IEEJ Trans. Electr. Electron. Eng.* **2019**, *14*, 383–393. [CrossRef]
10. Wang, S.; Bi, S.; Zhang, Y.-J.A. Demand Response Management for Profit Maximizing Energy Loads in Real-Time Electricity Market. *IEEE Trans. Power Syst.* **2018**, *33*, 6387–6396. [CrossRef]
11. Yang, H.; Li, M.; Jiang, Z.; Zhang, P. Multi-Time Scale Optimal Scheduling of Regional Integrated Energy Systems Considering Integrated Demand Response. *IEEE Access* **2020**, *8*, 5080–5090. [CrossRef]
12. Bahrami, S.; Sheikhi, A. From demand response in smart grid toward integrated demand response in smart energy hub. *IEEE Trans. Smart Grid* **2016**, *7*, 650–658. [CrossRef]
13. Zeng, B.; Zhao, D.; Singh, C.; Wang, J.; Chen, C. Holistic modeling framework of demand response considering multi-timescale uncertainties for capacity value estimation. *Appl. Energy* **2019**, *247*, 692–702. [CrossRef]
14. Zhang, J.; Domínguez-García, A.D. Evaluation of demand response resource aggregation system capacity under uncertainty. *IEEE Trans. Smart Grid* **2018**, *9*, 4577–4586. [CrossRef]
15. Xiong, J.; Sun, Y.; Wang, J.; Li, Z.; Xu, Z.; Zhai, S. Multi-stage equipment optimal configuration of park-level integrated energy system considering flexible loads. *Int. J. Electr. Power Energy Syst.* **2022**, *140*, 108050. [CrossRef]
16. Zeng, B.; Liu, Y.; Xu, F.; Liu, Y.; Sun, X.; Ye, X. Optimal demand response resource exploitation for efficient accommodation of renewable energy sources in multi-energy systems considering correlated uncertainties. *J. Clean. Prod.* **2021**, *288*, 125666. [CrossRef]
17. Liu, W.; Huang, Y.; Li, Z.; Yang, Y.; Yi, F. Optimal allocation for coupling device in an integrated energy system considering complex uncertainties of demand response. *Energy* **2020**, *198*, 117279. [CrossRef]
18. Zhang, K.; Shen, Y.Q. Interface capturing schemes based on sigmoid functions. *Comput. Fluids* **2024**, *280*, 106352. [CrossRef]
19. Pazouki, S.; Haghifam, M. Optimal planning and scheduling of energy hub in presence of wind, storage and demand response under uncertainty. *Int. J. Electr. Power Energy Syst.* **2016**, *80*, 219–239. [CrossRef]

symmetry MDPI

Article

An Adaptive Voltage Reference-Based Multi-Objective Optimal Control Method for the Power Flow Symmetry of Multi-Terminal DC Systems with the Large-Scale Integration of Offshore Wind Farms

Yuanshi Zhang [1,*], Yiwen Feng [1], Tongxin Xu [1], Yilei Li [2], Xinye Du [1], Chaoyang Yuan [1] and Hongrui Chen [1]

[1] School of Electrical Engineering, Southeast University, Nanjing 210096, China; 220233004@seu.edu.cn (Y.F.); 213212481@seu.edu.cn (T.X.); duxinye1018@163.com (X.D.); 220243119@seu.edu.cn (C.Y.); chrsky2002@163.com (H.C.)

[2] State Grid Nanjing Power Supply Company, Nanjing 210008, China; liyl1@js.sgcc.com.cn

* Correspondence: yuanshizhang@seu.edu.cn

Abstract: The optimization of the symmetry of MTDC systems after a contingency is crucial for the stable and economic operation of the MTDC systems. In this paper, a multi-objective optimal control method for the power flow symmetry of MTDC systems for the large-scale integration of offshore wind farms is proposed. A mirror relationship between the available headroom of DC lines and VSCs and their actual power flow distribution performance is established. A corresponding symmetry index is established for the MTDC network, and the multi-objective optimization problem is converted into a series of single-objective problems by the normal boundary intersection method, and solved by the original dyadic interior point method, so as to obtain the Pareto optimal solution with uniform distribution. The compromise optimal solution is decided according to the entropy weight double-basis point method, which provides decision-making guidance for the operators. The simulation results show that the normal boundary intersection method can solve the multi-objective dynamic optimal control problem of the VSC-HVDC system quickly and efficiently, and improve the symmetry of the power flow in an MTDC network.

Keywords: multi-terminal HVDC (MTDC) systems; voltage source converter; power flow symmetry; adaptive droop control; DC line power regulation

1. Introduction

In modern power systems, the concept of symmetry is pivotal in terms of optimizing power flow distribution and enhancing system stability [1,2]. Amidst the depletion of fossil fuel resources and the global focus on reducing greenhouse gas emissions, the past two decades have seen a significant rise in the use of sustainable energy sources in power generation [3–6]. Offshore wind power, with its high and consistent wind speeds, is expected to play a critical role in the transition to a low-carbon economy [7–13]. Two key technologies for integrating offshore wind farms into the existing AC grid include high-voltage AC systems and voltage source converter-based high-voltage DC (VSC-HVDC) grids. Among these, multi-terminal HVDC (MTDC) systems have emerged as the preferred solution due to their advantages in terms of the independent control of active and reactive power, undersea transmission capability, and flexible power flow management [1,14–19].

Classical control strategies for MTDC systems are typically classified into three categories: V-P control [20], voltage margin control [21], and DC voltage droop control [22–26].

V-P control adjusts DC voltage by selecting a primary VSC but faces significant limitations, especially if the DC voltage-controlled VSC fails [27]. Voltage margin control assigns control duties to slack buses but may lead to DC voltage oscillations [28,29]. To improve stability and reliability, distributed droop control is commonly used, enabling multiple VSCs to share voltage regulation and power distribution tasks [30]. The equivalent single-phase power flow model [31] for multiple converter stations connected to the AC grid is employed to simplify the analysis, focusing on the key dynamics of the DC grid and converter interactions, which are crucial for the multi-objective optimization of MTDC systems. The structure of the VSC-HVDC system is shown in Figure 1.

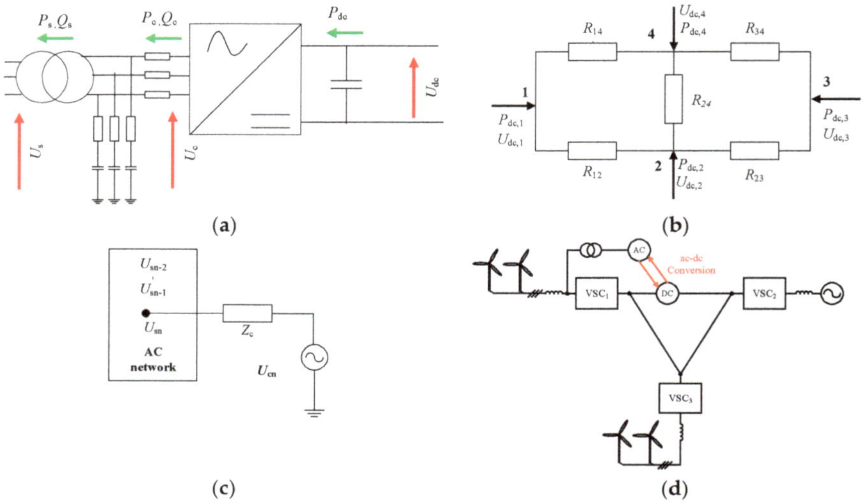

Figure 1. Structure of the VSC-HVDC system: (a) VSC-HVDC converter station; (b) multi-terminal DC network; (c) the equivalent single-phase power flow model for multiple converter stations con-nected to the AC grid; (d) an MTDC system with offshore wind farm integration.

In an MTDC system, effective power sharing among DC lines and VSC stations is essential to avoid the DC lines and converter overloading and to mitigate the impact on connected AC systems, particularly under disturbances such as large power variation in OWFs and VSC outages [32–36]. When the DC lines and VSCs in the MTDC system distribute the power flow proportionally according to their respective ratings or available headroom, the capacity configuration of the system will be fully utilized to enhance the symmetry of the MTDC system, otherwise it is likely to lead to power flow overloading in the DC lines as well as in the converters, which corresponds to a decrease in the symmetry of the system. Several approaches have been proposed to improve the power sharing of the MTDC system. For example, adaptive droop control has been explored for frequency support and active power sharing [37], and methods optimizing the desired power shares to meet grid requirements have also been suggested [38].

In addition to power sharing control, system operators must also regulate the voltages of DC nodes to promote enhancement of the symmetry of the power flow in MTDC systems, especially in systems characterized by long transmission distances and large power flows [39,40]. The greater the deviation of the system DC voltage, the less symmetrical the system power flow will be, which will lead to MTDC system instability [41,42]. The interplay between droop constants and line resistance can significantly impact both voltage regulation and power sharing, yet achieving accurate control remains difficult with traditional methods [33,34]. In [41], an analytical method was proposed to evaluate the effects of droop parameters on DC voltage variation and power sharing after a VSC outage. However, the study did not provide a precise method for accurate power sharing among the droop-controlled VSCs. Achieving precise voltage regulation at all DC nodes complicates the implementation of effective power sharing control [42]. Traditional optimal control methods may face significant challenges when attempting to simultaneously regulate DC line power distribution, active VSC power sharing, and voltage control at all DC nodes. These methods struggle to offer the necessary degrees of freedom for multi-objective optimization [43]. Additionally, due to the dynamic nature of the offshore wind farm (OWF)-integrated systems and the complexity of contingencies, DC line power, active VSC power, and voltage are prone to reaching their limits, making it difficult for these methods to find optimal solutions that adequately balance the multiple objectives [44–48].

To address these limitations, this paper introduces a novel normal boundary intersection (NBI)-based control method. The NBI approach excels at resolving trade-offs in multi-objective optimization problems by transforming them into a single-objective problem, allowing for the simultaneous optimization of multiple goals. Unlike conventional control methods, which often lack sufficient control freedom, the NBI-based method provides greater flexibility in handling power sharing, DC voltage regulation, and DC line power distribution in MTDC systems with offshore wind integration. Furthermore, NBI offers more efficient exploration of the solution space and provides a precise characterization of trade-offs between conflicting objectives, especially in the context of high-dimensional, nonlinear power systems. This ensures better performance under dynamic conditions, such as the variability and uncertainty of wind power generation, resulting in more robust and feasible solutions. The main contributions of this paper are as follows:

(1) This paper introduces a comprehensive approach to MTDC system control that balances multiple objectives to maintain the power flow symmetry of MTDC systems. This paper establishes a mirror relationship between the available headroom of DC lines and VSCs and their actual power flow distribution performance. A corresponding index is also established for the MTDC network to demonstrate the enhanced symmetry and stability of system power flow under the varying conditions of offshore wind integration.
(2) The proposed voltage reference-based multi-objective optimal control method allows proportional power sharing according to converter capacity, distributes DC line power proportionally, and regulates the DC voltage at all nodes to their nominal values, achieving three objectives simultaneously without any trade-offs. It is also effective in various contingencies such as power fluctuations and converter outages.
(3) The method employs the NBI framework, which enables precise control over DC line power distribution, VSC power sharing, and DC voltage regulation, overcoming the challenges posed by traditional methods such as control freedom limitations. This approach is particularly useful in scenarios where traditional techniques fail to provide analytic solutions.

2. Flexible Power Flow Control of MTDC Network Using Normal Boundary Intersection Method

2.1. *Adaptive Voltage Reference-Based Multi-Objective Optimal Control Framework*

The framework of the proposed multi-objective optimal control method is shown in Figure 2. The physical layer consists of the offshore wind farms (OWFs), wind farm VSCs (WFVSCs), AC grids, grid-side VSCs (GSVSCs), and the MTDC system. The node number of the proposed is defined as *#i*, *#j*, *#l* and *#k*. The control layers include the primary layer with power controllers and droop controllers. The voltage reference V^* is determined using the optimization method at the secondary layer. The WFVSCs operate under real power control, while the GSVSCs use adaptive droop control [17].

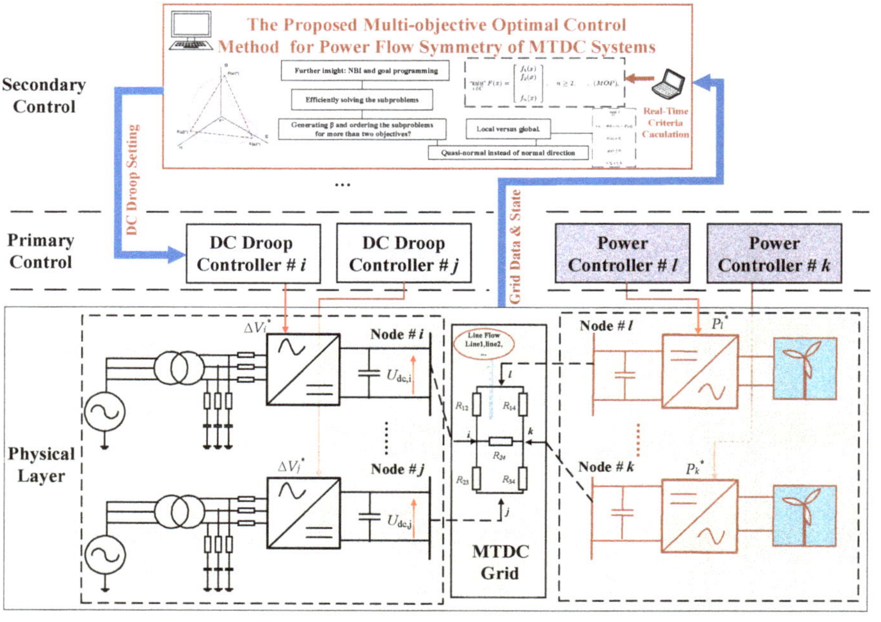

Figure 2. The framework of the proposed adaptive voltage reference-based multi-objective optimal control method.

To model the behavior of an MTDC system with offshore wind integration, the MMC station is represented using switching functions that capture the submodules (SMs), which include switches, a capacitor, and a discharging resistor. The SM-equivalent circuit, shown in Figure 3a, includes ideal diodes D1 and D2, with voltages V_{sp} and V_{sn} corresponding to the capacitor and diode conduction. The series connection of the SMs is simplified using a valve-equivalent circuit, where the total source voltage is the sum of the individual SM voltages. The capacitor voltage of each SM is discretized using the Forward Euler method, as given in Equation (1), with parameters such as the FPGA time step (TFPGA), SM capacitance (CSM), switching state (SW), arm current (i_{arm}), and discharging resistance (R_c). Importantly, the SM source voltages are computed in parallel on the FPGA, enabling real-time simulation of MMC valves with a large number of SMs.

$$v_{cap}(k) = v_{cap}(k-1) + \frac{T_{FPGA}}{C_{SM}}(SW \cdot i_{arm}(k-1) - \frac{v_{cap}(k-1)}{R_c}) \tag{1}$$

The FPGA-based equivalent circuit for a terminal MMC system is shown in Figure 3b. The valve model interfaces with the electrical network, where the arm current is measured from the network, and the source voltages are computed using the valve models. This setup allows for the accurate simulation of fast transients during faults with small time steps. The wind turbine model is based on steady-state power characteristics, with output power defined by Equation (2). Here, T_W represents turbine torque, ρ is the air density, R is the blade radius, and v_m is the wind velocity. The power conversion efficiency (Cp) depends on the blade pitch angle β and tip speed ratio λ, as shown in Equation (3), where

ω is the mechanical angular velocity of the blade. The wind farm's synchronous generator is connected to a rectifier and an inverter, modeled as voltage sources.

$$T_W = \frac{1}{2}\rho\pi R^3 v_m^2 C_p(\lambda, \beta)/\omega \tag{2}$$

$$\lambda = \frac{R\omega}{V_m} \tag{3}$$

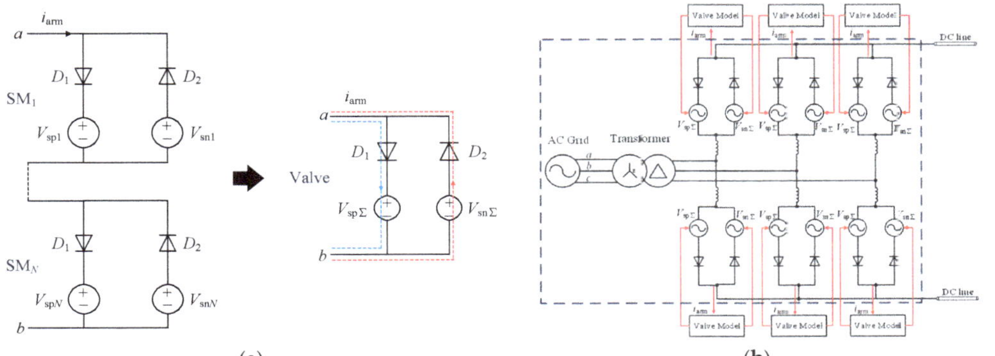

Figure 3. Equivalent circuit (**a**) MMC valve; (**b**) one-terminal MMC system.

2.2. Power Flow Symmetry of VSC-HVDC System

In the actual MTDC system, each VSC and DC line has its own rated value and available headroom, and the post-contingency power flow burden should be proportionally and symmetrically distributed based on the above parameters. As can be seen in Figure 4, this work establishes a mirror-symmetric relationship between the actual power sharing value and the available headroom. The bottom layer represents the available headroom of the VSC or DC line, and the upper layer represents the actual power distribution performance. The upper boundary is the mirror value in (4). It is calculated based on the symmetrical relationship between the available headroom and the actual power sharing performance. If the upper boundary is lower than the mirror value, it will cause the waste of system power capacity, which will lead to the power overload anyway, and only the symmetric allocation can promote the full utilization of the available headroom of the system and improve the rationality of system power sharing. In order to comprehensively evaluate the symmetrical distribution of DC line power, converter power, and DC voltage in the MTDC system, the following system power flow symmetry error index, which is called the Symmetry of Power Flow Error (SMPFE), is expressed as:

$$\begin{cases} SMPFE = \alpha(\mathbf{P}_L - \mathbf{P}_{L\_mirror})/\mathbf{P}_L^r + \beta(\mathbf{P}_{VSC} - \mathbf{P}_{VSC\_mirror})/P^r + \gamma(\mathbf{V} - \mathbf{V}_{mirror})/V^r \\ \mathbf{P}_{L_{mirror}} = \frac{\mathbf{P}_{mis} H_L H_L}{\sum_{j=1}^{m_L} H_{Lj}} + \mathbf{P}_{Lo} \\ \mathbf{H}_L = [H_{L1}, H_{L2}, \ldots, H_{LN_{dcl}}] \\ H_{Li} = P_{Li}^r - P_{Li} \\ \mathbf{P}_{VSC\_mirror} = \frac{\mathbf{P}_{mis}\mathbf{H}}{\sum_{j=1}^{m} H_j} + \mathbf{P}_o \\ \mathbf{H} = [H_1, H_2, \ldots, H_m] \\ \mathbf{H}_i = P_i^r - P_i \end{cases} \tag{4}$$

where α, β, and γ are the weighing factors. H_{Li} is the available headroom of the ith DC line ($i = 1, 2, \ldots, N_{dcl}$). H_i is the available headroom of the ith VSC station ($i = 1, 2, \ldots, m$). The subscript '_mirror' represents the ideal value of power distribution calculated according to the symmetrical principle of available headroom–actual allocation. It is worth noting that $\mathbf{V}_{mirror}$ is defined as the original voltage to achieve the goal of reducing the amount of voltage deviation in the MTDC grid.

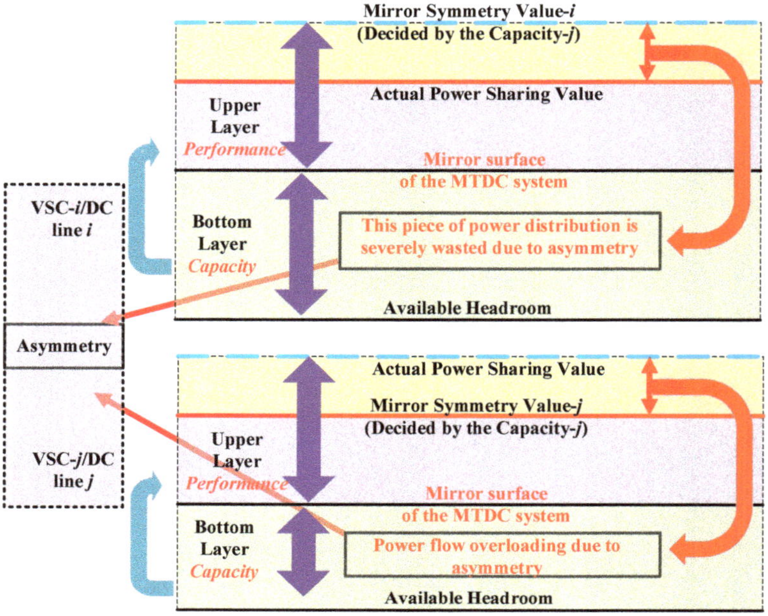

Figure 4. Diagram of the system's available headroom and the symmetrical distribution of power flow.

2.3. Multiple VSC-HVDC Control Objectives

This paper addresses three conflicting control objectives for the MTDC system: minimizing voltage deviation and ensuring proportional distribution of both active VSC power and the line power. These objectives enable the MTDC system to quickly and flexibly adjust power flow following contingencies, such as significant power fluctuations from offshore wind farms and VSC station outages.

In order to calculate the line power sharing accuracy of the proposed method, the overall objective to achieve the symmetry of the distribution of DC line power according to the power headroom can be formulated as follows:

$$f_1 = [(L_{51o} + \Delta P_{L51})/P_{r51} + (L_{54o} + \Delta P_{L54})/P_{r54}] + \sum_{i=1}^{N}(P_{Li}/P_{ri}) \tag{5}$$

For the MTDC network, the minimization of the total VSC power sharing error is established as the second objective:

$$f_2 = \frac{\Delta \boldsymbol{P} - \left(\frac{\boldsymbol{H}P_{mis}}{\sum_{j=1}^{m} H_j}\right)_1}{P^r} \tag{6}$$

In addition to power sharing control, system operators must also regulate the voltages of DC nodes by minimizing the total voltage deviation to improve the power flow symmetry of the MTDC system. For all nodes of the MTDC network, the total voltage deviation is used as the third objective, and it is calculated as:

$$f_3 = \frac{\boldsymbol{V} - \boldsymbol{V}_{ori1}}{V^r} \tag{7}$$

where P^r represents the power rating of the GSVSCs; $\boldsymbol{H}$ is the vector of the available headroom ($H = [H_1, \cdots H_m]$). $\boldsymbol{V}$ and $\boldsymbol{V}_{ori}$ are the post-contingency voltage and original voltage vectors, respectively. V_i^{min} and V_i^{max} are the lower and upper voltage boundaries of the ith VSC. P_{li} and P_{li}^r represent the actual and rated power of the ith DC line.

2.4. Multi-Objective Adaptive Droop Control Optimization Model

This paper presents a multi-objective optimization framework for MTDC networks, employing an adaptive voltage reference strategy. The proposed model aims to simultaneously reduce voltage deviations and minimize power-sharing discrepancies across VSCs and DC transmission lines. The droop characteristics of the MTDC system are examined as outlined below [49]:

$$P_i - P_i^* + R_i(V_i - V_i^*) = 0 \tag{8}$$

where P_i and P_i^* are the actual and reference active power of the ith VSC, R_i is the droop constant of the ith VSC, and V_i and V_i^* are the actual DC voltage and voltage reference.

P_i and P_i^* represent the actual and reference active power injected by the ith VSC, respectively, and R_i denotes the droop constant for the ith VSC.

The DC power of the ith DC node in (8) can be calculated as:

$$P_i = V_i \sum_{\substack{i=1 \\ i \neq j}}^{N_{dc}} G_{dc,ij}(V_i - V_j) \tag{9}$$

where N_{dc} indicates the number of DC nodes. The voltage and current upper and lower boundaries can be defined as follows:

$$\begin{cases} V_{i\min} \le V_i \le V_{i\max}, i = 1, 2, \cdots, N_{dc} \\ I_{dc,i\min} \le I_{dc,i} \le I_{dc,i\max}, i = 1, 2, \cdots, N_{dc} \end{cases} \tag{10}$$

where V_{imin} and V_{imax} are the lower and upper limits of the ith VSC. $I_{dc,i}$ represents the current amplitude at the ith DC node. The constraint of DC line power is shown in (11):

$$P_{l,i\min} \le P_{l,i} \le P_{l,i\max}, i = 1, 2, \cdots, N_{dcl} \tag{11}$$

where $P_{l,i}$ denotes the active power of the the ith DC line, and N_{dcl} denotes the total number of lines in the MTDC network.

The key control variable in this optimization model is the reference voltage V^* at the converter stations. As the adaptive droop control adjusts the reference voltage, the DC voltage across all nodes will also vary. Therefore, the DC voltage at each node can be considered to be a secondary control variable. The control variables x_c are defined as:

$$x_c = \left[V_1^*, V_2^*, \ldots, V_{N_{droop}}^*, V_1, V_2, \ldots, V_{Ndc}\right] \tag{12}$$

where N_{droop} represents the number of VSCs using droop control.

2.5. Normal Boundary Intersection Method

The proposed adaptive voltage reference-based multi-objective control optimization model, described in (2)–(8), is challenging to solve directly. While the weighted sum method is commonly used to construct a trade-off surface, it has two major limitations if the Pareto set is non-convex, points on the concave sections of the trade-off surface are overlooked, and, with uniform weights, the solutions are typically unevenly distributed across the criterion space. To address these issues, this paper employs the normal boundary intersection method, which achieves an evenly distributed set of points on the Pareto surface.

The multi-objective optimization model described by (2) to (8) is written in the following compact form [50]:

$$\begin{aligned} &min \quad Y = F(X) = [f_1(X), f_2(X), f_3(X)] \\ &s.t. \quad \begin{cases} g(V_{dc}, V_{ref}, P_{dc}, P_{ref}) = 0 \\ h(P_{dc}, P_L, \Delta V_{ref}, V_{dc}) \leq 0 \end{cases} \end{aligned} \tag{13}$$

where $f_1(X)$ is the overall objective of line power sharing, $f_2(X)$ is the converter power sharing error, $f_3(X)$ is the total voltage deviation, and g(x) and h(x) are the equality and inequality constraints, respectively.

The optimal solution $x_1^* \in C$ is obtained when only the minimum of total fuel consumption $f_1(X)$ is considered, which corresponds to the point as follows:

$$f^{1*}(f_1(x^{1*}), f_2(x^{1*}), f_3(x^{1*})) \tag{14}$$

Similarly, the optimal solutions x^{2*} and x^{3*} can be obtained by considering only the minimum of $f_2(X)$ and $f_3(X)$, respectively, corresponding to the points $f^{2*}(f_1(x^{2*}), f_2(x^{2*}), f_3(x^{2*}))$ and $f^{3*}(f_1(x^{3*}), f_2(x^{3*}), f_3(x^{3*}))$. In the coordinate space formed by each objective function, the points f^{1*}, f^{2*}, and f^{3*} form the endpoints of the Pareto front, and the plane defined by them is called the utopian plane, as shown in Figure 3.

The steps to form the Pareto frontier surface are described as follows:

(1) Specification of the objective function

To address the potential disparities in scale and magnitude between the objective functions, it is essential to normalize objective function (10) to ensure its values are confined within the range [0, 1], and the specification of the variables with an underscore "−". After specification, the variables are distinguished by an underscore "−", taking the *i*th objective function as an example:

$$\overline{f_i} = \frac{f_i - f_i^U}{f_i^N - f_i^U} \tag{15}$$

where $F^U = (f_1^U, f_2^U, f_3^U) = (f_1(x^{1*}), f_2(x^{2*}), f_3(x^{3*}))$ is the hypothetical optimum, known as the utopian point and is composed of minimum line power sharing error, minimum converter power distribution error, and minimum voltage deviation. F^N is the hypothetical worst point, called the bottom point, which consists of the maximum line power sharing error, the maximum converter power distribution error, and the maximum voltage deviation of the three objective function values corresponding to the single-objective optimal points x^{1*}, x^{2*}, and x^{3*}:

$$F^N = \left(f_1^N, f_2^N, f_3^N\right) \tag{16}$$

where $f_i^N = \max\{f_i(x^{1*}), f_i(x^{2*}), f_i(x^{3*})\}$.

(2) Generate uniformly distributed points on the utopian surface

Assume that the vector pointing from point f_1^* to point f_3^* is N_1, the vector pointing from point f_2^* to point f_3^* is N_2, and the vector pointing from point f_1^* to point f_3^* is N_3, as shown in Figure 5. N_k is divided into mk equal parts, so that the length of each equal part of the unit is $\delta_k = 1/m_k$, with k = 1, 2, 3. Any point on the utopian surface can be represented by a linear combination of the endpoints f_1^*, f_2^*, and f_3^*, taking the *j*th point A as an example:

$$p_j = \sum_{i=3}^{3} \beta_{ij} \overline{f}^{i*} \tag{17}$$

$$\begin{cases} \beta_{1j} = [0, 1, \ldots, m_1]\delta_1 \\ \beta_{2j} = [0, 1, \ldots, m_2']\delta_2 \\ \beta_{3j} = 1 - \beta_{1j} - \beta_{2j} \end{cases} \tag{18}$$

where $m_2' = \mathrm{I}[(1 - \beta_{1j})/\delta_2]$, I( ) is rounding the function. The value of the parameter β_{ij} determines the distribution of points on the utopian surface. When $\delta_1 = \delta_2 = 0.2$, the value of β_{ij} is taken as shown in Figure 6.

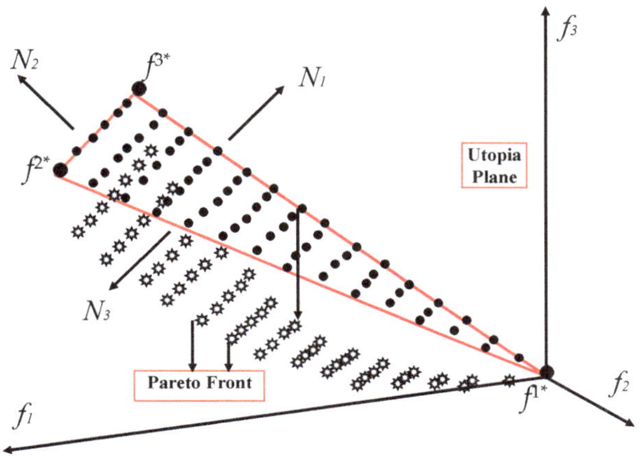

Figure 5. Utopia plane and Pareto front for the three-objective optimization model.

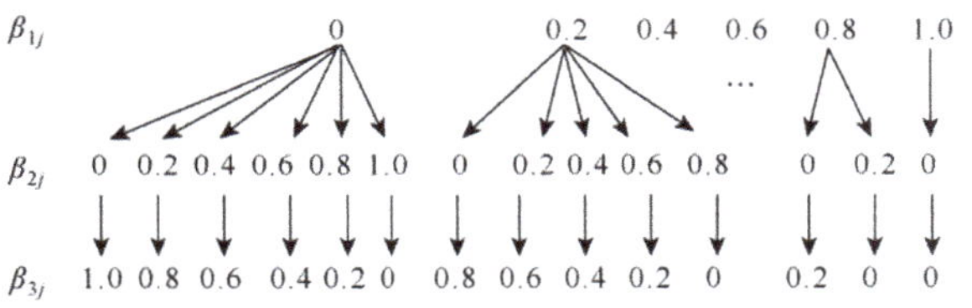

Figure 6. Values of vector β_{ij} used in the generation of points on the utopian plane.

p_j can also be written in the form of multiplication of the payment matrix $\overline{\Phi}$ and the equipartition vector $\boldsymbol{\beta}$, i.e.,

$$p_j = \overline{\Phi}\boldsymbol{\beta} = \begin{pmatrix} \overline{f_1}(x^{1*}) & \overline{f_1}(x^{2*}) & \overline{f_1}(x^{3*}) \\ \overline{f_2}(x^{1*}) & \overline{f_1}(x^{2*}) & \overline{f_1}(x^{3*}) \\ \overline{f_3}(x^{1*}) & \overline{f_1}(x^{2*}) & \overline{f_1}(x^{3*}) \end{pmatrix} \begin{pmatrix} \beta_{1j} \\ \beta_{2j} \\ \beta_{3j} \end{pmatrix} \tag{19}$$

(3) Find the Pareto optimal solution

The NBI method generates a uniformly distributed Pareto frontier by finding the intersection of the normal vector of the utopian surface with the boundary of the feasible domain in the objective function space. However, computing the normal vector for the utopian surface in 3D space is complex. Instead, a set of uniformly spaced parallel lines close to the normal direction can intersect with the utopian surface, and their intersections with the corresponding boundary of the feasible domain will also yield a uniformly distributed Pareto frontier. To simplify the calculation, the quasi-normal method is used, with the quasi-normal vector $\mathbf{n}$ given by (20), which represents the vector from point A on the utopian surface to the corresponding point B on the Pareto front, as shown in Figure 3.

$$\mathbf{n} = -\overline{\Phi}\mathbf{e} \tag{20}$$

where $\mathbf{e} = [1, 1...1]^T$. The form of the quasi-normal vector expressed in (17) is much simpler than that of the normal vector of the utopian surface, which helps to simplify the calculation. Moreover, the Pareto front defined by quasi-normal vectors remains unaffected when the objective functions are scaled by different coefficients. Therefore, the point B on the Pareto frontier can be determined using the following equation:

$$\overline{F}(x) = \overline{\Phi}\boldsymbol{\beta} + d\mathbf{n} = \overline{\Phi}(\boldsymbol{\beta} - d\mathbf{e}) \tag{21}$$

where d is the distance parameter. As d increases, the objective functions corresponding to the feasible solutions determined by $\overline{\Phi}\boldsymbol{\beta} + d\mathbf{n}$ are gradually improved. When d increases to the maximum value of d_{max}, each objective function reaches Pareto optimality. Therefore, given the vector of equidistant points $\boldsymbol{\beta}$, the multi-objective optimization problem in (10) can be converted into single-objective optimization problems by aiming to maximize the distance between points on the utopian surface and their corresponding points on the Pareto front, as follows:

$$\begin{cases} \min(-d) \\ \text{s.t. } \overline{F}(x) = \overline{\Phi}\boldsymbol{\beta} + d\mathbf{n} = \overline{\Phi}(\boldsymbol{\beta} - de) \\ g(x) = 0 \\ \underline{h}(x) \leq h(x) \leq \overline{h}(x) \end{cases} \tag{22}$$

By varying the values of $\boldsymbol{\beta}$, the multi-objective optimization problem is transformed into multiple single-objective optimization problems. These problems are addressed using the original pairwise interior point method, enabling the acquisition of a set of uniformly distributed Pareto-optimal solutions.

3. Multi-Objective Optimized Decision-Making

In engineering applications, each VSC has a single voltage reference value, and the scheduler must select a compromise optimal solution from the Pareto front. This paper employs the entropy weight two-basis point method, which incorporates subjective weight adjustments, to provide decision-making guidance for dispatchers [51]. First, objective weights are derived using the entropy weight method, considering the degree of difference among Pareto optimal solutions. These weights are then subjectively adjusted based on dispatcher experience or actual electricity demand. This approach not only reflects the subjective preferences of dispatchers but also objectively considers the significance of each objective function, providing a solid theoretical foundation for selecting the optimal point. A model is constructed using M optimal points on the Pareto front, and a comprehensive evaluation is carried out as follows [52]:

(1) Construction of an evaluation matrix R': for the three objective functions considered in this study, an evaluation matrix is formulated based on M Pareto optimal solutions.

$$R' = \begin{pmatrix} r'_{11} & r'_{12} & \cdots & r'_{1j} \ldots r'_{1M} \\ r'_{21} & r'_{22} & \cdots & r'_{2j} \ldots r'_{2M} \\ r'_{31} & r'_{32} & \cdots & r'_{3j} \ldots r'_{3M} \end{pmatrix} \tag{23}$$

where r'_{ij} is the value of the ith objective function corresponding to the jth Pareto optimal solution when i is taken as 1, 2, 3, respectively.

(2) Normalization of data: due to variations in scale and magnitude among the objective functions, the original data are standardized using the following normalization equation:

$$r_{ij} = \frac{\max\limits_j(r'_{ij}) - r'_{ij}}{\max\limits_j(r'_{ij}) - \min\limits_j(r'_{ij})} \tag{24}$$

where r_{ij} is the value of the ith objective function corresponding to the jth Pareto optimal solution after specification; $\min\limits_j(r'_{ij})$ are the maximum and minimum values of the ith row in R', respectively. The specified evaluation matrix R can be calculated as:

$$R = \begin{pmatrix} r_{11} & r_{12} & \cdots & r_{1j} \ \ldots r_{1M} \\ r_{21} & r_{22} & \cdots & r_{2j} \ \ldots r_{2M} \\ r_{31} & r_{32} & \cdots & r_{3j} \ \ldots r_{3M} \end{pmatrix} \tag{25}$$

(3) Calculation of entropy-based weights for each objective function: The entropy-based weight vector, denoted as $\alpha = (\alpha_1, \alpha_2, \alpha_3)^T$, quantifies the variation across solutions for each objective, representing the informational contribution of each objective function. The entropy weight for each objective (i = 1,2,3) is determined using the formula provided in (26):

$$\begin{cases} \alpha_i = \dfrac{1 - e_i}{\sum_{j=1}^{3}(1 - e_j)} \\ e_i = -\dfrac{\sum_{j=1}^{M}\left[\dfrac{r_{ij}}{\sum_{j=1}^{M} r_{ij}} ln(\dfrac{r_{ij}}{\sum_{j=1}^{M} r_{ij}}\right)]}{lnM} \end{cases} \tag{25}$$

(4) Using the subjective weights of the scheduler λ_i, i = 1, 2, 3, find the corrective weight coefficients $\omega_i = \frac{\alpha_i\lambda_i}{\sum_{i=1}^{3}\alpha_i\lambda_i}$. As can be seen, ω_i takes into account both the experience of the scheduler and the entropy weights, which reflect the degree of variability of different solutions on the Pareto frontier.

(5) Create a weighted specific evaluation matrix $\hat{R}$

$$\hat{R} = \begin{pmatrix} \omega_1 r_{11} & \omega_1 r_{12} & \cdots & \omega_1 r_{1j} \ \ldots \omega_1 r_{1M} \\ \omega_2 r_{21} & \omega_2 r_{22} & \cdots & \omega_2 r_{2j} \ \ldots \omega_2 r_{2M} \\ \omega_2 r_{31} & \omega_3 r_{32} & \cdots & \omega_3 r_{3j} \ \ldots \omega_3 r_{3M} \end{pmatrix} \tag{27}$$

It can be seen that the maximum and minimum values in the i-th row of $\hat{R}$ correspond to the most and least desirable scenarios of the i-th target, respectively.

(6) Determination of double-base points:

The ideal point is shown in (28):

$$\begin{cases} F^+ = (f_1^+, f_2^+, f_3^+) \\ f_1^+ = max(\hat{R}_{i1}, \hat{R}_{i2}, \ldots, \hat{R}_{in}) \end{cases} \tag{28}$$

The negative ideal point is calculated in (29):

$$\begin{cases} F^- = (f_1^-, f_2^-, f_3^-) \\ f_1^- = min(\hat{R}_{i1}, \hat{R}_{i2}, \ldots, \hat{R}_{in}) \end{cases} \tag{29}$$

(7) Evaluation of the relative closeness TJ_j for each Pareto optimal solution: the relative closeness of each Pareto optimal solution is calculated to measure its proximity to the ideal solution:

$$TJ_j = \frac{D_j^-}{D_j^+ + D_j^-} \tag{30}$$

where D_j^+ and D_j^- are the Euclidean distances from the jth solution to the positive and negative ideal points, respectively. It can be seen that the higher the relative closeness, the closer the solution is to the positive ideal point, so the Pareto optimal solution with the largest relative closeness is chosen as the compromise optimal solution. The algorithm to optimize the established SMPFE using the NBI method is proposed in Algorithm 1.

Algorithm 1: NBI-based for selecting the best droop setting for the multi-objective optimization of the VSC-HVDC system

1	Initialize objective functions $f_1(x)$, $f_2(x)$ and $f_3(x)$.
2	Solve for the optimal solution in the case of single-objective optimization x^{1*}, x^{2*} and x^{3*}.
3	Constitute Utopia plane and Pareto optimal surface for a three-objective case.
4	For i = 1 to M do:
5	Objective function specification processing $\overline{f_i} = \frac{f_i - f_i^U}{f_i^N - f_i^U}$.
6	Establishment of utopian points $F^U = (f_1(x^{1*}), f_2(x^{2*}), f_3(x^{3*}))$ and the hypothetical worst point(the bottom point) $F^N = (f_1^N, f_2^N, f_3^N)$.
7	Generate evenly distributed points on the utopian surface: $p_j = \overline{\Phi}\beta = \begin{pmatrix} \overline{f_1}(x^{1*}) & \overline{f_1}(x^{2*}) & \overline{f_1}(x^{3*}) \\ \overline{f_2}(x^{1*}) & \overline{f_1}(x^{2*}) & \overline{f_1}(x^{3*}) \\ \overline{f_3}(x^{1*}) & \overline{f_1}(x^{2*}) & \overline{f_1}(x^{3*}) \end{pmatrix} \begin{pmatrix} \beta_{1j} \\ \beta_{2j} \\ \beta_{3j} \end{pmatrix}$
8	Pareto optimal solution by quasi-normal vector method $\overline{F}(x) = \overline{\Phi}\beta + dn = \overline{\Phi}(\beta - de)$.
9	Converting a multi-objective optimization problem into a single-objective optimization problem with the objective of maximizing the distance between a point on the utopian surface and a point on the corresponding Pareto frontier: $\begin{cases} \min(-d) \\ \text{s.t. } \overline{F}(x) = \overline{\Phi}\boldsymbol{\beta} + d\mathbf{n} = \overline{\Phi}(\boldsymbol{\beta} - d\mathbf{e}) \\ g(x) = 0 \\ \underline{h}(x) \le h(x) \le \overline{h}(x) \end{cases}$
10	Obtain a series of uniformly distributed Pareto optimal points by solving a series of single-objective optimization problems.
11	end for

Algorithm 1: *Cont.*

For i=1 to M do:

Establishment of an evaluation matrix:

$$R' = \begin{pmatrix} r'_{11} & r'_{12} & \cdots & r'_{1j} \ldots r'_{1M} \\ r'_{21} & r'_{22} & \cdots & r'_{2j} \ldots r'_{2M} \\ r'_{31} & r'_{32} & \cdots & r'_{3j} \ldots r'_{3M} \end{pmatrix}$$

Specification processing of data:$r_{ij} = \frac{\max\limits_{j}(r'_{ij}) - r'_{ij}}{\max\limits_{j}(r'_{ij}) - \min\limits_{j}(r'_{ij})}$

Calculate the entropy weight of each objective function:

$$\begin{cases} \alpha_i = \dfrac{1 - e_i}{\sum_{j=1}^{3}(1 - e_j)} \\ e_i = -\dfrac{\sum_{j=1}^{M}\left[\dfrac{r_{ij}}{\sum_{j=1}^{M} r_{ij}} ln\left(\dfrac{r_{ij}}{\sum_{j=1}^{M} r_{ij}}\right)\right]}{lnM} \end{cases}$$

Use of subjective weights of dispatchers: $\omega_i = \frac{\alpha_i \lambda_i}{\sum_{i=1}^{3} \alpha_i \lambda_i}$

Create a weighted specific evaluation matrix $\overset{\Lambda}{R}$.

Establishment of double base points $F^{+} = \left(f_1^{+}, f_2^{+}, f_3^{+}\right)$ and $F^{-} = \left(f_1^{-}, f_2^{-}, f_3^{-}\right)$.

Calculate the relative closeness of each Pareto optimal solution and select the Pareto optimal solution with the largest relative closeness as the compromise optimal solution:

$TJ_j = \frac{D_j^{-}}{D_j^{+} + D_j^{-}}$

end for

4. Case Studies

A five-terminal VSC-based multi-terminal DC (MTDC) system integrating offshore wind farms is used to validate the proposed autonomous control strategies for converter power sharing and DC pilot voltage regulation, as shown in Figure 7. The MTDC network is modeled in MATLAB/Simulink using the SimPowerSystems Blockset with OPAL-RT's RT-LAB libraries. Two offshore wind farms are connected to VSC-3 and VSC-5, and the other VSCs are linked to three different AC systems. The wind farm converters use a real power regulation scheme, while the grid-side converters employ a DC voltage droop control strategy. Each VSC is modeled with the average value representation of a modular multilevel converter (MMC). DC cables are modeled using the RT-LAB/ARTEMIS Blockset, with cable and VSC parameters provided in Tables 1–3. VSCs 1 and 2 use MMCs with half-bridge structure, while the other VSCs use MMCs with a full-bridge structure. Converter loss coefficients are provided in Table 4, and DC voltage limits are set between 640 kV and 600 kV. Initially, WFVSC-3′s power output shifts from −750 MW to −400 MW due to wind power variation.

Table 1. DC cable rated power and length.

DC Cable	1 to 2	1 to 3	1 to 4	1 to 5	2 to 3	3 to 4	4 to 5
Length (km)	160	400	250	320	320	320	500
Rated Power $P_{l,i}^{r}$ (MW)	430	430	430	430	430	430	430

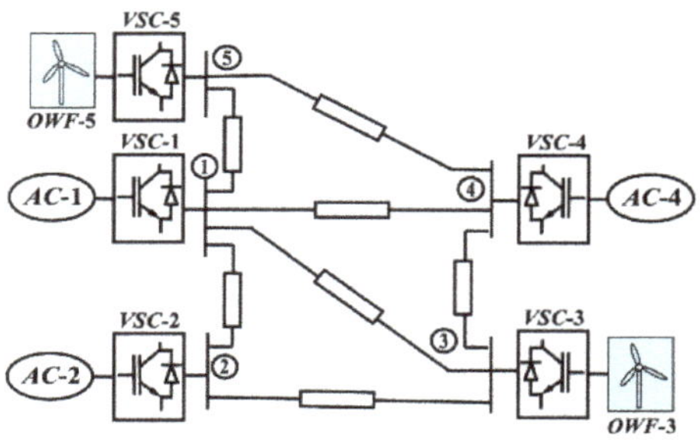

Figure 7. Diagram of the test system.

Table 2. DC cable parameter.

Parameter	R(Ω/km)	L(mH/km)	C(μF/km)
Value	0.0200	0.1463	0.2662

Table 3. VSC station parameter.

VSC No.	1	2	3	4	5
Nominal Voltage V_i^r(kV)	630	630	630	630	630
Rated Power P_i^r(MW)	400	400	800	400	900
Droop Coefficient k_i(in P.U.))	0.08	0.06	0	0.04	0
Power Reference $P_i{}^*$(MW)	350	500	−750	400	−500

Table 4. VSC converter loss coefficients.

Type	a_{dc}	b_{dc}	c_{dc}	
Half-bridge Full-bridge	8.80	4.00 6.70	0.47 0.96	$\times 10^{-3}$

The proposed control method is validated through its application in two scenarios: an increase in wind farm generation (Case 1) and a GSVSC outage (Case 2).

4.1. Case1: Power Increase in OWF Generation

In Case A, the proposed control method effectively balances DC line and VSC power while maintaining the DC voltage during increased wind power generation. At t = 0.5 s, the active power of VSC-5 increases from −500 MW to −750 MW. The dynamic optimal scheduling Pareto frontier, showing power fluctuations due to offshore wind, is presented in Figure 8. The control method is activated at t = 1.5 s. The results for the DC line powers, DC voltages, VSC voltages, DC powers, and SMPFE under the proposed control system are shown in Figure 9, Figure 10, Figure 11, and Figure 12, respectively.

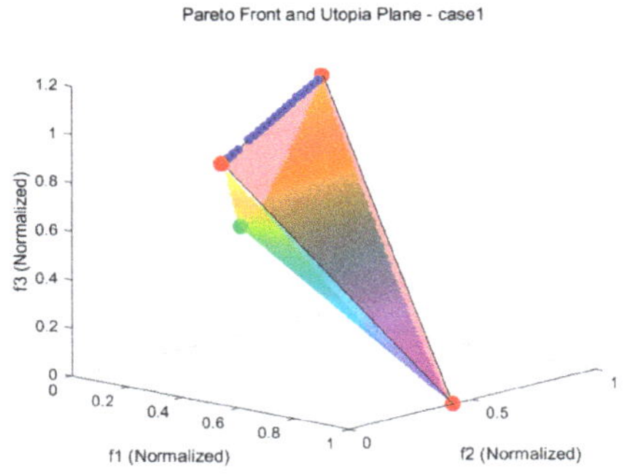

Figure 8. Pareto front and utopian plane under power increase at VSC-5.

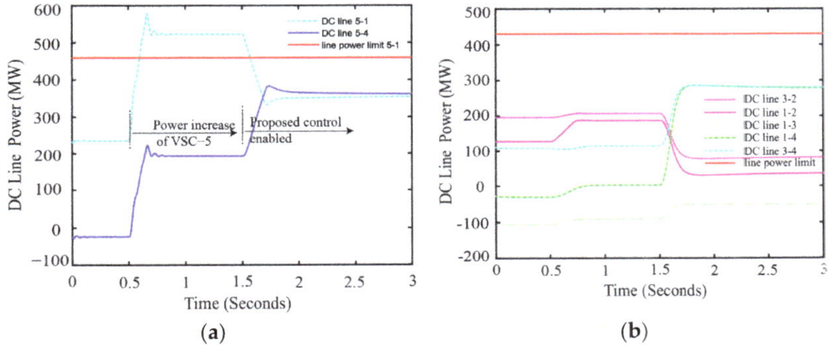

Figure 9. DC line power under increase at VSC-5: (**a**) powers of DC lines 5-1 and line 5-4; (**b**) powers of DC lines 3-2, 1-2, 1-3, 1-4, and 3-4.

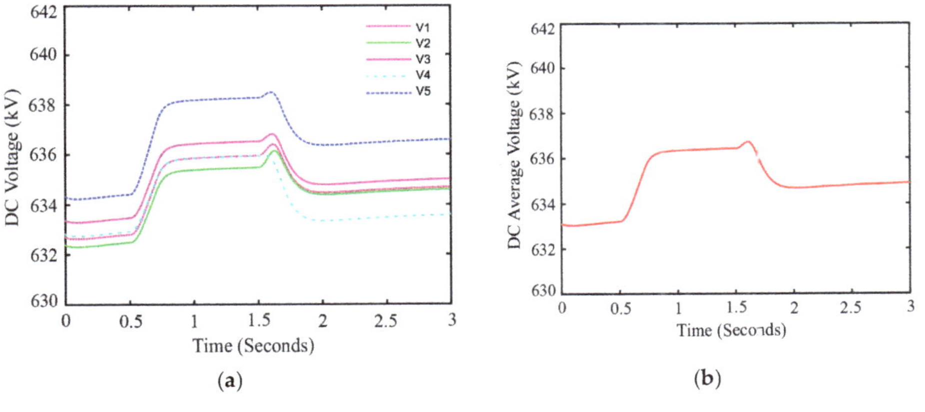

Figure 10. DC voltage and DC average voltage under power increase at VSC-5 (**a**) DC voltages; (**b**) DC average voltage.

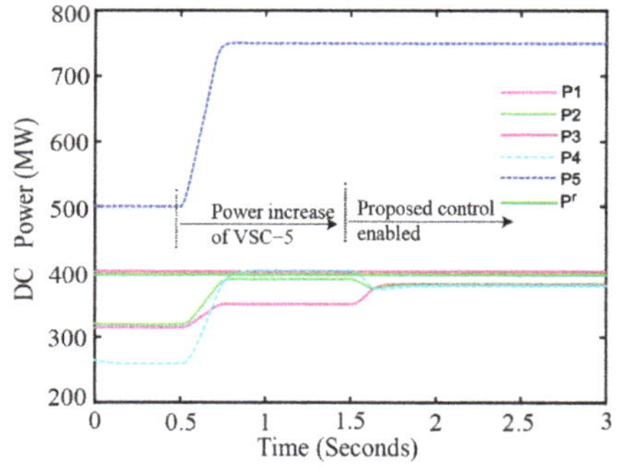

Figure 11. DC powers under power increase at VSC-5.

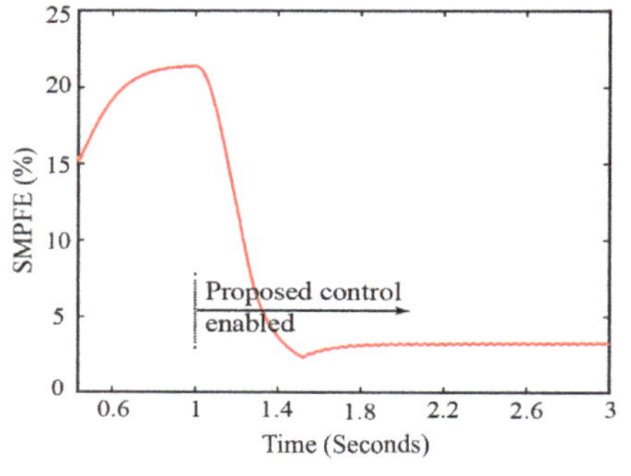

Figure 12. SMPFE of Case A.

4.2. GSVSC Outage

In Case B, the proposed control strategy is applied to achieve proportional power sharing and DC voltage regulation across all nodes after a grid-side VSC (GSVSC) outage. Specifically, VSC-2 suffers an outage at t = 0.5 s, and the control is activated at t = 1s. The pareto front and utopian plane under VSC-2 outage is shown in Figure 13. Figures 14–17 present the results, including DC line power flows, DC voltage levels, average VSC voltages, DC power distribution, and the computed SMPFE under the proposed method.

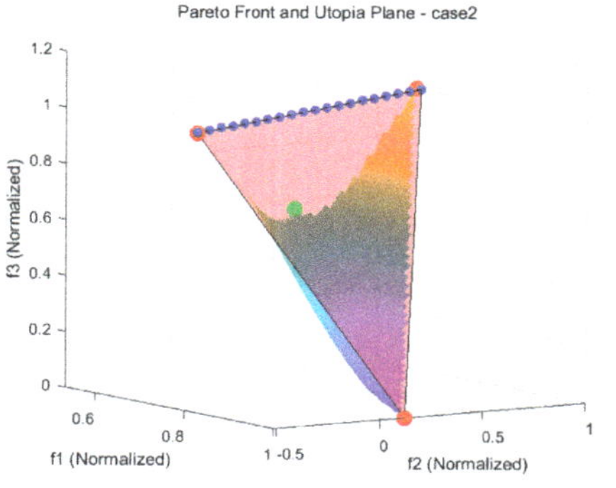

Figure 13. Pareto front and utopian plane under VSC-2 outage.

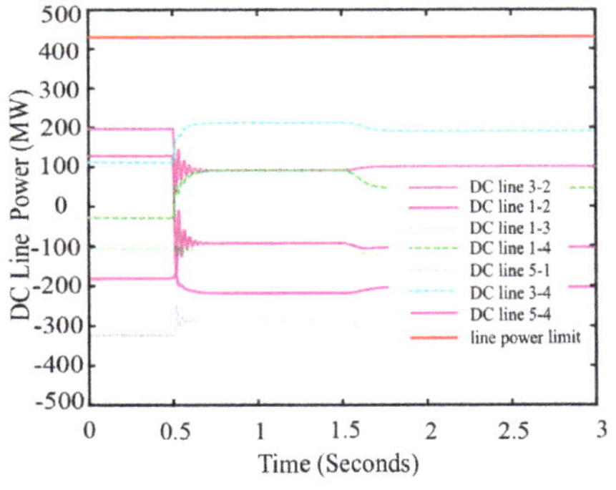

Figure 14. DC line powers under VSC-2 outage.

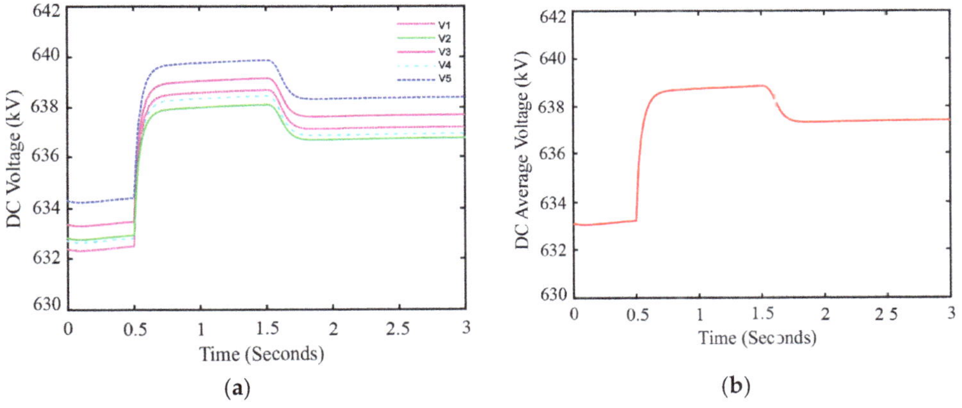

Figure 15. DC voltage and DC average voltage under VSC-2 outage: (**a**) DC voltages; (**b**) DC average voltage.

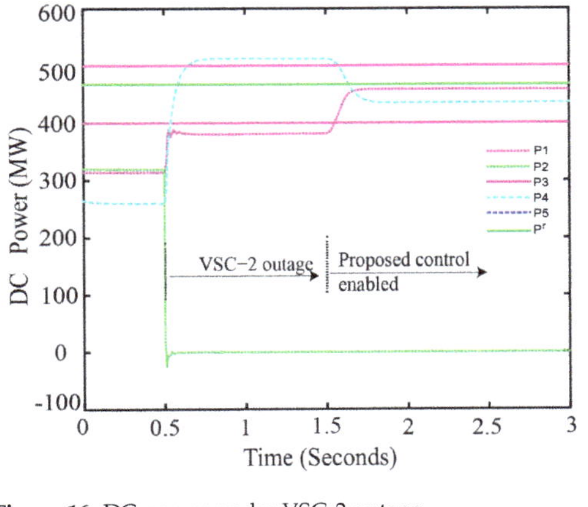

Figure 16. DC power under VSC-2 outage.

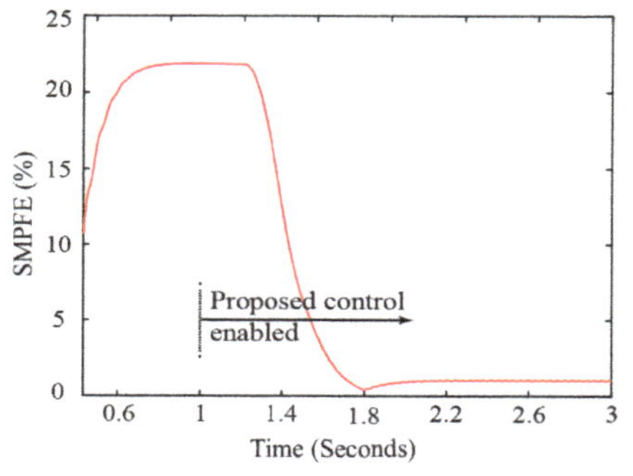

Figure 17. SMPFE of Case B.

4.3. Analysis of Compromise Optimal Solution of Entropy Weight Two-Basis Point Method

The subjective weight λ can be selected according to the dispatcher's preference. In this paper, assuming $\lambda = (\frac{1}{3}, \frac{1}{3}, \frac{1}{3})^T$, the modified weight coefficients ω_i of each objective function under Case 1 and 2 determined by the entropy weight two-base point method are shown in Table 5. It can be seen that among the three objective functions, the difference between the points of the Pareto optimal solution caused by the symmetric distribution of the line power flow is the largest; then, this objective plays the largest role in the comprehensive evaluation, so the compromise optimal solution is the largest. In this case, the selection of the compromise optimal solution will favor the point with the smaller error in the symmetrical distribution of line power flow.

Table 5. The modified weight coefficients ω_i of each objective function.

Case	VSC Power ω_1	Line Power ω_2	Voltage Deviation ω_3
1	0.2175	0.5649	0.2176
2	0.5492	0.2285	0.2223

The Pareto optimal solutions are ranked, and the solution with the highest relative closeness is chosen as the compromise optimal solution for decision-making. The location of this solution on the Pareto front is depicted in Figure 11, with the corresponding values of each objective function presented in Table 6. The values of the three objective functions were established as Symmetric Error Value 1 (*SME1*), Symmetric Error Value 2 (*SME2*), and Symmetric Error Value 3 (*SME3*), respectively.

Table 6. =The values of each objective function at the compromise optimal solution.

Case	VSC Power (*SME1%*)	Line Power (*SME2%*)	Voltage Deviation (*SME3%*)
1	1.8214	17.1932	1.6557
2	10.8443	3.0794	3.8345

For Case1, the *SME1* corresponding to the compromise optimal solution is only 0.174% higher than the minimal *SME1* of line power, while the SME2 and SME3 are 5.663% and 1.832% higher than the single-objective optimization value, respectively. For Case2, the *SME1* corresponding to the compromise optimal solution is 1.174% higher than the minimal *SME1* under the single-optimization case, while the *SME2* and *SME3* are only 1.124% and

2.012% higher than the single-objective optimization value, respectively. It can be seen that the compromise optimal solution is obviously closer to the ideal utopian point, and it is a higher quality optimization solution on the Pareto frontier.

5. Discussion

The dynamic simulation results before and after activating the proposed NBI-based control are presented in Figures 9–11. In Figure 9a, following a power increase at VSC-5, DC line 5-1's power rises from 318.1 MW to 461.6 MW, exceeding its 430 MW limit, while DC line 5-4 increases from 188.7 MW to 296.6 MW, remaining within its capacity. After activating the control at t= 1.5 s, the power of both lines is nearly proportionally shared, with power sharing errors of 0.55% and 0.70%, ensuring all DC line powers stay within their limits, as shown in Figure 9a,b. Figures 10 and 11 illustrate the DC voltage and power profiles, respectively, with all converter DC powers remaining within rated values.

The proposed control significantly improves the ability to share the power burden of droop-controlled VSCs, distribute the power of DC lines, and regulate DC voltage, as shown in Figures 9–11. Specifically, Figure 11 highlights that with the active power increase in VSC-5, traditional droop control causes an imbalanced power distribution among VSCs-1, 2, and 4, with VSC-1 having more headroom, while VSC-4 exceeds its 400 MW limit and VSC-2 nearly reaches its limit. After activating the NBI-based control at t = 1.5 s, the active powers of the VSCs are shared proportionally to their headroom, ensuring all stay within their limits. Figure 10 shows that the DC voltage rises with the active power change in VSC-5, approaching 640 kV. However, after activating the proposed control, the voltage increase is mitigated, and the DC voltages across all nodes are adjusted closer to their original values. The optimization of the three objectives is illustrated in Figure 14, showing a significant decrease in the SMPFE.

Figure 14 demonstrates that after activating the proposed control method, all DC line powers remain within their limits, with each line distributing power according to available headroom, enhancing power flow symmetry. Figure 15a,b show a noticeable voltage increase when VSC-2 fails at t= 0.5 s. After the control is activated at t =1 s, the voltage rise is mitigated, and the DC voltages across all nodes are adjusted closer to their original steady-state values. Figure 16 highlights that, under the VSC-2 failure, VSC-4 becomes overloaded while VSC-1 retains significant headroom. Upon activation of the NBI-based method, the active power imbalance from the VSC-2 outage is proportionally redistributed between VSCs 1 and 4 based on the available headroom. Similarly to Case A, the method successfully achieves proportional and symmetrical power distribution, active power sharing, and voltage control, with a minimal SMPFE value after the disturbance. As shown in Figure 17, the SMPFE initially peaks at 21.87%, but drops by about 18.2% following the activation of the control. With the reduction in the established SMPFE, the symmetry of the power flow distribution of the MTDC system is significantly improved, and the power of the VSC and DC lines can be proportionally and symmetrically distributed in accordance with reasonable rules, making full use of their available headroom, reducing the risk of the power flow overloading the HVDC system after a contingency, and promoting safe and symmetrical operation of the system.

6. Conclusions

This paper presents a novel control strategy for achieving power flow symmetry in converter power sharing, line power distribution, and DC voltage regulation in DC grids with large-scale offshore wind integration. The proposed method is validated using a multi-objective optimization model and the voltage droop control scheme. Dynamic simulations, including disturbances such as power fluctuations in the wind farm VSC and grid-side VSC outages, demonstrate the effectiveness of the approach. The NBI method is used to generate uniformly distributed Pareto fronts in three-dimensional space, providing dispatchers with more precise decision-making data. Results show that the method enables proportional and symmetrical power sharing and line power distribution according to the available headroom, enhancing the symmetry of power flow in MTDC systems. Crucially, this method functions independently, eliminating the need for a centralized controller or a global power flow solution. Future work will explore the sensitivity of DC voltage reference adjustments in relation to DC line power distribution, VSC power sharing and voltage regulation, as well as expanding the simulation to include additional contingency scenarios, such as DC line disconnection and topology changes, to further assess the robustness of the proposed method.

Author Contributions: Conceptualization, Y.Z. and Y.F.; methodology, Y.F.; software, Y.Z.; writing—original draft preparation, Y.Z. and Y.F.; writing—review and editing, Y.Z., Y.F., T.X., Y.L. and X.D.; visualization, C.Y. and H.C., supervision, Y.Z. and Y.F. All authors have read and agreed to the published version of the manuscript.

Funding: This work was supported by the National Natural Science Foundation of China under Grant 52207083.

Data Availability Statement: Data are available upon reasonable request to the corresponding author.

Conflicts of Interest: Author Yilei Li was employed by the company State Grid Nanjing Power Supply Company. The remaining authors declare that the research was conducted in the absence of any commercial or financial relationships that could be construed as a potential conflict of interest.

Nomenclature

The main symbols appearing in this paper are defined below.

MTDC	Multi-Terminal DC
VSC	Voltage Source Converter
OWF	Offshore Wind Farm
VSC-HVDC	Voltage Source Converter-Based High-Voltage Direct Current
NBI	Normal Boundary Intersection
SMPFE	Symmetry of Power Flow Error

References

1. Wu, M.; Ma, D.; Xiong, K.; Yuan, L. Deep Reinforcement Learning for Load Frequency Control in Isolated Microgrids: A Knowledge Aggregation Approach with Emphasis on Power Symmetry and Balance. *Symmetry* **2024**, *16*, 322. [CrossRef]
2. Wu, D.; Wu, L.; Wen, T.; Li, L. Microgrid Operation Optimization Method Considering Power-to-Gas Equipment: An Improved Gazelle Optimization Algorithm. *Symmetry* **2024**, *1*, 83.
3. Chen, X.; Hu, Q.; Zhang, Y.; Song, M.; Wu, Z.; Xiong, J. POMDP-Based Dispatch Scheme for Residential Distributed Energy Resources Under Customer Fatigue Consideration. *IEEE Trans. Smart Grid* **2024**, *15*, 5665–5677.
4. Zhang, Y.; Zou, B.; Jin, X.; Luo, Y.; Song, M.; Ye, Y.; Hu, Q.; Chen, Q.; Zambroni, A.C. Mitigating power grid impact from proactive data center workload shifts: A coordinated scheduling strategy integrating synergistic traffic-data-power networks. *Appl. Energy* **2025**, *377*, 124697.
5. Zou, B.; Zhang, Y.; Chen, Q.; Hu, Q.; Hu, X.; Shi, J.; Li, Z.; Wang, Q. Low-carbon Economic Schedule of the H_2 DRI-EAF Steel Plant Integrated with a Power-to-hydrogen System Driven by Blue Hydrogen and Green Hydrogen. *IET Renew. Power Gen.* **2024**, *18*, 3839–3854.
6. Qian, T.; Liang, Z.; Chen, S.; Hu, Q.; Wu, Z. A Tri-Level Demand Response Framework for EVCS Flexibility Enhancement in Coupled Power and Transportation Networks. *IEEE Trans. Smart Grid* **2024**, *16*, 598–611.
7. Zhang, Y.; Qian, W.; Ye, Y.; Li, Y.; Tang, Y.; Long, Y.; Duan, M. A novel non-intrusive load monitoring method based on ResNet-seq2seq networks for energy disaggregation of distributed energy resources integrated with residential houses. *Appl. Energy* **2023**, *349*, 121703.

8. Hu, J.; Liu, X.; Shahidehpour, M.; Xia, S. Optimal Operation of Energy Hubs with Large-Scale Distributed Energy Resources for Distribution Network Congestion Management. *IEEE Trans. Sustain. Energy* **2021**, *12*, 1755–1765. [CrossRef]
9. Xiao, D.; Lin, Z.; Chen, H.; Hua, W.; Yan, J. Windfall profit-aware stochastic scheduling strategy for industrial virtual power plant with integrated risk-seeking/averse preferences. *Appl. Energy* **2024**, *357*, 122460. [CrossRef]
10. Qian, T.; Liang, Z.; Shao, C.; Guo, Z.; Hu, Q.; Wu, Z. Unsupervised learning for efficiently distributing EVs charging loads and traffic flows in coupled power and transportation systems. *Appl. Energy* **2025**, *377*, 124476. [CrossRef]
11. Popat, M.; Wu, B.; Liu, F.; Zargari, N. Coordinated Control of Cascaded Current-source Converter Based Offshore Wind Farm. *IEEE Trans. Sustain. Energy* **2012**, *3*, 557–565.
12. Xiao, D.; Chen, H.; Cai, W.; Wei, C.; Zhao, Z. Integrated Risk Measurement and Control for Stochastic Energy Trading of a Wind Storage System in Electricity Markets. *Prot. Control Mod. Power Syst.* **2023**, *8*, 60.
13. Zhang, F.; Bieber, L.; Zhang, Y.; Li, W.; Wang, L. A Multi-Port DC Power Flow Controller Integrated with MMC Stations for Offshore Meshed Multi-Terminal HVDC Grids. *IEEE Trans. Sustain. Energy* **2023**, *14*, 1676–1691. [CrossRef]
14. Chang, Y.; Cai, X. Hybrid Topology of a Diode-Rectifier-Based HVDC System for Offshore Wind Farms. *IEEE J. Emerg. Sel. Top. Power Electron.* **2019**, *7*, 2116–2128. [CrossRef]
15. Zhang, Y.; Ge, Y.; Wang, S.; Pan, W.; Feng, Y.; Qiu, P. Research on the Optimal Operation of a Prosumer Micro Energy Network Centred on Data Centres. *IET Renew. Power Gen.* **2024**, *18*, 3869–3889.
16. Castro, L.M.; Acha, E. On the dynamic modeling of marine VSC-HVDC power grids including offshore wind farms. *IEEE Trans. Sustain. Energy* **2020**, *11*, 2889–2900.
17. Kou, P.; Liang, D.; Wu, Z.; Ze, Q.; Gao, L. Frequency Support From a DC-Grid Offshore Wind Farm Connected Through an HVDC Link: A Communication-Free Approach. *IEEE Trans. Energy Convers.* **2018**, *33*, 1297–1310.
18. Nami, A.; Rodriguez-Amenedo, J.L.; Arnaltes, S.; Cardiel-Alvarez, M.A.; Baraciarte, R.A. Control of the Parallel Operation of DR-HVDC and VSC-HVDC for Offshore Wind Power Transmission. *IEEE Trans. Power Deliv.* **2022**, *37*, 1682–1691.
19. Wang, L.; Thi, M.S.-N. Stability Analysis of Four PMSG-Based Offshore Wind Farms Fed to an SG-Based Power System Through an LCC-HVDC Link. *IEEE Trans. Ind. Electron.* **2013**, *60*, 2392–2400.
20. Li, Z.; Yang, L.; Xu, Y. A Dynamics-Constrained Method for Distributed Frequency Regulation in Low-Inertia Power Systems. *Appl. Energy* **2023**, *344*, 121256.
21. Wang, W.; Shi, X.; Wu, G.; Cao, Y. Interaction Between Grid-Forming Converters with AC Grids and Damping Improvement Based on Loop Shaping. *IEEE Trans. Power Syst.* **2023**, *39*, 1905–1917.
22. Abdelwahed, M.A.; El-Saadany, E.F. Power Sharing Control Strategy of Multiterminal VSC-HVDC Transmission Systems Utilizing Adaptive Voltage Droop. *IEEE Trans. Sustain. Energy* **2016**, *8*, 605–615.
23. Liu, H.; Wang, Y.; Wang, M.; Guo, Y.; Wang, J.; Shen, G.; Liao, G. Small-Signal Analysis of DC Microgrid and Multi-Objective Optimization Segmented Droop Control Suitable for Economic Dispatch. *J. Mod. Power Syst. Clean. Energy* **2020**, *8*, 564–572. [CrossRef]
24. Wang, Y.; Wen, W.; Wang, C.; Liu, H.; Zhan, X.; Xiao, X. Adaptive Voltage Droop Method of Multiterminal VSC-HVDC Systems for DC Voltage Deviation and Power Sharing. *IEEE Trans. Power Deliv.* **2018**, *34*, 169–176.
25. Wang, W.; Li, Y.; Cao, Y.; Häger, U.; Rehtanz, C. Adaptive droop control of VSC-MTDC system for frequency support and power sharing. *IEEE Trans. Power Syst.* **2017**, *33*, 1264–1274.
26. Vennelaganti, S.G.; Chaudhuri, N.R. Stability Criterion for Inertial and Primary Frequency Droop Control in MTDC Grids with Implications on Ratio-based Frequency Support. *IEEE Trans. Power Syst.* **2020**, *35*, 3541–3551.
27. Zhang, Y.; Wang, L.; Li, W. Autonomous DC Line Power Flow Regulation Using Adaptive Droop Control in HVDC Grid. *IEEE Trans. Power Deliv.* **2020**, *36*, 3550–3560.
28. Haileselassie, T.M.; Uhlen, K. Impact of DC Line Voltage Drops on Power Flow of MTDC Using Droop Control. *IEEE Trans. Power Syst.* **2012**, *27*, 1441–1449.
29. Shi, N.; Cheng, R.; Liu, L.; Wang, Z.; Zhang, Q.; Reno, M.J. Data-Driven Affinely Adjustable Robust Volt/VAr Control. *IEEE Trans. Smart Grid* **2024**, *15*, 247–259.
30. Wang, Y.; Qiu, F.; Liu, G.; Lei, M.; Yang, C.; Wang, C. Adaptive Reference Power Based Voltage Droop Control for VSC-MTDC Systems. *J. Mod. Power Syst. Clean. Energy* **2023**, *11*, 381–388.
31. Davari, M.; Mohamed, Y.A.-R.I. Robust Droop and DC-Bus Voltage Control for Effective Stabilization and Power Sharing in VSC Multiterminal DC Grids. *IEEE Trans. Power Electron.* **2018**, *33*, 4373–4395. [CrossRef]
32. Wang, P.; Deng, N.; Zhang, X.-P. Droop Control for a Multi-Line Current Flow Controller in Meshed Multi-Terminal HVDC Grid Under Large DC Disturbances. *IEEE Power Energy Technol. Syst. J.* **2018**, *5*, 35–46.
33. Dong, H.; Xu, Z.; Song, P.; Tang, G.; Xu, Q.; Sun, L. Optimized Power Redistribution of Offshore Wind Farms Integrated VSC-MTDC Transmissions After Onshore Converter Outage. *IEEE Trans. Ind. Electron.* **2017**, *64*, 8948–8958. [CrossRef]
34. Sun, K.; Xiao, H.; Pan, J.; Liu, Y. VSC-HVDC Interties for Urban Power Grid Enhancement. *IEEE Trans. Power Syst.* **2021**, *36*, 4745–4753. [CrossRef]

35. Adeuyi, O.D.; Cheah-Mane, M.; Liang, J.; Jenkins, N. Fast Frequency Response from Offshore Multiterminal VSC–HVDC Schemes. *IEEE Trans. Power Deliv.* **2017**, *32*, 2442–2452.
36. Raza, A.; Dianguo, X.; Yuchao, L.; Xunwen, S.; Williams, B.W.; Cecati, C. Coordinated Operation and Control of VSC Based Multiterminal High Voltage DC Transmission Systems. *IEEE Trans. Sustain. Energy* **2016**, *7*, 364–373. [CrossRef]
37. Ambia, M.N.; Meng, K.; Xiao, W.; Al-Durra, A.; Dong, Z.Y. Adaptive Droop Control of Multi-Terminal HVDC Network for Frequency Regulation and Power Sharing. *IEEE Trans. Power Syst.* **2021**, *36*, 566–578.
38. Zhang, Y.; Shotorbani, A.M.; Wang, L.; Li, W. A Combined Hierarchical and Autonomous DC Grid Control for Proportional Power Sharing with Minimized Voltage Variation and Transmission Loss. *IEEE Trans. Power Deliv.* **2022**, *37*, 3213–3224.
39. Rouzbehi, K.; Miranian, A.; Luna, A.; Rodriguez, P. DC Voltage Control and Power Sharing in Multiterminal DC Grids Based on Optimal DC Power Flow and Voltage-Droop Strategy. *IEEE J. Emerg. Sel. Top. Power Electron.* **2014**, *2*, 1171–1180. [CrossRef]
40. Xiao, L.; Xu, Z.; An, T.; Bian, Z. Improved Analytical Model for the Study of Steady State Performance of Droop-Controlled VSC-MTDC Systems. *IEEE Trans. Power Syst.* **2017**, *32*, 2083–2093. [CrossRef]
41. Zhang, Y.; Meng, X.; Shotorbani, A.M.; Wang, L. Minimization of AC-DC Grid Transmission Loss and DC Voltage Deviation Using Adaptive Droop Control and Improved AC-DC Power Flow Algorithm. *IEEE Trans. Power Syst.* **2021**, *36*, 744–756. [CrossRef]
42. Wang, Z.; He, J.; Xu, Y.; Zhang, F. Distributed Control of VSC-MTDC Systems Considering Tradeoff Between Voltage Regulation and Power Sharing. *IEEE Trans. Power Syst.* **2020**, *35*, 1812–1821. [CrossRef]
43. Zhang, Y.; Shotorbani, A.M.; Wang, L.; Li, W. Distributed Voltage Regulation and Automatic Power Sharing in Multi-Terminal HVDC Grids. *IEEE Trans. Power Syst.* **2020**, *35*, 3739–3752. [CrossRef]
44. Li, B.; Li, Q.; Wang, Y.; Wen, W.; Li, B.; Xu, L. A Novel Method to Determine Droop Coefficients of DC Voltage Control for VSC-MTDC System. *IEEE Trans. Power Deliv.* **2020**, *35*, 2196–2211. [CrossRef]
45. Zhang, F.; Gu, W.; Zhang, Y.; Wang, L.; Li, W. General Linearized Model of Voltage Source Converter with Fixed Nodal Admittance Matrix. *IEEE Trans. Power Electron.* **2024**, *39*, 12143–12148. [CrossRef]
46. Gu, Z.; Pan, G.; Gu, W.; Qiu, H.; Lu, S. Robust optimization of scale and revenue for integrated power-to-hydrogen systems within energy, ancillary services, and hydrogen markets. *IEEE Trans. Power Syst.* **2023**, *39*, 5008–5023. [CrossRef]
47. Pan, G.; Gu, W.; Gu, Z.; Lin, J.; Zhou, S.; Wu, Z.; Lu, S. Feasibility of Scaling up the Cost-Competitive and Clean Electrolytic Hydrogen Supply in China. *Engineering* **2024**, *39*, 154–165. [CrossRef]
48. Kirakosyan, A.; El-Saadany, E.F.; Moursi, M.S.E.; Al Hosani, K. DC Voltage Regulation and Frequency Support in Pilot Voltage Droop-Controlled Multiterminal HVdc Systems. *IEEE Trans. Power Deliv.* **2018**, *33*, 1153–1164. [CrossRef]
49. Zhang, Y.; Qian, W.; Shao, J.; Zhang, F.; Wang, L.; Hu, Q.; Li, W. Adaptive Voltage Reference Based Controls of Converter Power Sharing and Pilot Voltage in HVDC System for Large-Scale Offshore Wind Integration. *IEEE Open J. Power Energy* **2024**, *11*, 55–67. [CrossRef]
50. Zhang, C.; Xu, Y.; Zhang, R. Multi-Objective Adaptive Robust Voltage/VAR Control for High-PV Penetrated Distribution Networks. *IEEE Trans. Smart Grid* **2020**, *11*, 5288–5300. [CrossRef]
51. Pan, C.; Shao, C.; Hu, B.; Xie, K.; Li, C.; Ding, J. Modeling the Reserve Capacity of Wind Power and the Inherent Decision-Dependent Uncertainty in the Power System Economic Dispatch. *IEEE Trans. Power Syst.* **2023**, *38*, 4404–4417. [CrossRef]
52. Roman, C.; Rosehart, W. Evenly Distributed Pareto Points in Multi-Objective Optimal Power Flow. *IEEE Trans. Power Syst.* **2006**, *21*, 1011–1012. [CrossRef]

symmetry

MDPI

Article

Stability Control Method Utilizing Grid-Forming Converters for Active Symmetry in the Elastic Balance Region of the Distribution Grid

Zhipeng Lv [1], Bingjian Jia [2], Zhenhao Song [1], Hao Li [1], Shan Zhou [1] and Zhizhou Li [1,*]

1 State Grid Shanghai Energy Interconnection Research Institute Co., Ltd., Shanghai 201210, China
2 Xinjiang Information Industry Co., Ltd., Urumqi 830011, China
* Correspondence: zhizhoulee@126.com

Abstract: The development of the elastic balance area within the distribution network places greater demands on the interaction between sources and loads, which impacts the stability of the power system. While achieving symmetry in active power is essential for stable operation, it is challenging to attain perfection due to various disruptions that can exacerbate frequency and voltage instability. Additionally, due to the inherent resonance characteristics of LCL filters and the time-varying nature of weak grid line impedance, grid-connected inverters may interact with the grid, potentially leading to oscillation issues. A grid-forming inverter control method that incorporates resonance suppression is proposed to address these challenges. First, a control model for the grid-forming inverter based on the Virtual Synchronous Generator (VSG) is established, enabling the system to exhibit inertia and damping characteristics. Considering the interaction between the VSG grid-connected system and the weak grid, sequence impedance models of the VSG system, which feature voltage and current double loops within the $\alpha\beta$ coordinate system, are developed using harmonic linearization techniques. By combining the impedance analysis method, the stability of the system under weak grid conditions is evaluated using the Nyquist criterion. The validity of the analysis is confirmed through simulations. Finally, in order to ensure the effectiveness and correctness of the simulation, an experimental prototype of an NPC three-level LCL grid-forming inverter is built, and the experimental results have verified that the system has good elastic support capability and resonance suppression capability in the elastic region.

Keywords: elastic balance region; active symmetry; grid-forming inverter; impedance modeling; stability analysis

1. Introduction

With the advancement of new power systems, distributed energy sources such as photovoltaic and wind power generation, along with flexible loads, have been extensively integrated into the distribution system [1,2]. The randomness and volatility of outputs from distributed power sources, coupled with the extensive involvement of flexible loads in demand response, will significantly alter the operating characteristics of the power grid. Active power symmetry plays a vital role in this context. It refers to the balanced distribution and exchange of active power among different components within the power grid, including power sources, grids, loads, and energy storage. This change will shift the operating mode from a deterministic power balance—where power sources follow load demands—to a dynamic balance characterized by interactions between power

https://doi.org/10.3390/sym17020263

sources and loads. As a result, balancing power and energy between these entities will become more challenging. Additionally, the safety and stability concerns related to large-scale interactions between power sources and loads are becoming increasingly prominent. The low inertia and weak damping characteristics of these systems exacerbate stability concerns [3,4]. Grid-forming inverters have gained significant attention because they behave like synchronous voltage sources and can provide some inertia support to the electrical grid. The most common control strategies for these inverters include droop control and Virtual Synchronous Generator control, both of which adjust power output in response to grid demand. Additionally, matching control and virtual oscillator control can enhance grid performance. Matching control optimizes the compatibility between the inverter and the grid, while virtual oscillator control introduces nonlinear oscillation mechanisms to support grid stability [5–7].

To improve the quality of grid-connected electrical power, an LCL-type filter needs to be connected in series with the inverter output side. However, there are significant resonance issues associated with this configuration [8]. Currently, LCL resonance suppression methods can be categorized into two primary approaches: (1) Passive damping methods involve inserting or paralleling resistors in the LCL filter to suppress resonance [9,10]. While passive damping does not introduce additional control loops to the system, it incurs considerable power losses. (2) Active damping methods mainly include the pre-filtering method and the state variable feedback method [11,12]. Reference [13] employs a notch filter in the system control loop to suppress LCL resonance using pole-zero cancelation, However, this method is sensitive to system parameters. Adaptive algorithms have been used in References [14,15] to reduce the parameter sensitivity issue, though this significantly increases the complexity of system control. Due to various drawbacks associated with the pre-filtering method, researchers have increasingly focused on the state variable feedback method. Commonly used state variables include inverter-side current, capacitor voltage, or current and grid current. When employing the grid current feedback method, the active damping controller utilizes a second-order differential component, which can effectively suppress LCL resonance [16,17]. However, these methods primarily target grid-connected inverters under conventional single-current loop control and are not readily applicable to VSG-based grid-connected systems.

Regarding system stability, the premise for VSG to actively underpin weak grids is that the VSG can function reliably when connected to the grid due to the VSG control strategy simulating the motion equation of synchronous generators; when the load changes, VSG can detect the changes and make corresponding adjustments, so that the inverter can provide good elastic support for the active frequency and reactive voltage stability of the system when facing interference in the elastic balance region [18]. References [19,20] established the state-space equation of a multi-machine parallel system using the state-space modeling method and analyzed the static stability of the grid-connected system. However, these studies primarily focus on traditional VSG control strategies, with limited research on the influence of LCL filters on the stability of VSG grid-connected systems.

Addressing the aforementioned issues, a grid-forming inverter control method with resonance suppression is discussed. In this method, VSG technology is used to make the grid-forming inverter have similar inertia and damping characteristics as the synchronous machine, ensuring that the grid provides elastic support for frequency and voltage while reducing the harmonic distortion rate of grid current. Additionally, the harmonic linearization method is used to build the sequential impedance model of the inverter system on the grid side. The stability of the system is then analyzed from an impedance perspective.

2. Topology and Control Strategies of VSG

Figure 1 illustrates the topology and control structure of the VSG-based grid-forming inverter under weak grid conditions. V_{dc} is the input DC voltage; C_1 and C_2 are the DC voltage divider capacitors; Point O is the midpoint of the DC side; v_{ia}, v_{ib}, and v_{ic} are the inverter output bridge arm voltages; C is the filtering capacitor; L_1 and L_2 are, respectively, the filter inductors on the inverter side and the grid side; i_{1a}, i_{1b}, and i_{1c} are the inverter-side three-phase currents; i_{2a}, i_{2b}, and i_{2c} are the grid-side currents; e_g is the grid voltage; and L_g is the grid inductor. By applying an appropriate control strategy, the voltage waveform output by the inverter can be controlled and, after filtering through the LCL filter, the output voltage is integrated into the grid.

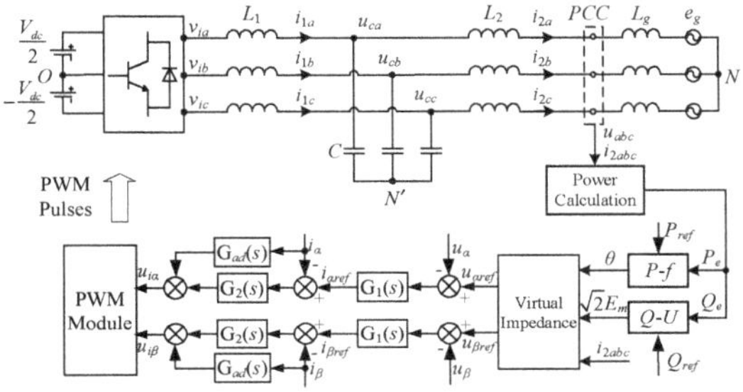

Figure 1. Overall control block diagram of the grid-connected system based on VSG.

2.1. *P-f* Control

The *P-f* loop of the VSG primarily exhibits inertia and damping characteristics by emulating the rotor dynamics of conventional synchronous generators. Combining with the synchronous generator's second-order model [21,22], the P-f equation for the VSG is represented as follows (1):

$$\begin{cases} J\frac{\mathrm{d}\omega}{\mathrm{d}t} = \frac{P_{ref}}{\omega_n} - \frac{P_e}{\omega_n} - D_p(\omega - \omega_n) \\ \frac{\mathrm{d}\theta}{\mathrm{d}t} = \omega \end{cases} \tag{1}$$

where, *J*-the VSG's virtual inertia, P_{ref} is the given active power, ω and ω_n refer to the output and rated angular frequencies of the VSG, θ represents the phase angle, and P_e is the instantaneous active power.

2.2. *Q-U* Control

In the VSG, the Q-U loop primarily simulates the excitation controller of synchronous generators [21,22]. The Q-U equation can be expressed as (2)

$$\sqrt{2}E_r = \frac{1}{K_q s}[\sqrt{2}(U_n - U)D_q + Q_{ref} - Q_e] \tag{2}$$

where K_q is the excitation constant, U is the effective value of the output voltage, D_q is the Q-U Loop droop constant, U_n is the rated effective voltage, Q_{ref} is the reference value for reactive power, Q_e is the instantaneous reactive power, and E_r represents the virtual internal voltage's effective value.

The internal voltage of the VSG is jointly defined by the outer loop control, which is expressed as Equation (3):

$$\begin{cases} e_a = \sqrt{2}E_r \sin\theta \\ e_b = \sqrt{2}E_r \sin(\theta - \frac{2}{3}\pi) \\ e_c = \sqrt{2}E_r \sin(\theta + \frac{2}{3}\pi) \end{cases} \tag{3}$$

2.3. Virtual Impedance Control

According to the electrical characteristics of the stator, the internal voltage E_q, stator current I, and terminal voltage U of a synchronous generator exhibit the following relationship:

$$U = E_q - I(r_a + j\omega L_d) \quad (4)$$

where r_a is the armature resistance of the synchronous generator and L_d is the synchronous inductance.

To achieve the stator electrical characteristics shown in Equation (4) for the grid-tied system, virtual impedance is employed in the VSG control strategy and transformed into the αβ coordinate system as shown in Equation (5):

$$\begin{cases} u_{\alpha ref} = e_\alpha - R_v i_\alpha + \omega L_v i_\beta \\ u_{\beta \mathrm{ref}} = e_\beta - R_v i_\beta - \omega L_v i_\alpha \end{cases} \quad (5)$$

where R_v denotes the virtual resistor, L_v denotes the virtual inductor; e_α and e_β are the αβ axis components of the voltages e_a, e_b, and e_c within VSG; i_α and i_β are the αβ axis components of the system grid-side currents; and the reference components of the voltage output from the virtual impedance loop in the αβ axis are $u_{\alpha ref}$ and $u_{\beta ref}$.

The voltage reference value obtained through the power outer-loop control and virtual impedance control in Sections 2.1–2.3 can be directly used as the modulation signal for SPWM, allowing the grid-connected system to exhibit inertia and damping characteristics as those of a synchronous generator. Nevertheless, since the relatively decelerated dynamic response of the outer loop control, the grid-tied power quality cannot be fully ensured. Therefore, it is necessary to add an inner-loop controller to accelerate the system's dynamic response and improve power quality.

2.4. VSG Voltage Loop Control

The quasi-PR controller not only retains the advantages of the PR (Proportional-Resonant) controller but also reduces the impact of frequency deviation on the AC side. Therefore, this paper selects a voltage loop design based on the quasi-PR controller. The expression for the quasi-PR controller is shown in Equation (6).

$$G_{PR}(s) = K_p + \frac{2K_r\omega_i s}{s^2 + 2\omega_i s + \omega_o^2} \quad (6)$$

where K_p denotes the proportional factor, K_r denotes the resonant factor, and ω_i is the fundamental resonant bandwidth. In this paper, ω_i = πrad/s and ω_o is the angular frequency reference value.

The voltage loop expression of the VSG based on the quasi-PR controller $G_1(s)$ is as follows:

$$\begin{cases} i_{\alpha ref} = G_1(s)(u_{\alpha ref} - u_\alpha) \\ i_{\beta ref} = G_1(s)(u_{\beta ref} - u_\beta) \end{cases} \quad (7)$$

where Kp1 is the proportional coefficient of $G_1(s)$, K_{r1} is the resonant coefficient of $G_1(s)$, $u_{\alpha ref}$ and $u_{\beta ref}$ are the voltage loop αβ axis reference components output by the virtual impedance, u_α and u_β represent the output voltages of the αβ axis components at the PCC of the system, and $i_{\alpha ref}$ and $i_{\beta ref}$ are the current loop αβ axis reference components output by the voltage loop.

2.5. VSG Current Loop Control

After obtaining the reference value of the current loop controlled by the VSG voltage loop, it is necessary to adjust the grid side current of the grid-connected system through the current loop. Based on the proportional controller $G_2(s)$, the expression of the VSG current loop can be expressed as (8)

$$\begin{cases} u_{i\alpha} = G_2(s)(i_{\alpha ref} - i_\alpha) \\ u_{i\beta} = G_2(s)(i_{\beta ref} - i_\beta) \end{cases} \tag{8}$$

where K_{p2} is the proportional coefficient of $G_2(s)$. The outputs of the current loop, $u_{i\alpha}$ and $u_{i\beta}$, serve as the αβ axis components of the inverter's modulation wave; $i_{\alpha ref}$ and $i_{\beta ref}$ serve as the reference components of the current loop's αβ axis, outputted by the voltage loop; and i_α and i_β represent the current of the αβ axis components at the grid-side in the system.

To address the resonance issues introduced by the LCL filter, an active damping component based on grid-side feedback is incorporated into the current control loop.

$$G_{ad}(s) = k_{ad}s^2 \tag{9}$$

where k_{ad} represents the active damping coefficient.

3. Impedance Modeling of Grid-Forming Inverters Based on VSG Under Weak Grid Conditions

Considering that the three-phase grid-connected system in this paper does not control the dq coordinate system, and the advantages of harmonic linearization modeling method are clear physical meaning, convenient measurement and verification of sequential impedance model, and relatively simple stability criteria, in order to create the impedance model for the three-phase grid-connected system, this paper uses the harmonic linearization modeling technique.

3.1. Sequence Impedance Modeling

In the time domain, after injecting a small signal perturbation on the grid side, the expression for the output voltage $u_a(t)$ of the system and the grid-side current $i_{2a}(t)$ are as follows:

$$\begin{aligned} u_a(t) = \; & V_1 \cos(2\pi f_1 t) + V_p \cos(2\pi f_p t + \varphi_{vp}) \\ & + V_n \cos(2\pi f_n t + \varphi_{vn}) \end{aligned} \tag{10}$$

$$\begin{aligned} i_{2a}(t) = \; & I_1 \cos(2\pi f_1 t + \varphi_{i1}) + I_p \cos(2\pi f_p t + \varphi_{ip}) \\ & + I_n \cos(2\pi f_n t + \varphi_{in}) \end{aligned} \tag{11}$$

where V_1, V_p, and V_n represent the amplitudes of the disturbance voltages of each sequence, respectively; I_1, I_p, and I_n represent the corresponding response current amplitudes; f_1, f_p, and f_n represent the corresponding perturbation frequencies; φ_{vp} and φ_{vn} are the initial phase angles of the positive-sequence and negative-sequence disturbance voltages, respectively; φ_{i1}, φ_{ip} and φ_{in} are the initial phase angles of the corresponding response currents.

Equations (10) and (11) can be transformed into the frequency domain as follows:

$$\begin{cases} u_a[f] = \begin{cases} \mathbf{V}_1, f = \pm f_1 \\ \mathbf{V}_p, f = \pm f_p \\ \mathbf{V}_n, f = \pm f_n \end{cases} \\ i_{2a}[f] = \begin{cases} \mathbf{I}_1, f = \pm f_1 \\ \mathbf{I}_p, f = \pm f_p \\ \mathbf{I}_n, f = \pm f_n \end{cases} \end{cases} \tag{12}$$

where $V_1 = V_1/2$; $V_p = (V_p/2)e \pm j\varphi_{vp}$; $V_n = (V_n/2)e \pm j\varphi_{vn}$; $I_1 = (I_1/2)e \pm j\varphi_{i1}$; $I_p = (I_p/2)e \pm j\varphi_{ip}$; $I_n = (I_n/2)e \pm j\varphi_{in}$.

Since the grid-connected system studied in this paper is three-phase symmetric, Equation (12) can be transformed into the αβ coordinate system.

$$\begin{cases} u_\alpha[f] = \begin{cases} \boldsymbol{V}_1, f = \pm f_1 \\ \boldsymbol{V}_p, f = \pm f_p \\ \boldsymbol{V}_n, f = \pm f_n \end{cases} \\ u_\beta[f] = \begin{cases} \mp j\boldsymbol{V}_1, f = \pm f_1 \\ \mp j\boldsymbol{V}_p, f = \pm f_p \\ \pm j\boldsymbol{V}_n, f = \pm f_n \end{cases} \end{cases} \tag{13}$$

$$\begin{cases} i_{2\alpha}[f] = \begin{cases} \boldsymbol{I}_1, f = \pm f_1 \\ \boldsymbol{I}_p, f = \pm f_p \\ \boldsymbol{I}_n, f = \pm f_n \end{cases} \\ i_{2\beta}[f] = \begin{cases} \mp j\boldsymbol{I}_1, f = \pm f_1 \\ \mp j\boldsymbol{I}_p, f = \pm f_p \\ \pm j\boldsymbol{I}_n, f = \pm f_n \end{cases} \end{cases} \tag{14}$$

According to the fundamental principles of the modeling method of harmonic linearization, the expression of the function between the disturbance voltage and the response current can be obtained.

$$\begin{cases} Z_{vp} = -\frac{V_p}{I_p} \\ Z_{vp} = -\frac{V_n}{I_n} \end{cases} \tag{15}$$

where $Z_{vp}(s)$ and $Z_{vn}(s)$ represent the positive and negative sequence impedance of the VSG system. By combining the above three layers of functional relationships, the functional relationship between $V_p(V_n)$ and $I_p(I_n)$ can be solved, and the sequence impedance model of the VSG system can be obtained. Based on the operating principle of the system shown in Figure 1, the small signal flow chart of the system as shown in Figure 2.

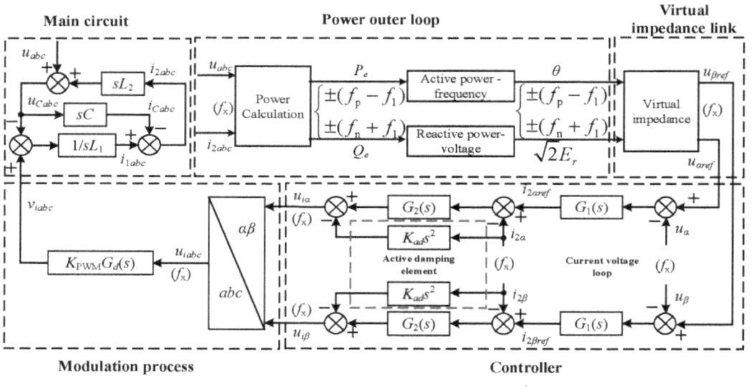

Figure 2. Small signal flowchart of the VSG system.

Figure 2 illustrates the small signal flow diagram of the VSG system, where the main circuit is composed of filters, and the control loop consists of the VSG power loop, virtual impedance segment, voltage–current loop, and modulation segment. In the figure, the subscript x = p, n represents the positive and negative sequence disturbance component at frequency f. Based on Figure 2, the impedance modeling of the VSG system can be divided into five parts. Below, the impedance models for these five components are established.

3.2. *Modeling of the Main Circuit for the VSG Grid-Connected Systems*

Injecting small signal voltage disturbances in positive and negative sequences with frequencies f_p and f_n on the grid side will generate current response signals with frequencies f_p and f_n, respectively. Taking phase a as an example, Figure 3 represents the harmonic small signal equivalent model of the main circuit. In the figure, the subscript x = p indicates the positive sequence disturbance component at frequency f_p, while x = n indicates the negative sequence disturbance component at frequency f_n.

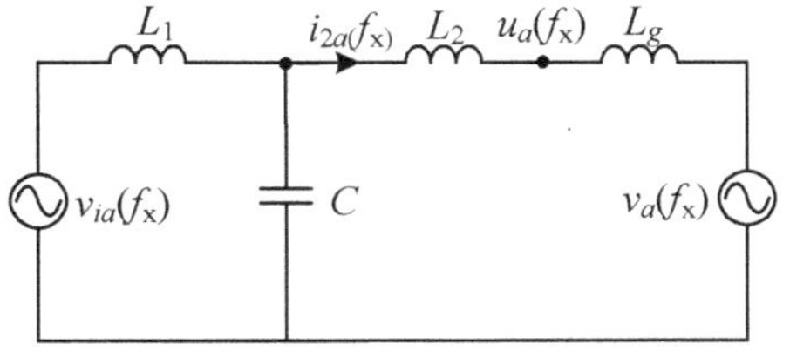

Figure 3. Equivalent model of harmonic small signal in phase A main circuit.

Based on Figure 3, the corresponding main circuit small signal expression is (16)

$$v_{ia}[f] = [L_1 L_2 C s^3 + (L_1 + L_2)s] i_{2a}[f] + (L_1 C s^2 + 1) u_a[f] \tag{16}$$

3.3. *VSG Power Outer Loop Modeling*

It is known that the expression for the output power of the system in the αβ coordinate system can be expressed as (17)

$$\begin{cases} P_e = 1.5G_f(s)(u_\alpha i_{2\alpha} + u_\beta i_{2\beta}) \\ Q_e = 1.5G_f(s)(u_\beta i_{2\alpha} - u_\alpha i_{2\beta}) \end{cases} \tag{17}$$

where $G_f(s) = \omega l/(s + \omega l)$ represents the transfer function of a low-pass filter and ωl represents the cutoff frequency of the low-pass filter.

By substituting Equations (14) and (15) into Equation (18), and combining the frequency domain convolution theorem while neglecting the effects of second-order small-signal disturbance terms, the expressions for P_e and Q_e of the system in the frequency domain can be expressed as (18)

$$P_e = \begin{cases} 3G_f(s)(\boldsymbol{V}_1\boldsymbol{I}_1^* + \boldsymbol{I}_1\boldsymbol{V}_1^*),\ f = \mathrm{dc} \\ 3G_f(s)(\boldsymbol{I}_\mathrm{p}\boldsymbol{V}_1^* + \boldsymbol{V}_\mathrm{p}\boldsymbol{I}_1^*), f = \pm(f_\mathrm{p} - f_1) \\ 3G_f(s)(\boldsymbol{V}_1\boldsymbol{I}_\mathrm{n} + \boldsymbol{V}_\mathrm{n}\boldsymbol{I}_1),\ f = \pm(f_\mathrm{n} + f_1) \end{cases} \tag{18}$$

$$Q_e = \begin{cases} 3\mathrm{j}G_f(s)(-\boldsymbol{V}_1\boldsymbol{I}_1^* + \boldsymbol{I}_1\boldsymbol{V}_1^*), f = \mathrm{dc} \\ \pm 3\mathrm{j}G_f(s)(\boldsymbol{I}_\mathrm{p}\boldsymbol{V}_1^* - \boldsymbol{V}_\mathrm{p}\boldsymbol{I}_1^*), f = \pm(f_\mathrm{p} - f_1) \\ \pm 3\mathrm{j}G_f(s)(-\boldsymbol{V}_1\boldsymbol{I}_\mathrm{n} + \boldsymbol{V}_\mathrm{n}\boldsymbol{I}_1), f = \pm(f_\mathrm{n} + f_1) \end{cases} \tag{19}$$

The expression for θ can be obtained from the VSG active power controller as (20)

$$\theta(s) = M(s)(P_{ref} - P_e + D_p\omega_n^2) \tag{20}$$

In the above formula, $M(s) = 1/(J_{\omega ns2} + D_{\omega ns})$.

By substituting Equation (18) into Equation (20), the expressions about θ can be expressed as (21)

$$\theta[f] = \begin{cases} -3G_f(s)M(s)(\boldsymbol{I}_\text{p}\boldsymbol{V}_1^* + \boldsymbol{V}_\text{p}\boldsymbol{I}_1^*), & f = \pm(f_\text{p} - f_1) \\ -3G_f(s)M(s)(\boldsymbol{V}_1\boldsymbol{I}_\text{n} + \boldsymbol{V}_\text{n}\boldsymbol{I}_1), & f = \pm(f_\text{n} + f_1) \end{cases} \tag{21}$$

Regarding the output θ of the VSG active power-frequency loop, the corresponding phase angle perturbation Δθ can be introduced by voltage perturbations, which can be expressed as (22)

$$\Delta\theta[f] = \begin{cases} -3G_f(s)M(s)(\boldsymbol{I}_\text{p}\boldsymbol{V}_1^* + \boldsymbol{V}_\text{p}\boldsymbol{I}_1^*), f = \pm(f_\text{p} - f_1) \\ -3G_f(s)M(s)(\boldsymbol{V}_1\boldsymbol{I}_\text{n} + \boldsymbol{V}_\text{n}\boldsymbol{I}_1), f = \pm(f_\text{n} + f_1) \end{cases} \tag{22}$$

The phase angle disturbance Δθ yields θ = $θ_1$ + Δθ, where $θ_1$ is the fundamental wave angle of the grid-connected system, obtaining Equation (23).

$$\begin{aligned} \cos\theta[f] &= \cos(\theta_1[f] + \Delta\theta[f]) \\ &\approx \cos\theta_1[f] - \Delta\theta[f] * \sin\theta_1[f] \end{aligned} \tag{23}$$

where $\theta = \theta_1 + \Delta\theta$, with φ_1 denoting the power angle of the VSG, $\varphi_1 \approx \text{arsin}[\text{PN}\omega_{nLv}/(V_N V_1)]$.

In the frequency domain, the value of θ_1 can be calculated as (24)

$$\begin{cases} \cos\theta_1[f] = e^{\pm \text{j}\varphi_1}/2, & f = \pm f_1 \\ \sin\theta_1[f] = -e^{\pm \text{j}(\varphi_1 + \pi/2)}/2, & f = \pm f_1 \end{cases} \tag{24}$$

By substituting Equations (22) and (24) into Equation (23), we can obtain Equation (25).

$$\cos\theta[f] = \begin{cases} \frac{1}{2}e^{\pm \text{j}\varphi_1}, & f = \pm f_1 \\ -\frac{3}{2}e^{\pm \text{j}\varphi_\text{vir}}G_f(s \mp \text{j}2\pi f_1)M(s \mp \text{j}2\pi f_1) \bullet \\ (\boldsymbol{I}_\text{p}\boldsymbol{V}_1^* + \boldsymbol{V}_\text{p}\boldsymbol{I}_1^*), & f = \pm f_\text{p} \\ -\frac{3}{2}e^{\mp \text{j}\varphi_\text{vir}}G_f(s \pm \text{j}2\pi f_1)M(s \pm \text{j}2\pi f_1) \bullet \\ (\boldsymbol{V}_1\boldsymbol{I}_\text{n} + \boldsymbol{V}_\text{n}\boldsymbol{I}_1), & f = \pm f_\text{n} \end{cases} \tag{25}$$

where $\varphi_{vir} = \varphi_1 + \pi/2$.

From the reactive power–voltage controller Equation (2) of the VSG, the internal voltage magnitude $\sqrt{2}E_m$ in the frequency domain is given by

$$\sqrt{2}E_r[f] = \begin{cases} V_1, & f = \text{dc} \\ \frac{-2D_q \boldsymbol{V}_\text{p} \mp 3\text{j}G_f(s)(\boldsymbol{I}_\text{p}\boldsymbol{V}_1^* - \boldsymbol{V}_\text{p}\boldsymbol{I}_1^*)}{K_q s}, & f = \pm(f_\text{p} - f_1) \\ \frac{-2D_q \boldsymbol{V}_\text{n} \pm 3\text{j}G_f(s)(\boldsymbol{V}_1\boldsymbol{I}_\text{n} - \boldsymbol{V}_\text{n}\boldsymbol{I}_1)}{K_q s}, & f = \pm(f_\text{n} + f_1) \end{cases} \tag{26}$$

By combining Equations (25) and (26) with the VSG power outer loop inner voltage Equation (3), we can neglect the effects of second-order small-signal disturbance terms and transform them into the αβ coordinate system. Considering that the system is three-phase balanced and symmetrical, $e_\alpha[f] = e_a[f]$, the expression for e_α is obtained as (27)

$$e_\alpha[f] = \begin{cases} H_1(s)\boldsymbol{V}_\mathrm{p} + H_2(s)(\boldsymbol{I}_\mathrm{p}\boldsymbol{V}_1^* - \boldsymbol{V}_\mathrm{p}\boldsymbol{I}_1^*) + \\ H_3(s)(\boldsymbol{I}_\mathrm{p}\boldsymbol{V}_1^* + \boldsymbol{V}_\mathrm{p}\boldsymbol{I}_1^*), f = \pm f_\mathrm{p} \\ H_1^*(s)\boldsymbol{V}_\mathrm{n} - H_2^*(s)(\boldsymbol{V}_1\boldsymbol{I}_\mathrm{n} - \boldsymbol{V}_\mathrm{n}\boldsymbol{I}_1) + \\ H_3^*(s)(\boldsymbol{V}_1\boldsymbol{I}_\mathrm{n} + \boldsymbol{V}_\mathrm{n}\boldsymbol{I}_1), f = \pm f_\mathrm{n} \end{cases} \quad (27)$$

where

$$\begin{cases} H_1(s) = -D_q e^{\pm \mathrm{j}\varphi_1} / K_q(s \mp \mathrm{j}2\pi f_1) \\ H_2(s) = \mp 1.5\mathrm{j}G_f(s \mp \mathrm{j}2\pi f_1)e^{\pm \mathrm{j}\varphi_1} / K_q(s \mp \mathrm{j}2\pi f_1) \\ H_3(s) = -1.5V_1 e^{\pm \mathrm{j}\varphi_{vir}} G_f(s \mp \mathrm{j}2\pi f_1) M(s \mp \mathrm{j}2\pi f_1) \end{cases} \quad (28)$$

3.4. Virtual Impedance Modeling of the VSG

As shown in Figure 2, the output of the VSG power outer loop, after passing through the virtual impedance stage, becomes the reference quantity for the output voltage loop. Combining Equation (5), the α-axis reference component $u_{\alpha ref}$ of the voltage loop is given by (29)

$$u_{\alpha ref}[f] = \begin{cases} H_1(s)\boldsymbol{V}_\mathrm{p} + H_2(s)(\boldsymbol{I}_\mathrm{p}\boldsymbol{V}_1^* - \boldsymbol{V}_\mathrm{p}\boldsymbol{I}_1^*) + H_3(s)(\boldsymbol{I}_\mathrm{p}\boldsymbol{V}_1^* + \boldsymbol{V}_\mathrm{p}\boldsymbol{I}_1^*) - \\ R_v\boldsymbol{I}_\mathrm{p} \mp \mathrm{j}\omega_n L_v \boldsymbol{I}_\mathrm{p}, \qquad f = \pm f_\mathrm{p} \\ H_1^*(s)\boldsymbol{V}_\mathrm{n} - H_2^*(s)(\boldsymbol{V}_1\boldsymbol{I}_\mathrm{n} - \boldsymbol{V}_\mathrm{n}\boldsymbol{I}_1)] + H_3^*(s)(\boldsymbol{V}_1\boldsymbol{I}_\mathrm{n} + \boldsymbol{V}_\mathrm{n}\boldsymbol{I}_1) - \\ R_v\boldsymbol{I}_\mathrm{n} \pm \mathrm{j}\omega_n L_v \boldsymbol{I}_\mathrm{n}, \qquad f = \pm f_\mathrm{n} \end{cases} \quad (29)$$

3.5. VSG Inner Loop Modeling

Using the voltage loop reference obtained from Equation (29) and the voltage outer loop control principle in Figure 2, the α-axis current loop reference component $i_{\alpha ref}$ is given by

$$i_{\alpha ref}[f] = \begin{cases} G_1(s)(u_{\alpha ref}[f] - \boldsymbol{V}_\mathrm{p}), \ f = \pm f_\mathrm{p} \\ G_1(s)(u_{\alpha ref}[f] - \boldsymbol{V}_\mathrm{n}), f = \pm f_\mathrm{n} \end{cases} \quad (30)$$

From the small-signal flowchart of the VSG system depicted in Figure 2, it can be known that the α-axis modulation wave of the current loop output is denoted as $u_{i\alpha}[f]$:

$$u_{i\alpha}[f] = \begin{cases} G_2\Big(s\Big)(i_{\alpha ref}[f] - \boldsymbol{I}_\mathrm{p}) - K_{ad}s^2\boldsymbol{I}_\mathrm{p}, \ f = \pm f_\mathrm{p} \\ G_2\Big(s\Big)(i_{\beta ref}[f] - \boldsymbol{I}_\mathrm{n}) - K_{ad}s^2\boldsymbol{I}_\mathrm{n}, f = \pm f_\mathrm{n} \end{cases} \quad (31)$$

3.6. Modulation Link Modeling

Given that the grid-connected system is balanced in three phases, $u_{ia}[f] = u_{i\alpha}[f]$. The PWM modulation signal in the αβ coordinate system is derived from the output of the inner loop. By transforming it into the abc coordinate system, the response signal of the voltage of the bridge arm can be obtained as (32)

$$v_{ia}[f] = K_\mathrm{PWM} u_{ia}[f] G_d(s) \quad (32)$$

By combining Equations (16), (31), and (32), the sequence impedances $Z_{v\mathrm{p}}$ and $Z_{v\mathrm{n}}$ of the grid-forming inverter can be determined as follows:

$$Z_{v\mathrm{p}}(s) = \frac{-0.5V_1K_1(s)[H_2(s) + H_3(s)] + (R_v \pm \mathrm{j}\omega_n L_v)K_1(s) + K_2(s) + L_1L_2Cs^3 + (L_1 + L_2)s}{0.5I_1K_1(s)[H_2(s) - H_3(s)] + K_1(s)[1 - H_1(s)] + L_1Cs^2 + 1} \quad (33)$$

$$Z_{vn}(s) = \frac{0.5V_1K_1(s)[H_2^*(s) - H_3^*(s)] + [R_v \mp j\omega_n L_v]K_1(s) + K_2(s) + L_1L_2Cs^3 + (L_1 + L_2)s}{-0.5I_1K_1(s)[H_2^*(s) + H_3^*(s)] + K_1(s)[1 - H_1^*(s)] + L_1Cs^2 + 1} \quad (34)$$

where

$$\begin{aligned} K_1(s) &= K_{\text{PWM}}G_1(s)G_2(s)G_d(s) \\ K_2(s) &= K_{\text{PWM}}K_{ad}s^2G_d(s) + K_{\text{PWM}}G_2(s)G_d(s) \end{aligned} \quad (35)$$

$$\begin{aligned} K_1(s) &= K_{\text{PWM}}G_1(s)G_2(s)G_d(s) \\ K_2(s) &= K_{\text{PWM}}K_{ad}s^2G_d(s) + K_{\text{PWM}}G_2(s)G_d(s) \end{aligned} \quad (36)$$

4. Stability Analysis

To further analyze the stability of the VSG system, this paper analyzes the impedance proportion of grid-forming inverters and weak grids based on the Nyquist criterion. Given that the VSG system exhibits symmetrical characteristics in three-phase balance during grid-connected operation, the positive and negative sequence impedances are decoupled in the system. Therefore, the impedance proportion applied to determine the stability of the grid-connected system is calculated using Equation (36).

$$\begin{cases} H_{vp}(s) = \frac{Z_{vp}(s)}{Z_{gp}(s)} \\ H_{vn}(s) = \frac{Z_{vn}(s)}{Z_{gn}(s)} \end{cases} \quad (37)$$

Here, H_{vp}(s) and H_{vn}(s) represent the positive and negative sequence impedance proportions, respectively, between the grid-forming inverter and the weak grid. Z_{gp}(s) and Z_{gn}(s) represent the positive and negative sequence impedances of the grid, respectively, and Z_g(s) = Z_{gp}(s) = Z_{gn}(s) = sL_g.

Figure 4 shows the Nyquist diagrams of the grid-connected system's $H_{vp}(s)$ and $H_{vn}(s)$ under different weak grid conditions, and Table 1 shows the analysis of the Nyquist diagram in Figure 4. Among them, the red solid line represents the system's positive sequence impedance ratio $H_{vp}(s)$, and the blue dashed line represents the system's negative sequence impedance ratio $H_{vn}(s)$. When the difference between the number of positive crossings N+ and the number of negative crossings N- of the open-loop frequency characteristic curve of the system in the section of the negative real axis $(-1, -\infty)$ multiplied by 2 is equal to the number of right poles of the open-loop system, the closed-loop system is stable. Both $H_{vp}(s)$ and $H_{vn}(s)$ have two poles in the right half-plane. Therefore, when the grid-connected system is stable, the Nyquist curves of both $H_{vp}(s)$ and $H_{vn}(s)$ should cross the negative real axis to the left of the point $(-1, j0)$ positively once. It can be seen from Figure 4 that as the grid becomes weaker, the Nyquist curves of $H_{vp}(s)$ and $H_{vn}(s)$ each cross the negative real axis to the left of the point $(-1, j0)$ positively once. From this, it can be concluded that the grid-connected system with the active damping link added can operate stably under weak grid conditions.

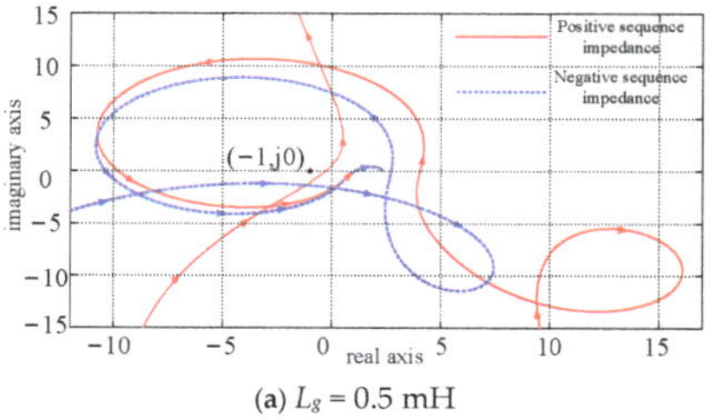

(a) L_g = 0.5 mH

Figure 4. *Cont.*

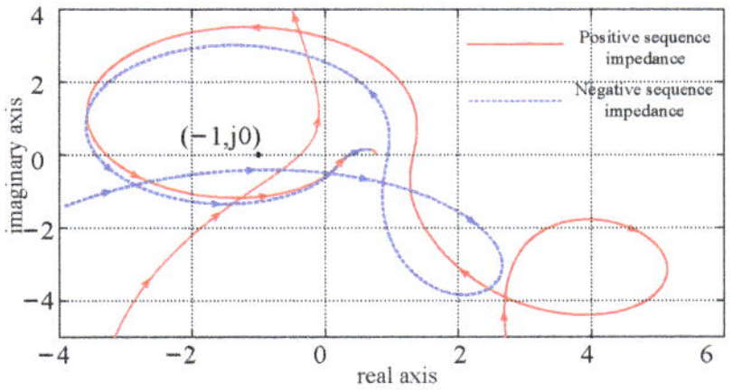

(**b**) L_g = 1.5 mH

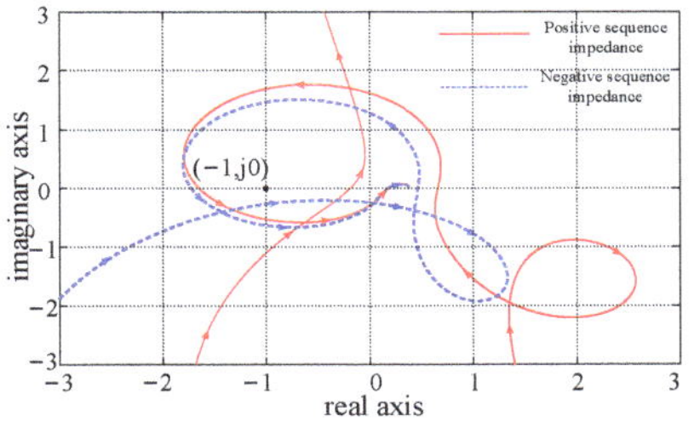

(**c**) L_g = 3 mH

Figure 4. Nyquist plots of $H_{vp}(s)$ and $H_{vn}(s)$ under different weak grid conditions.

Table 1. Analysis of the Nyqist diagram in Figure 4.

L_g/mH	N + (H_{vp}(s))	N − (H_{vn}(s))	2[(N+) − (N−)]	Stability
0.5	1	1	2	stable
1.5	1	1	2	stable
3	1	1	2	stable

5. Simulation Analysis

To assess the efficacy of the proposed VSG control strategy for the grid-forming inverter, a simulation model based on VSG is constructed using MATLAB/Simulink 2023a under weak grid conditions. The results of this simulation are then analyzed, with the parameters outlined in Table 2.

Table 2. Simulation model parameters for the system.

Parameter	Value	Parameter	Value
L_1	2 mH	L_2	1 mH
C	50 µF	L_g	0~3 mH
V_{dc}	300 V	U_g	76.7 V
P_{ref}	2000 W	Q_{ref}	0 Var
J	0.5	D_p	5
K_q	5.8	D_q	368
R_v	0.1Ω	L_v	3 mH
K_{p1}	0.01	K_{r1}	2.1
K_{p2}	0.06	K_{ad}	6*10^{-9}
f_s	10 kHz	f_{sw}	10 kHz

5.1. *Verification of the Dynamic Performance of the VSG*

5.1.1. Participate in the Grid Frequency Modulation Capability Verification

(1) Simulation of participation in the grid frequency modulation capability verification 1: At the initial moment of 0 s, the grid-connected system operates at rated conditions with P_{ref} = 0 W and Q_{ref} = 0 Var. At 0.4 s, a sudden drop in the grid angular frequency of

0.2 π rad/s is introduced, and, at 0.75 s, the grid angular frequency is restored to its rated state. The output power, output voltage, and grid-side current of the grid-connected system are illustrated in Figure 5.

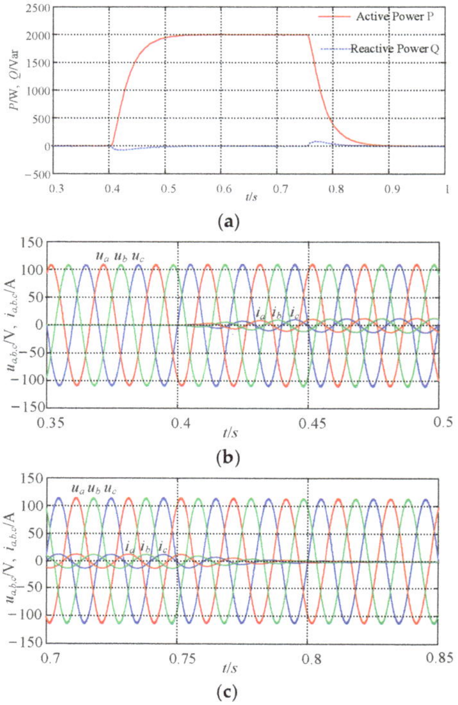

Figure 5. Simulation 1 of participation in the frequency regulation of the power grid. (**a**) Waveform graph of P and Q of the system during variations in grid frequency. (**b**) Waveform graph of output voltage and current of the system during sudden drops in grid frequency. (**c**) Waveform graph of output voltage and current of the system during recovery of grid frequency.

In Figure 5, at 0 s, the system operates at rated conditions, with the output P and Q tracking the given power values of P_{ref} and Q_{ref}. Following a sudden drop in the grid angular frequency of 0.2 π rad/s at 0.4 s, the output P of the system increases to 2000 W, accompanied by changes in the grid-side current. At 0.75 s, when the grid angular frequency is restored to its rated state, the output P returns to the set value, and the grid-side current also reverts to its initial condition. During the fluctuations in grid frequency, the VSG can adjust its active output through primary frequency control, ensuring the stability of the system. The variations in output voltage and grid-side current are consistent with changes in the system's active power, indicating that the VSG control strategy proposed in this paper is effective in participating in grid frequency regulation.

(2) Simulation of participation in the grid frequency modulation capability verification 2: At the initial moment of 0 s, the grid-connected system operates at rated conditions with P_{ref} = 0 W and Q_{ref} = 0 Var. At 0.54 s, a sudden increase in the grid angular frequency of 0.2 π rad/s is introduced, and, at 0.75 s, the grid angular frequency is restored to its rated state. The output power, output voltage, and grid-side current of the grid-connected system are illustrated in Figure 6.

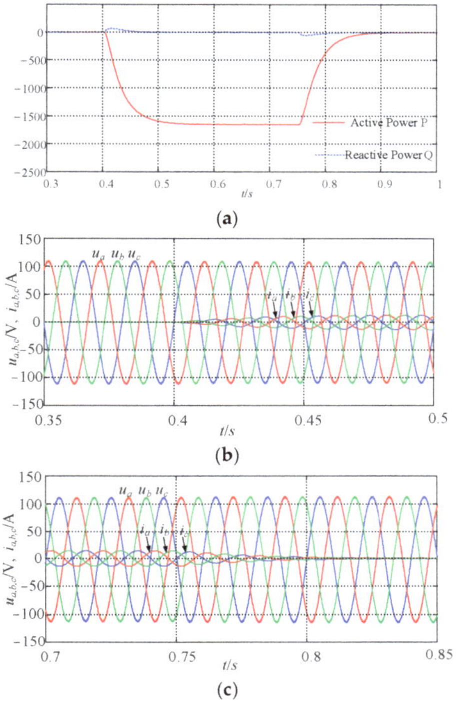

Figure 6. Simulation 2 of participation in the frequency regulation of the power grid. (**a**) Waveform graph of P and Q of the system during variations in grid frequency. (**b**) Waveform graph of output voltage and current of the system during sudden drops in grid frequency. (**c**) Waveform graph of output voltage and current of the system during recovery of grid frequency.

As shown in Figure 6, at 0 s, the grid-connected system operates at rated conditions, with the output P and Q tracking the specified power values of P_{ref} and Q_{ref}. Following a sudden increase in the grid angular frequency of 0.2 π rad/s at 0.4 s, the output P of the grid-connected system decreases to −2000 W, accompanied by corresponding changes in the grid-side current. At 0.75 s, when the grid angular frequency is restored to its rated state, the output P returns to the set value, and the grid-side current also reverts to its initial condition. During fluctuations in grid frequency, the VSG can adjust its active output through primary frequency control, ensuring the stability of the system. The variations in output voltage and grid-side current are consistent with changes in the system's active power, further demonstrating that the VSG control strategy proposed in this paper effectively participates in grid frequency regulation.

(3) Validation of the ability to participate in frequency regulation under unbalanced grid conditions: In Figure 7, in order to verify the system's frequency support capability under an unbalanced power grid, at 0 s, the grid-connected system operates at rated conditions, with the output P and Q tracking the specified power values of P_{ref} and Q_{ref}. Following a sudden increase in the grid angular frequency of 0.15 π rad/s at 0.4 s, the output P of the grid-connected system decreases to 15,000 W, accompanied by corresponding changes in the grid-side current. At 0.75 s, when the grid angular frequency is restored to its rated state, the output P returns to the set value, and the grid-side current also reverts to its initial condition. It can be seen that under the conditions of an unbalanced power grid, during fluctuations in grid frequency, the VSG can adjust its active output through primary

frequency control, ensuring the stability of the system. The variations in output voltage and grid-side current are consistent with changes in the system's active power, indicating that the VSG control strategy proposed in this paper can also effectively participate in grid frequency regulation under the conditions of a three-phase unbalance in the power grid.

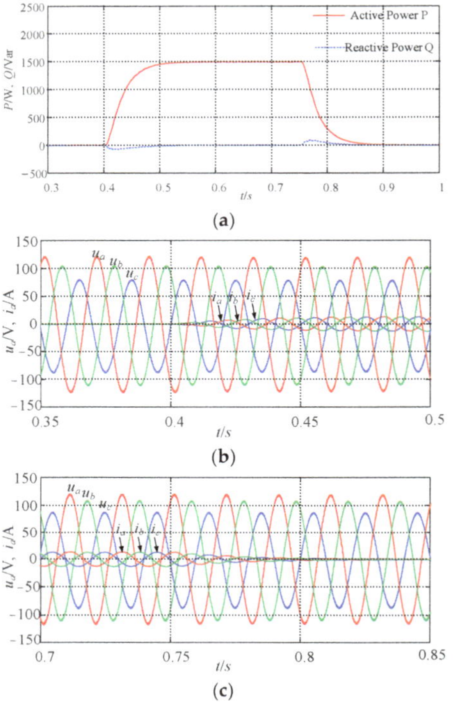

Figure 7. Validation of the ability to participate in frequency regulation under unbalanced grid conditions. (**a**) Waveform graph of P and Q of the system during variations in grid frequency. (**b**) Waveform graph of output voltage and current of the system during sudden drops in grid frequency. (**c**) Waveform graph of output voltage and current of the system during recovery of grid frequency.

5.1.2. Participate in the Grid Voltage Modulation Capability Verification

(1) Simulation of participation in the grid voltage modulation capability verification 1: At the initial moment of 0 s, the grid-connected system operates at rated conditions with P_{ref} = 0 W and Q_{ref} = 0 Var. At 0.4 s, a sudden drop of 2.5% in the grid voltage amplitude is introduced, and, at 0.75 s, the grid voltage amplitude returns to the set value. The output power, output voltage, and grid-side current of the grid-connected system are illustrated in Figure 7 Simulation 1 of Participation in the Voltage Regulation of the Power Grid.

From Figure 7, at 0 s, the grid-connected system operates at rated conditions, with the output P and Q tracking the specified values of P_{ref} and Q_{ref}. Following a sudden drop of 2.5% in the grid voltage amplitude at 0.4 s, the output Q of the system increases to 1000 Var, accompanied by corresponding changes in the grid-side current. At 0.75 s, when the grid voltage is restored to its rated state, the output Q returns to the set value, and the grid-side current also reverts to its initial condition. During fluctuations in grid voltage amplitude, the VSG can adjust its reactive output through voltage regulation to ensure the stability of the system. The variations in output voltage and grid-side current are consistent with changes in the system's Q, indicating that the VSG control strategy proposed in this paper effectively participates in grid voltage regulation.

(2) Simulation of participation in the grid voltage modulation capability verification 2: At the initial moment of 0 s, the grid-connected system operates at rated conditions with P_{ref} = 0 W and Q_{ref} = 0 Var. At 0.4 s, a sudden increase of 2.5% in the grid voltage amplitude is introduced, and, at 0.75 s, the grid voltage amplitude returns to the set value. The output power, output voltage, and grid-side current of the grid-connected system are illustrated in Figure 8 Simulation 2 of Participation in the Voltage Regulation of the Power Grid.

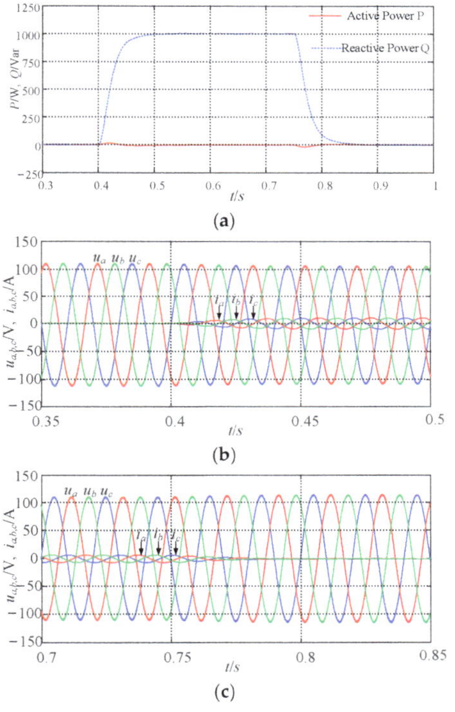

Figure 8. Simulation 1 of participation in the voltage regulation of the power grid. (**a**) Waveform graph of P and Q of the system during variations in grid voltage. (**b**) Waveform graph of output voltage and current of the system during sudden drops in grid voltage. (**c**) Waveform graph of output voltage and current of the system during recovery of grid voltage.

As shown in Figure 9, at 0 s, the grid-connected system operates at rated conditions, with the output P and Q tracking the specified values of P_{ref} and Q_{ref}. Following a sudden increase of 2.5% in the grid voltage amplitude at 0.4 s, the output Q of the system decreases to −1000 Var, accompanied by corresponding changes in the grid-side current. At 0.75 s, when the grid voltage amplitude is restored to its rated state, the output Q returns to the set value, and the grid-side current also reverts to its initial condition. During fluctuations in grid voltage amplitude, the VSG can adjust its reactive output through voltage control to ensure the stability of the system. The variations in output voltage and grid-side current are consistent with changes in the system's Q, further demonstrating that the VSG control strategy proposed in this paper effectively participates in grid voltage regulation.

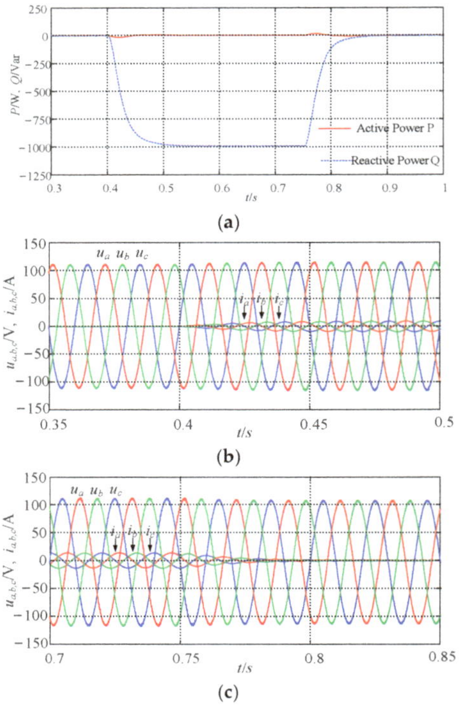

Figure 9. Simulation 2 of participation in the voltage regulation of the power grid. (a) Waveform graph of P and Q of the system during variations in grid voltage. (b) Waveform graph of output voltage and current of the system during sudden increases in grid voltage. (c) Waveform graph of output voltage and current of the system during recovery of grid voltage.

(3) Validation of participation in voltage regulation capability under unbalanced grid conditions: In Figure 10, to verify the system's voltage support capability under unbalanced grid conditions, at 0 s, the grid-connected system operates at rated conditions, with the output P and Q tracking the specified values of P_{ref} and Q_{ref}. Following a sudden drop of 1.875% in the grid voltage amplitude at 0.4 s, the output Q of the system increases to 750 Var, accompanied by corresponding changes in the grid-side current. At 0.75 s, when the grid voltage is restored to its rated state, the output Q returns to the set value, and the grid-side current also reverts to its initial condition. It can be seen that under the conditions of an unbalanced grid, during fluctuations in grid voltage amplitude, the VSG can adjust its reactive output through voltage regulation to ensure the stability of the system. The variations in output voltage and grid-side current are consistent with changes in the system's Q, indicating that the VSG control strategy proposed in this paper can effectively participate in grid voltage regulation under conditions of three-phase unbalance in the grid.

From the above analysis, it can be seen that adopting the VSG control strategy can enable the system to have good frequency and voltage active support capabilities in the elastic region. In the case of a weak grid, the system will have good elastic support capabilities.

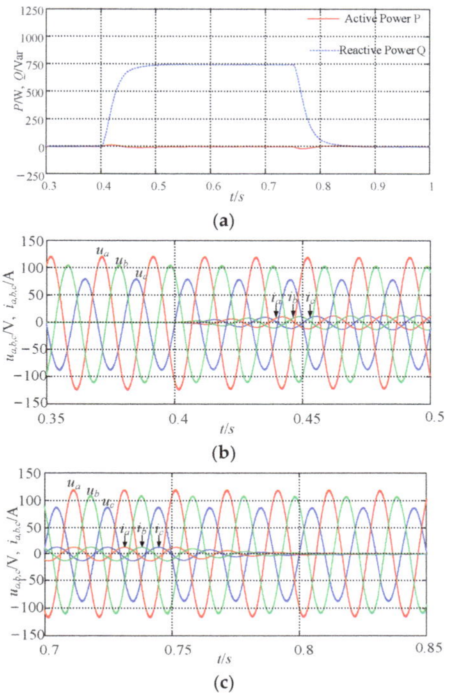

Figure 10. Validation of participation in voltage regulation capability under unbalanced grid conditions. (**a**) Waveform graph of P and Q of the system during variations in grid voltage. (**b**) Waveform graph of output voltage and current of the system during sudden increases in grid voltage. (**c**) Waveform graph of output voltage and current of the system during recovery of grid voltage.

5.2. Verification of Steady-State Performance of the VSG

Figures 11–13 present the simulation waveforms of the grid-forming inverter equipped with an additional active damping control strategy when the grid-connected system operates at rated conditions, with P_{ref} = 2000 W and Q_{ref} = 0 Var. The simulations consider grid line impedances of 0 mH, 1.5 MH, and 3 mH, respectively. Here, u_i (i = a, b, c) are the three-phase grid voltage waveforms, while i_n(n = a, b, c) denotes the three-phase grid current waveforms.

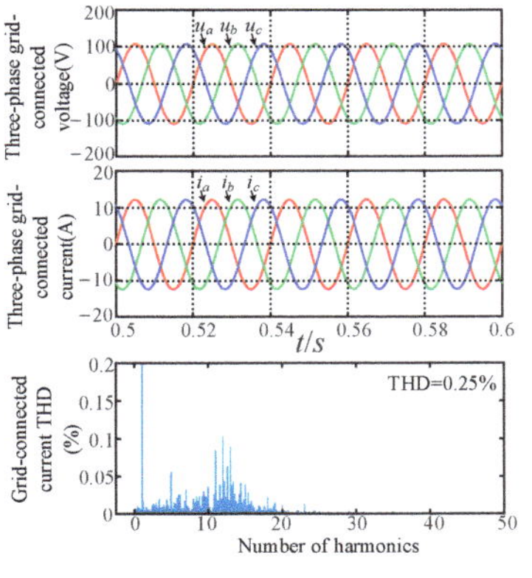

Figure 11. Simulation waveform when Lg = 0 mH.

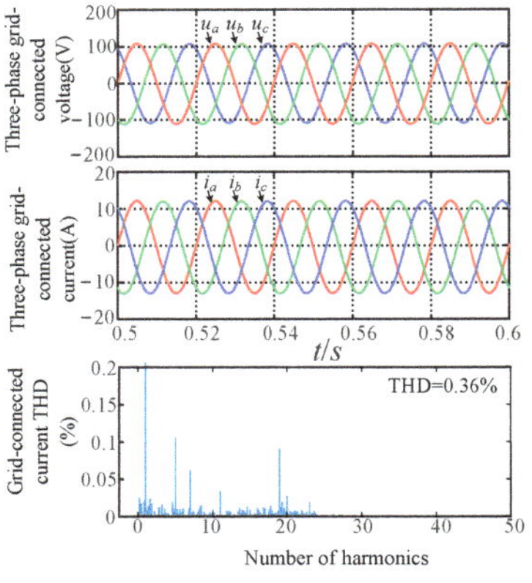

Figure 12. Simulation waveform when Lg = 1.5 mH.

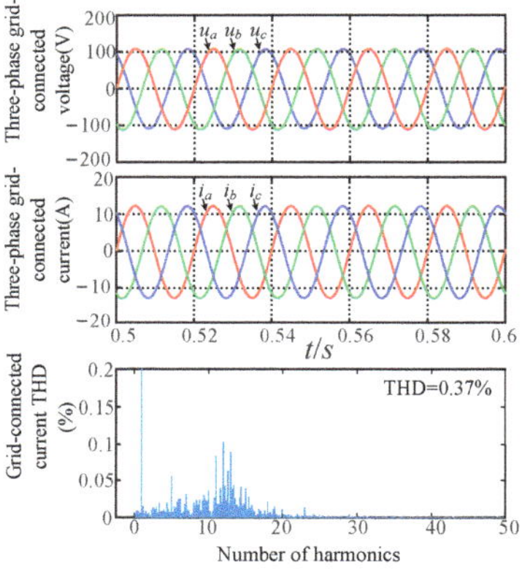

Figure 13. Simulation waveform when Lg = 3 mH.

From Figures 11–13 and Table 3, the grid-forming inverter with VSG control operates stably under all three grid conditions. The three-phase grid currents are sine-balanced and exhibit relatively low Total Harmonic Distortion (THD). Specifically, when L_g = 0 mH, the THD is 0.25%; when L_g = 1.5 mH, the THD is 0.36%; and when L_g = 3 mH, the THD is 0.37%. These values meet the grid connection requirements.

Table 3. Analysis of steady-state performance simulation results of VSG.

L_g/mH	THD
0	0.25%
1.5	0.36%
3	0.37%

5.3. IEEE Equivalent Power System Model Validation

In order to verify the effectiveness and reliability of the designed system and method operating in a large power grid, this section establishes an IEEE power system equivalent model of the inverter based on the scheme designed in this paper. The specific topology diagram is shown in Figure 14, and the specific parameters are shown in Table 4.

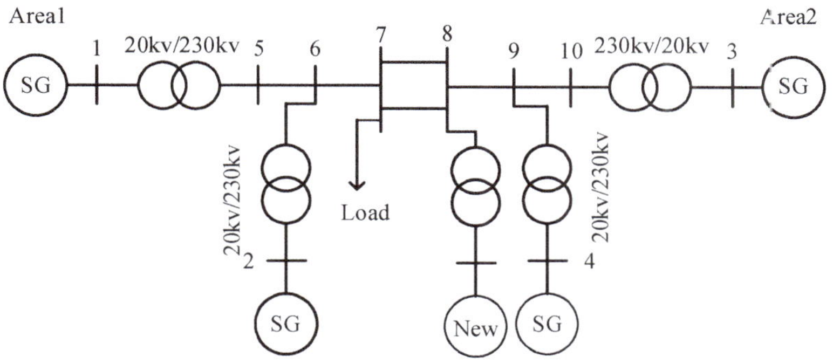

Figure 14. The topology of IEEE power system model with grid-connected inverters.

Table 4. Simulation parameter settings.

Parameters	Numeric Value
Zone 1 generator capacity (MW)	198
Zone 2 generator capacity (MW)	198
Rated power output of the PV array (KW)	500
The number of photovoltaic power station units	150
Voltage side capacitor DC voltage (V)	1500

5.3.1. Verification of System Frequency Regulation Capability

When the system reaches a steady state, at t = 5 s, there is a sudden increase in active load in the power grid, and the corresponding frequency fluctuation curve and inverter output active power change are shown in Figure 15. After the sudden change occurs, the inverter starts to increase power to suppress the frequency drop. Finally, at t = 12 s, the inverter increases active power by 10 MW, and the frequency stabilizes at 49.925 Hz, achieving active support for the power grid frequency.

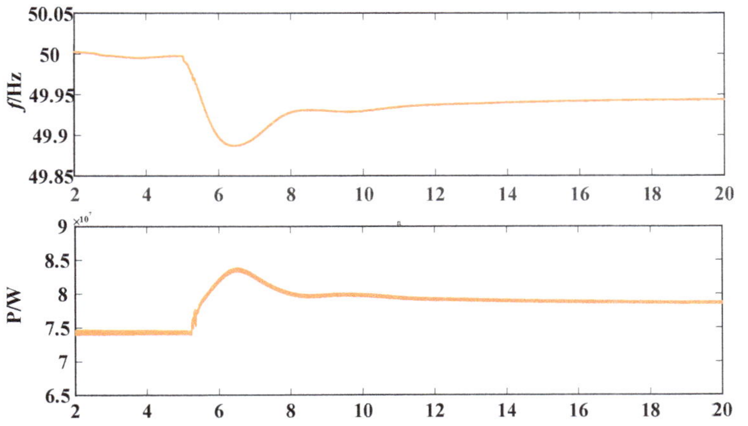

Figure 15. Waveform diagram of grid frequency and active power during load burst.

5.3.2. Verification of System Voltage Regulation Capability

When the system reaches a steady state, at t = 5 s, there is a sudden increase in reactive load in the power grid, and the corresponding voltage fluctuation curve and inverter output reactive power change are shown in Figure 16. After the sudden change occurs, the inverter starts to increase reactive power to suppress the voltage drop. Finally, at t = 18 s, the inverter increases reactive power by 25 MVar, and the voltage stabilizes at 0.97 p.u, achieving reactive support for the power grid voltage.

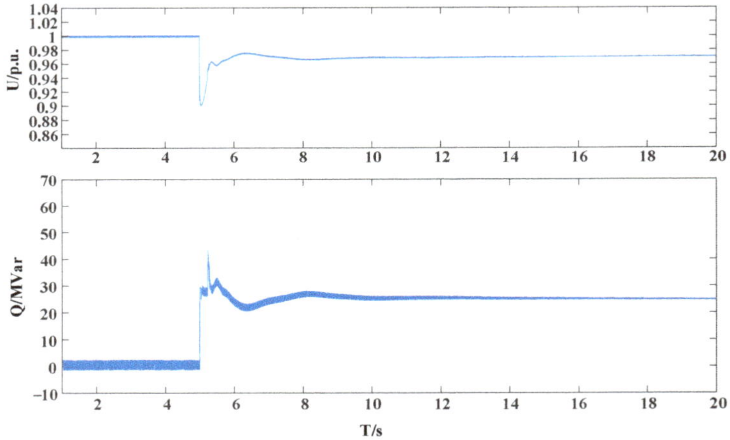

Figure 16. Waveform diagram of grid voltage and reactive power during load burst.

In summary, after equating the system proposed in this article to the IEEE power system equivalent model, in the large-scale power grid system, when facing active and reactive load disturbances, the system will actively adjust the output of active and reactive power according to the sudden changes in power grid load to cope with changes in system frequency and voltage, thereby achieving active support for regional elastic power grids.

5.4. Experimental Validation

In order to ensure the effectiveness and correctness of the simulation, an experimental prototype of an NPC three-level LCL grid-forming inverter is built, and the corresponding experimental parameters are consistent with those of the simulation in the experimental platform.

Figure 17 shows the comparative experimental waveforms of the grid-connected system when the grid impedances are L_g = 0 mH, 1.5 mH, and 3 mH with and without the active damping loop. From Figure 17a,c,e, it can be observed that when the system does not include the active damping loop, the THD of the grid current changes little with increasing grid impedance and does not meet the grid connection requirements, exhibiting significant resonance phenomena. In contrast, Figure 17b,d,f show that after incorporating the active damping loop, the resonance phenomena of the grid-connected inverter are effectively suppressed compared to the case without it. At this point, for the three different grid impedance scenarios, the THD of the grid current is 1.518%, 1.7664%, and 1.790%, respectively, which shows significant improvement compared to the THD values under the same conditions without the active damping loop. All these values meet the grid connection requirements, indicating that the system has good current quality.

According to the comprehensive analysis, the resonance phenomenon of the grid-forming inverter is effectively suppressed under weak grid conditions by adopting the VSG control with the active damping loop proposed in this paper, and the system remains stable, has good current quality, and meets the grid-connected standard.

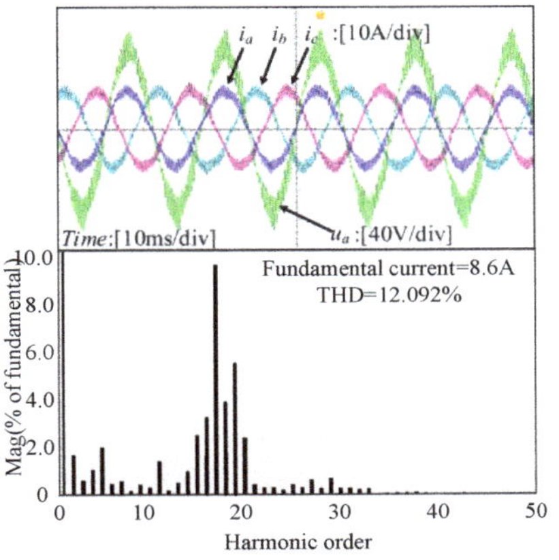

(**a**) L_g = 0 mH without the active damping loop.

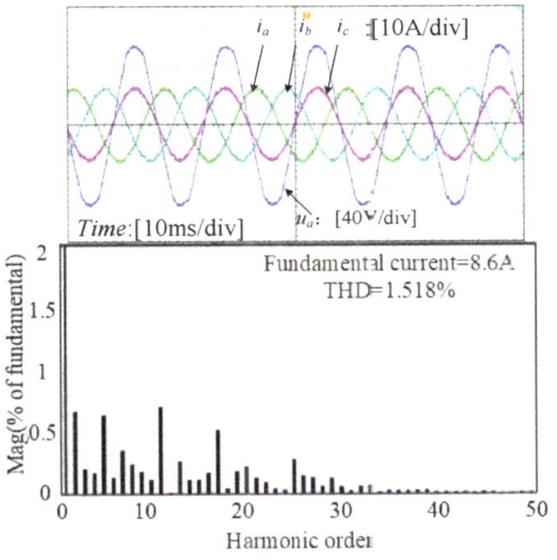

(**b**) L_g = 0 mH with active damping loop.

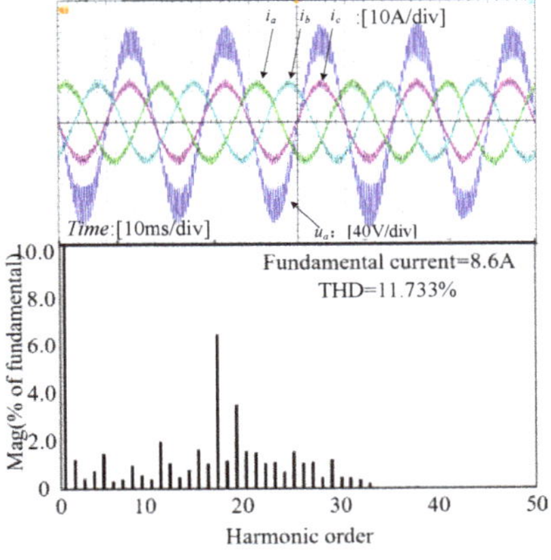

(**c**) L_g = 1.5 mH without the active damping loop.

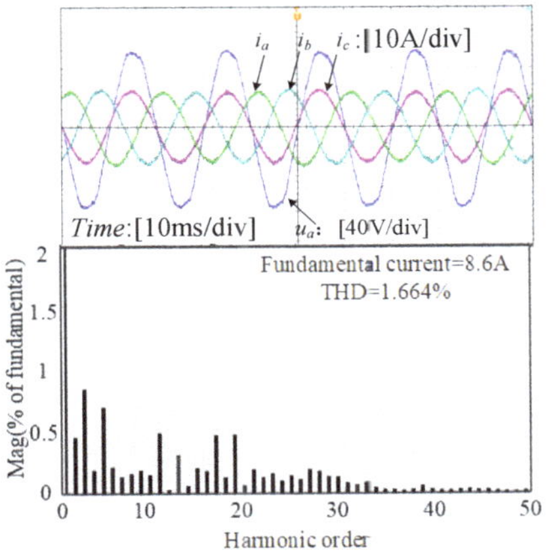

(**d**) L_g = 1.5 mH with the active camping loop.

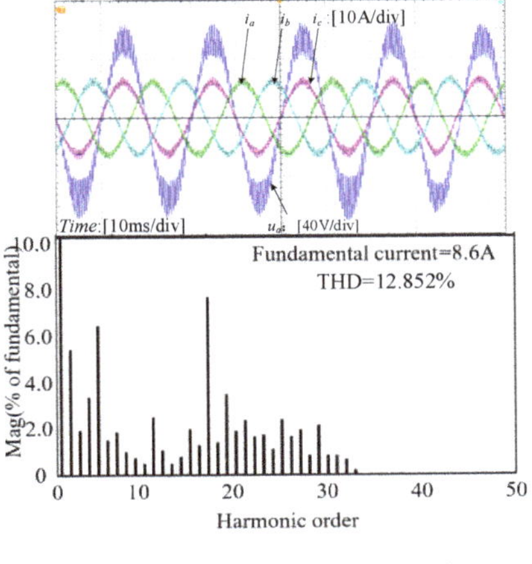

(**e**) L_g = 3 mH without the active damping loop.

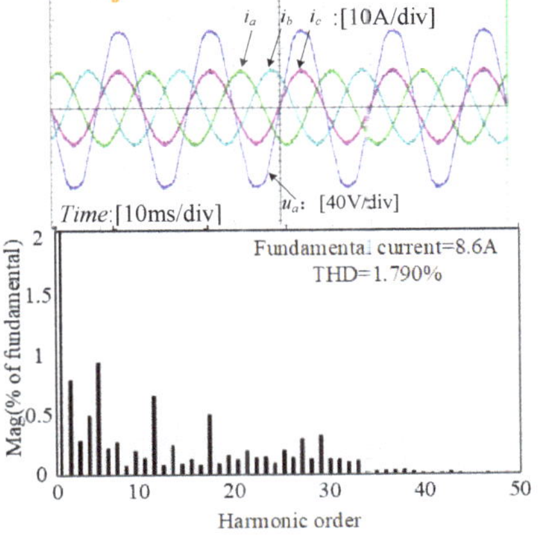

(**f**) L_g = 3 mH with the active damping loop.

Figure 17. Experimental waveforms of the grid-connected system under different impedance conditions.

6. Conclusions

This paper examines the grid-side components of power systems that incorporate new energy generation sources and considers the frequency and voltage support capabilities required by elastic regional power grids. A control system based on Virtual Synchronous Generator (VSG) technology is designed. A small-signal sequential impedance model of the VSG system in the $\alpha\beta$ coordinate frame is established using harmonic linearization methods. The stability of the system is analyzed from an impedance perspective, leading to the following conclusions, which were verified by comparative experiments.

(1) The control system based on VSG technology effectively implements grid-side management for power systems utilizing wind and photovoltaic energy. It provides essential elastic support for voltage and frequency stabilization, effectively suppresses resonance, and avoids causing oscillation, thereby enhancing power quality.

(2) The grid-connected system employing VSG technology demonstrated stable operation under varying grid impedance conditions, indicating its robustness and adaptability across different operational environments, thereby providing a resilient area for the stable operation of the power grid.

Author Contributions: Conceptualization, Z.L. (Zhipeng Lv) and H.L.; methodology, B.J.; software, H.L.; validation, Z.S., H.L. and S.Z.; formal analysis, Z.L. (Zhipeng Lv); data curation, Z.S.; writing—original draft preparation, Z.S.; writing—review and editing, Z.L. (Zhizhou Li); visualization, Z.L. (Zhizhou Li); supervision, S.Z.; project administration, Z.L. (Zhizhou Li); All authors have read and agreed to the published version of the manuscript.

Funding: This research received no external funding.

Data Availability Statement: The original contributions presented in the study are included in the article, further inquiries can be directed to the corresponding author.

Conflicts of Interest: Author Zhipeng Lv, Zhenhao Song, Hao Li, Shan Zhou and Zhizhou Li was employed by the company State Grid Shanghai Energy Interconnection Research Institute Co., Ltd. Author Bingjian Jia was employed by the company Xinjiang Information Industry Co., Ltd. Authors declare that the research was conducted in the absence of any commercial or financial relationships that could be construed as a potential conflict of interest.

References

1. Yang, X.; Song, Y.; Wang, G.; Wang, W. A Comprehensive Review on the Development of Sustainable Energy Strategy and Implementation in China. *IEEE Trans. Sustain. Energy* **2010**, *1*, 57–65. [CrossRef]
2. Blaabjerg, F.; Teodorescu, R.; Liserre, M.; Timbus, A.V. Overview of Control and Grid Synchronization for Distributed Power Generation Systems. *IEEE Trans. Ind. Electron.* **2006**, *53*, 1398–1409. [CrossRef]
3. Lasseter, R.H.; Chen, Z.; Pattabiraman, D. Grid-Forming Inverters: A Critical Asset for the Power Grid. *IEEE J. Emerg. Sel. Top. Power Electron.* **2020**, *8*, 925–935. [CrossRef]
4. Elmelegi, A.; Mohamed, E.A.; Aly, M.; Ahmed, E.M.; Al-Attar, M.; Elbaksawi, O. Optimized Tilt Fractional Order Cooperative Controllers for Preserving Frequency Stability in Renewable Energy-Based Power Systems. *IEEE Access* **2020**, *9*, 8261–8277. [CrossRef]
5. Wang, J.; Ma, K. Inertia and Grid Impedance Emulation of Power Grid for Stability Test of Grid-Forming Converter. *IEEE Trans. Power Electron.* **2023**, *38*, 2469–2480. [CrossRef]
6. Du, W.; Chen, Z.; Kevin, P.; Schneider, K.; Lasseter, R.H.; Nandanoori, S.P.; Tuffner, F.K.; Kundu, S. A Comparative Study of Two Widely Used Grid-Forming Droop Controls on Microgrid Small-Signal Stability. *IEEE J. Emerg. Sel. Top. Power Electron.* **2022**, *8*, 963–975. [CrossRef]
7. Rathnayake, D.B.; Akrami, M.; Phurailatpam, C.; Me, S.P.; Hadavi, S.; Jayasinghe, G.; Zabihi, S.; Bahrani, B. Grid Forming Inverter Modeling, Control, and Applications. *IEEE Access* **2021**, *9*, 114781–114807. [CrossRef]
8. Liu, T.; Liu, J.; Liu, Z.; Liu, Z. A Study of Virtual Resistor-Based Active Damping Alternatives for LCL Resonance in Grid-Connected Voltage Source Inverters. *IEEE Trans. Power Electron.* **2020**, *35*, 247–262. [CrossRef]

9. Zhao, H.; Wu, W.; Orabi, M.; Koutroulis, E.; Chung, H.S.-H.; Blaabjerg, F. A Passive Damping Design Method of Three-phase Asymmetric-LCL Power filter for Grid-Tied Voltage Source Inverter. In Proceedings of the 2023 IEEE Energy Conversion Congress and Exposition (ECCE), Nashville, TN, USA, 29 October–2 November 2023; pp. 94–98.
10. Sharma, H.R.; Suryawanshi, H.M.; Chaturvedi, P.; Govind, D. Implementation of Passive Damping Technique in LCL Filter for Three-Phase Grid Connected Inverter. In Proceedings of the 2022 IEEE 2nd International Conference on Sustainable Energy and Future Electric Transportation (SeFeT), Hyderabad, India, 4–6 August 2022; pp. 1–6.
11. Fonsêca, T.Q.; Ribeiro, R.L.A.; Rocha, T.d.O.A.; Costa, F.B.; Guerrero, J.M. Voltage Grid Supporting by Using Variable Structure Adaptive Virtual Impedance for LCL-Voltage Source Converter DG Converters. *IEEE Trans. Ind. Electron.* **2020**, *67*, 9326–9336. [CrossRef]
12. Beres, R.N.; Wang, X.; Liserre, M.; Blaabjerg, F.; Bak, C.L. A Review of Passive Power Filters for Three-Phase Grid-Connected Voltage-Source Converters. *IEEE J. Emerg. Sel. Top. Power Electron.* **2016**, *4*, 54–69. [CrossRef]
13. Yao, W.; Yang, Y.; Zhang, X.; Blaabjerg, F.; Loh, P.C. Design and analysis of robust active damping for LCL filters using digital notch filters. *IEEE Trans. Power Electron.* **2016**, *32*, 2360–2375. [CrossRef]
14. Peña-Alzola, R.; Liserre, M.; Blaabjerg, F.; Ordonez, M.; Kerekes, T. A Self-commissioning notch filter for active damping in three phase LCL-filter based grid-tied converter. *IEEE Trans. Power Electron.* **2014**, *29*, 6754–6761. [CrossRef]
15. Han, Y.; Li, Z.; Yang, P.; Wang, C.; Xu, L.; Guerrero, J.M. Analysis and Design of Improved Weighted Average Current Control Strategy for LCL-Type Grid-Connected Inverters. *IEEE Trans. Energy Convers.* **2017**, *32*, 941–952. [CrossRef]
16. Wang, X.; Wang, Y.; Yang, H.; Dai, H. Grid-Connected Current Double Loop Feedback Suppression Strategy of Parallel Inverter in the Weak Grid. In Proceedings of the 2023 IEEE 2nd International Power Electronics and Application Symposium (PEAS), Xi'an, China, 10–13 November 2023; pp. 2023–2028.
17. Xu, J.; Xie, S.; Tang, T. Active Damping-Based Control for Grid-Connected LCL-Filtered Inverter with Injected Grid Current Feedback Only. *IEEE Trans. Ind. Electron.* **2014**, *61*, 4746–4758. [CrossRef]
18. Wu, W.; Chen, Y.; Zhou, L.; Luo, A.; Zhou, X.; He, Z.; Yang, L.; Xie, Z.; Liu, J.; Zhang, M. Sequence Impedance Modeling and Stability Comparative Analysis of Voltage-Controlled VSGs and Current-Controlled VSGs. *IEEE Trans. Ind. Electron.* **2018**, *66*, 6460–6472. [CrossRef]
19. Wei, Y.; Zhang, H.; Song, Q.; Sun, K. Control strategy for parallel-operated virtual synchronous generators. In Proceedings of the 2016 IEEE 8th International Power Electronics and Motion Control Conference (IPEMC 2016–ECCE Asia), Hefei, China, 22–26 May 2016.
20. Liu, J.; Miura, Y.; Bevran, H.; Ise, T. Enhanced virtual synchronous generator control for parallel inverters in microgrids. *IEEE Trans. Smart Grid* **2017**, *8*, 2268–2277. [CrossRef]
21. Wang, Q.; Zhou, D.; Yin, S.; Lei, Y.; He, T. Improved Adaptive Inertia and Damping Coefficient Control Strategy of VSG Based on Optimal Damping Ratio. In Proceedings of the 2022 International Power Electronics Conference (IPEC-Himeji 2022- ECCE Asia), Himeji, Japan, 15–19 May 2022; pp. 102–107.
22. Ibrahim, N.F.; Mahmoud, K.; Lehtonen, M.; Darwish, M.M.F. Comparative Analysis of Three-Phase PV Grid Connected Inverter Current Control Schemes in Unbalanced Grid Conditions. *IEEE Access* **2023**, *11*, 42204–42221. [CrossRef]

symmetry

MDPI

Article

Quantitative State Evaluation Method for Relay Protection Equipment Based on Improved Conformer Optimized by Two-Stage APO

Yanhong Li [1,2], Min Zhang [1], Shaofan Zhang [1] and Yifan Zhou [3,*]

[1] Guangzhou Power Supply Bureau of Guangdong Power Grid Co., Ltd., Guangzhou 510620, China; liyanhong@sjtu.edu.cn (Y.L.); zhangmin@gzps.corp.csg (M.Z.); zhangshaofan@gzps.corp.csg (S.Z.)
[2] School of Electrical Engineering, Shanghai Jiao Tong University, Shanghai 200240, China
[3] NR Engineering Co., Ltd., 69 Suyuan Avenue, Nanjing 211102, China
* Correspondence: zhouyifan_1984@163.com

Abstract: State evaluation of relay protection equipment constitutes a crucial component in ensuring the stable, secure, and symmetric operation of power systems. Current methodologies predominantly encompass fuzzy-rule-based control systems and data-driven machine learning approaches. The former relies on manual experience for designing fuzzy rules and membership functions and exhibits limitations in high-dimensional data integration and analysis. The latter predominantly formulates state evaluation as a classification task, which demonstrates its ineffectiveness in identifying equipment at boundary states and faces challenges in model parameter selection. To address these limitations, this paper proposes a quantitative state evaluation method for relay protection equipment based on a two-stage artificial protozoa optimizer (two-stage APO) optimized improved Conformer (two-stage APO-IConf) model. First, we modify the Conformer architecture by replacing pre-layer normalization (Pre-LN) in residual networks with post-batch normalization (post-BN) and introducing dynamic weighting coefficients to adaptively regulate the connection strengths between the first and second feed-forward network layers, thereby enhancing the capability of the model to fit relay protection state evaluation data. Subsequently, an improved APO algorithm with two-stage optimization is developed, integrating good point set initialization and elitism preservation strategies to achieve dynamic equilibrium between global exploration and local exploitation in the Conformer hyperparameter space. Experimental validation using operational data from a substation demonstrates that the proposed model achieves a RMSE of 0.5064 and a MAE of 0.2893, representing error reductions of 33.6% and 35.0% compared to the baseline Conformer, and 9.1% and 15.2% error reductions over the improved Conformer, respectively. This methodology can provide a quantitative state evaluation and guidance for developing maintenance strategies for substations.

Keywords: relay protection equipment; quantitative state evaluation; power system symmetry; two-stage APO algorithm; improved conformer model

1. Introduction

In the development of new power systems, the symmetry equilibrium of grid structures faces critical challenges due to the large-scale integration of high-penetration renewable energy [1,2], power electronic devices [3], and heterogeneous loads [4], leading to escalating topological complexity and heightened risks for secure and stable operation.

In January 2022, China's National Development and Reform Commission and National Energy Administration issued the 14th Five-Year Plan for Modern Energy Systems, explicitly proposing the promotion of the construction of new power systems, accelerating the digital transformation of power systems, and enhancing grid intelligence [5]. During the development of digital twin models for smart substation protection systems, the state evaluation of relay protection equipment is critically important. As a safety barrier for power grids, the health status of relay protection equipment directly affects the maintenance of the operational symmetry and dynamic stability of power systems. When asymmetric faults occur, negative- and zero-sequence components emerge in the system, potentially leading to severe accidents. Healthy relay protection devices can correctly monitor these components and facilitate the restoration of the grid symmetry. Therefore, an accurate status evaluation and timely detection of potential faults in these devices are critical for preserving the symmetric characteristics of power grids.

Traditional methods for evaluating the state of relay protection equipment often rely on manual periodic inspections, which depend heavily on workers' empirical judgments, leading to inefficiency, subjectivity, and an inability to meet modern grid demands for real-time and precise evaluations. To improve maintenance efficiency and reduce resource waste, early researchers developed mathematical models for reliability assessment based on the failure rates of relay protection systems and introduced stochastic failure indicators [6] for model correction. However, these early approaches could only predict binary failure states (functional or failed) without quantifying specific health conditions.

Advancements in information technology and equipment upgrades have enabled substations to collect and analyze multi-source data, including operational telemetry, historical maintenance records, and familial device data, leading to increasingly diversified evaluation criteria. To characterize device health, researchers typically define four states: normal, attention, abnormal, and critical [7–9]. Subsequent studies incorporated analog/digital signals and power module data, while references [10–12] expanded the considerations to channel testing results, operational lifespan, fault alarms, environmental temperature, and voltage levels. These works gradually established small-scale expert rule libraries, where the device status is scored based on predefined rules for each indicator and aggregated for final evaluation. Reference [13] specifically addressed the temperature impacts by proposing an overheating analysis framework. However, these studies only partially incorporated the aforementioned factors and failed to establish a comprehensive evaluation system. To integrate multi-dimensional state features, reference [14] adopted entropy-based weight assignment, while references [15,16] utilized the fuzzy analytical hierarchy process (FAHP) to determine indicator weights and reference [17] combined entropy and FAHP method. Reference [18] applied Bayesian theory, and reference [19] integrated Bayesian networks with FAHP to achieve more objective and reasonable weight allocations. Based on FAHP-derived weights, reference [20] implemented fuzzy control methods with various membership functions for comprehensive fuzzy evaluations of relay protection devices.

The aforementioned methods, expert rule libraries, FAHP, and fuzzy control, overemphasize subjective human judgment. For instance, FAHP pairwise comparison matrices and fuzzy control membership functions rely on expert-defined rules, neglecting data-driven statistical patterns. Furthermore, fuzzy control struggles with high-dimensional data due to the complex rule limitations. Recent advances in intelligent technologies [21] have shifted the focus to data-driven and AI-based evaluation methods. Reference [22] demonstrated the feasibility of AI for power system fault analysis. Reference [23] employed back propagation (BP) neural networks for health assessment but showed poor accuracy for degraded devices. Reference [24] enhanced evaluation precision using least-squares support vector machines (LSSVM) and Bayesian network decision trees (BNDT). Addressing data imbalance, reference [8] achieved over 96% accuracy by combining generative adversarial networks (GAN) for data augmentation with random forest classifiers. Reference [25] validated the superiority of data-driven methods using CNN-BiGRU models with scenario-simulated sample-balancing. However, due to constraints in maintenance, most substations currently still rely on the four-state qualitative classification (normal, attention, abnormal, and critical) to assess equipment status, implementing uniform maintenance for devices deemed abnormal and critical. Driven by the need to build smart grids and develop digital twin models, substation operators now seek a more detailed understanding of equipment conditions rather than merely determining whether maintenance is required. The use of continuous numerical values to precisely reflect equipment conditions is a good approach. This would enable preventive maintenance strategies for devices in critical transitional states between attention and abnormalities.

Current mainstream data-driven approaches [8,23–25] primarily formulate state evaluation as classification tasks, where neural networks directly assign discrete labels like "normal" or "attention". Although these methods achieve high accuracy, they fail to quantify the boundary states. For example, two devices labeled "attention" with scores of 89 and 80 cannot be differentiated, limiting precise maintenance guidance. Conventional models (e.g., MLP and CNN) have limitations in extracting complex patterns from high-dimensional heterogeneous data sources. This paper aims to address the limitations stated above. Recent advances in Conformer models, which integrate the multi-head self-attention (MHSA) mechanism of Transformers with depthwise separable convolutional layers of CNNs, have demonstrated superior capabilities in synergistic global-local feature modeling across domains, such as speech recognition. Inspired by these developments, this study proposes an improved Conformer architecture specifically designed for the multi-dimensional and heterogeneous nature of relay protection device state data, which encompasses fundamental condition evaluation data, operational condition evaluation data, maintenance condition evaluation data, and ancillary factor evaluation data. Owing to the synergistic design of Transformer and CNN, this architecture efficiently captures both local characteristics and global features of the device state data, thereby improving the accuracy of the state quantification evaluation. Furthermore, the proposed framework incorporates a two-stage APO strategy to learn discriminative multi-dimensional features from device state data, enabling precise state quantification rather than categorical labels. Granular quantification provides critical guidance for formulating data-driven maintenance strategies.

The differences between the method proposed in this paper and conventional methods are shown in Table 1.

The principal contributions of this study are summarized as follows:

(1) **Architectural improvement of the Conformer model:** An improved Conformer architecture tailored for the quantitative state evaluation of relay-protection equipment is proposed. By replacing pre-layer normalization with post-batch normalization in residual networks and introducing dynamic weighting coefficients to adaptively regulate the connectivity between the first and second feed-forward network segments, the model significantly enhances its capability to fit relay protection state evaluation data, achieving superior fitting accuracy compared to traditional machine learning models.

(2) **Enhancement of the optimization algorithm:** A two-stage APO algorithm is developed, incorporating good point set initialization and elitism preservation strategies. This innovation synergistically strengthens global search and local optimization while mitigating stochastic interference.

(3) **Model-algorithm co-optimization:** The two-stage APO algorithm is integrated to optimize the hyperparameters of the improved Conformer model. By harmonizing the APO's global search capabilities with Conformer's hybrid architectural features,

a dynamic equilibrium between exploitation and exploration is achieved in the hyperparameter space, effectively resolving the parameter selection challenges in relay protection state evaluation models.

(4) **Experimental validation:** Experimental validation conducted on the PyCharm platform demonstrates the efficacy of the proposed methodology. The results show a RMSE of 0.5064 and a MAE of 0.2893. These outcomes can provide a quantitative state evaluation and guidance for developing maintenance strategies for substations.

Table 1. Comparative analysis of the proposed method and previous methods.

Category	Fuzzy Control Method [15,26–29]	Traditional Machine Learning Models [23–25,28,29]	Proposed Method
Model Architecture	Expert rule base and membership functions	Single architecture	Conformer architecture and two-stage APO algorithm
Data Dimensionality Handling	Medium	High	High
Complex Feature Extraction	Low	Moderate	Strong
Methodological Objectivity	Low	High	High
Computational Complexity	Low	Medium	Relatively High
State Boundary Identification	Sensitive to classification thresholds	Sensitive to classification thresholds	Support quantifiable continuous scoring
Evaluation Metrics	Classification accuracy and F1-score	Classification accuracy and F1-score	RMSE and MAE

The rest of this paper is organized as follows: Section 2 elaborates on the quantitative state evaluation methodology for relay protection equipment, including the state indicator system and evaluation workflow; Section 3 details the structural improvements of the Conformer model and the operational principles of the two-stage APO algorithm; Section 4 describes the model training procedures and performance evaluation metrics; Section 5 validates the effectiveness of the proposed method through practical case studies; Section 6 concludes the research.

2. Quantitative State Evaluation Methodology for Relay Protection Equipment

2.1. State Evaluation Indicator System

Currently, state evaluation data for relay protection equipment primarily originate from three sources: offline file imports, online acquisition via relay protection master stations, and manual entry. Offline imported files mainly include equipment ledger files that contain device models, commissioning dates, and other metadata. Relay protection master stations can acquire data online, such as analog signal verification results, real-time operational anomalies, alerts, homologous data comparison results, device communication rates, and trip operation records. For data that are unobtainable through online or offline means, manual collection and entry are employed, including familial defect records, secondary circuit insulation defects, periodic inspection completion status, corrective action implementation status, spare parts inventory status, and equipment production status (discontinued/active).

Based on the available data, the state evaluation indicator system categorizes data into four classes: fundamental condition evaluation data, operational condition evaluation data, maintenance condition evaluation data, and ancillary factor evaluation data. The data are organized as listed in Table 2.

Table 2. State evaluation indicator system for relay protection equipment.

Category	Subcategory	Evaluation Metrics
Fundamental Condition Evaluation	Basic information	Commissioning date, Standard version compliance, Device model, Device name/ID, Record generation timestamp
Operational Condition Evaluation	Real-Time Analog Verification	Amplitude/phase of analog signals, Zero-sequence current over-limit alarms, Differential current alarms (line protection), Differential current alarms (transformer protection), Current/voltage sampling over-limit alarms, Device temperature alarms, DC voltage over-limit alarms
	Real-Time Operational Alerts	Tier-2 operational anomaly alarms, Tier-3 operational anomaly alarms
	Operational Parameter Verification	Active setting zone verification status, Setting value deviations from baseline, Soft plate discrepancies, Hard plate discrepancies, Clock synchronization alarms
	Homologous Data Comparison	Data inconsistency alarms from homologous sources
	Online Risk Identification	Channel anomaly frequency (fiber-optic/high-frequency channels), CT-N wire disconnections, PT-N wire disconnections, Analog sampling disconnections, Single-bit alarms, Frequent Tier II-V signal alarms
	Process-Level Secondary Virtual Circuits	GOOSE link disconnections, Optical port anomalies (signal strength), Network port frame loss/CRC errors
	Device Communication	Device communication rate
	Familial Defects	Same-model defect frequency (current cycle), Same-model defect frequency (lifecycle)
	Secondary Circuit Insulation	Insulation defect frequency (current cycle), Insulation defect frequency (lifecycle)
	Action History Validation	Incorrect protection operations, Incorrect breaker operations, Incorrect fault zone identification records
	Misoperation/Refusal Records	Misoperation/refusal counts
Maintenance Condition Evaluation	Periodic Inspection Status	Periodic inspection completion status
	Corrective Action Status	Unimplemented corrective actions, Historical corrective action frequency
Ancillary Factor Evaluation	Manufacturer Support Factors	Production discontinuation status, Spare parts inventory status

In the evaluation metric parameters listed above, most of the recorded data pertain to the occurrence frequencies. Under basic information, the commissioning date and record generation timestamp document the equipment's operational start date and the record's creation date, respectively, which are converted into the device's actual operational duration. The standard version compliance adopts a binary yes/no value to indicate whether the device is included in the standardized equipment list. The device model and device name/ID are textual fields in which identical models or names share uniform encoding to differentiate the device identities.

In real-time analog verification, the system logs the count of analog signal over-limit alarms (amplitude/phase) during an inspection cycle, including zero-sequence current over-limit alarms, differential current alarms for line/transformer protection, current/voltage sampling over-limit alarms, device temperature alarms, and DC voltage over-limit alarms.

Real-time operational alerts track tier-2 and tier-3 operational anomaly alarm quantities within the evaluation period.

For operational parameter verification, the active setting zone verification status reflects whether the current setting zone number operates normally or not. Setting deviation values from baseline quantifies entries deviating from predefined benchmarks, while the remaining metrics count discrepancy alarms for soft plates, hard plates, and clock synchronization.

Homologous data comparisons arise from multiple measurement points on the same device. Data inconsistency alarms are triggered when faulty measurement points or device malfunctions cause discrepancies between homologous data sources.

Online risk identification records the frequencies of anomalies in the communication channels (fiber-optic/high-frequency), CT-N/PT-N wire disconnections, analog sampling disconnections, single-bit alarms, and frequent tier II-V signal alarms.

Process-level secondary virtual circuits document failure counts in data transmission links such as GOOSE link disconnections, optical port signal anomalies, and network port frame loss/CRC errors.

The device communication rate is expressed as a percentage. For legacy devices lacking sampling capabilities, this value defaults to 100% based on expert consensus.

Familial defects quantify failures using same-model device defect frequencies (current cycle or lifecycle).

The secondary circuit insulation counts insulation defect alarms in secondary circuits.

Action history validation and misoperation/refusal records quantify historical errors through incorrect protection/breaker operations, fault zone misidentifications, and misoperation/refusal events.

The periodic inspection status uses yes/no to indicate the completion of scheduled inspections. Corrective action status evaluates historical corrective measures based on their implementation frequency. Manufacturer support factors indicate whether the device is discontinued and spare parts are available.

Due to space constraints of the table, a single indicator may encompass multiple types of captured data. For instance, metrics involving "alarm counts" can include both the number of monitoring points that trigger alarms within an evaluation cycle and the total number of alarm occurrences across all monitoring points.

In the table above, the equipment operation time data are converted into floating-point years; status information such as maintenance status and spare parts status are quantized into binary states 0 or 1; continuous variables like device communication rate are converted into decimals; and abnormal alarm information is quantized into analog limit-exceeding counts and abnormal alarm counts, respectively. Furthermore, analog parameters are acquired and validated by analog signal monitoring modules, which detect over-limit conditions and generate the corresponding over-limit alarm counts. This approach significantly reduces the amount of analog data required to evaluate equipment status. Normalize the alarm count data to the range of 0–1 to eliminate differences in feature scales, facilitating the subsequent model learning of features.

It should be noted that the indicators in this paper do not directly incorporate the action time data of relay protection devices. If devices fail to operate correctly or timely due to faults, relevant failure alarm information (e.g., misoperation counts and refusal counts) can be collected. These records are then used to train the models. Once trained, the model can immediately generate quantified health scores for devices based on the collected data, including both online data collection and automatic offline data import. The time required for scoring is negligible.

2.2. Quantitative State Evaluation Workflow

The quantitative state evaluation process for relay protection equipment based on the aforementioned indicator system operates as follows.

Grid experts initially scored the equipment within a range of 0–100 using existing technical specifications and scoring rule databases. However, because these specifications and rules are predominantly derived from engineering experience with inherent subjectivity and single rules cannot accurately evaluate diverse equipment states, field technicians must physically inspect the equipment and iteratively adjust scores through expert consultation to obtain final ratings.

The two-stage APO-IConf model learns expert scoring logic and field adjustment patterns from historical data patterns, enabling direct score generation on newly acquired inspection data. This approach replaces multi-round empirical adjustments by maintenance personnel, thereby enhancing evaluation objectivity and efficiency. Compared to direct classification methods, quantifying equipment scores allows technicians to precisely discern device conditions, particularly for equipment near critical state boundaries, facilitating proactive maintenance planning to prevent severe failures.

3. Two-Stage APO-IConf-Based Quantitative State Evaluation Method for Relay Protection Equipment

3.1. Standard Conformer Encoder Architecture

The Conformer model was initially proposed by Google Research in 2020 [30] to address the limitations of Transformer models in capturing local features. Designed to synergize the Transformer's global modeling capabilities with CNN's local feature extraction strengths, the Conformer architecture enhances model performance. The standard Conformer model comprises both encoder and decoder layers. However, in quantitative state evaluation tasks for relay protection equipment, where text/audio outputs are unnecessary, the decoder layer is omitted, retaining only the encoder structure, as shown in Figure 1.

The feed-forward network (FFN) in the Conformer model inherits the core design principles of traditional Transformers while enhancing the representational capacity through modular enhancements and parameter optimization. Each Conformer encoder block contains two FFN modules positioned before and after the multi-head self-attention layer and the convolutional module, respectively. This architectural design facilitates a two-phase non-linear transformation process that enables a progressive feature refinement. The FFN employs swish activation functions, which demonstrate marginally superior performance compared to ReLU functions [31].

The front-end FFN performs preliminary feature enhancement on the input to supply richer contextual information for the self-attention layer, while the rear-end FFN processes the local features output by the convolutional module through global integration and recalibration. Applying identical 0.5 coefficients to both front-end FFN and rear-end FFN modules enhances accuracy when stacking multiple encoder layers while maintaining implementation balance and reducing model complexity.

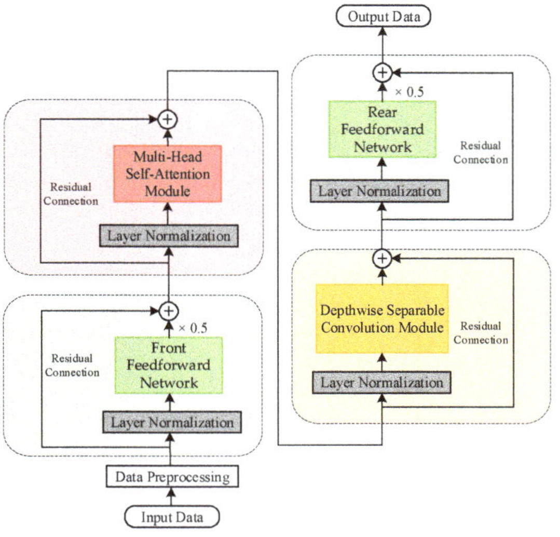

Figure 1. Standard Conformer encoder architecture.

The multi-head self-attention mechanism module employs relative positional encoding, which enhances the learning and generalization capabilities for variable-length input sequences. The input sequence undergoes linear transformations to generate three vector sets: Query (Q), Key (K), and Value (V). Relative positional encoding replaces absolute positional encoding [32] with its mathematical formulation given in Equation (1). Here, indices i and j denote positional coordinates of the target (Q) and contextual (K) elements respectively, while d_k represents the dimensionality of the K vectors.

$$P_{i,j} = \frac{i-j}{\sqrt{d_k}} \tag{1}$$

By partitioning the Q, K, and V vectors into h independent heads, where the vector dimensionality of the model is defined as d_{model}, each head attains a dimensionality of d_{model}/h (with d_{model} and h being appropriately chosen to ensure integer dimensionality). This configuration enabled parallel feature learning across distinct subspaces.

$$Q_i = QW_i^Q, \quad K_i = KW_i^K, \quad V_i = VW_i^V \tag{2}$$

where W_i^Q, W_i^K, W_i^V are learnable parameter matrices. Each head performs scaled dot-product attention, which incorporates relative positional encoding [33].

$$\text{Attention}(Q_i, K_i, V_i) = \text{Softmax}\left(\frac{Q_i K_i^T}{\sqrt{d_k}} + P\right) V_i \tag{3}$$

The relative positional offset P enhances the perception of local temporal relationships. The outputs from all heads are concatenated and linearly transformed to produce the final multi-head attention output, where W_i^Q denotes the fusion weight matrix.

$$\begin{aligned} \text{MultiHead}(Q,K,V) &= \text{Concat}(\text{head}_1,\ldots,\text{head}_h)W^O \\ &= \sum_{i=1}^{h} \text{Attention}(Q_i,K_i,V_i)W_i^O \end{aligned} \tag{4}$$

The convolutional layer in the Conformer employs a gated depthwise separable convolution, which differs from standard CNN layers through the combination of depthwise and pointwise convolutions. This architecture offers several advantages, including a reduced parameter count and enhanced computational efficiency. The workflow can be summarized as follows, with a schematic diagram of the Conformer convolutional layer provided in Figure 2. The depthwise separable convolution process begins with pointwise convolution to reduce the channel dimensions for parameter efficiency, followed by GLU to selectively filter features. Depthwise convolution processes spatial features per channel independently. Batch normalization stabilizes training, while the swish activation enhances non-linear capabilities with smoother gradients. A second pointwise convolution adjusts the channel dimensions, and dropout regularizes the model by randomly deactivating the neurons.

Figure 2. Schematic of the Conformer convolutional layer.

Depthwise convolution independently performs single-channel convolution on each input channel, extracting spatial features without altering the channel count, thereby significantly reducing the parameter quantity. Pointwise convolution employs 1×1 convolutional kernels primarily for adjusting channel numbers or cross-channel feature fusion. By linearly combining information across different channels, it modifies the feature map depth while preserving the spatial resolution while maintaining high computational efficiency. Figure 3 shows a schematic of the depthwise and pointwise convolutions. Where H represents the convolution height, W denotes the convolution width, C_{in} is the number of input channels, K specifies the convolution kernel size, and C_{out} indicates the number of output channels.

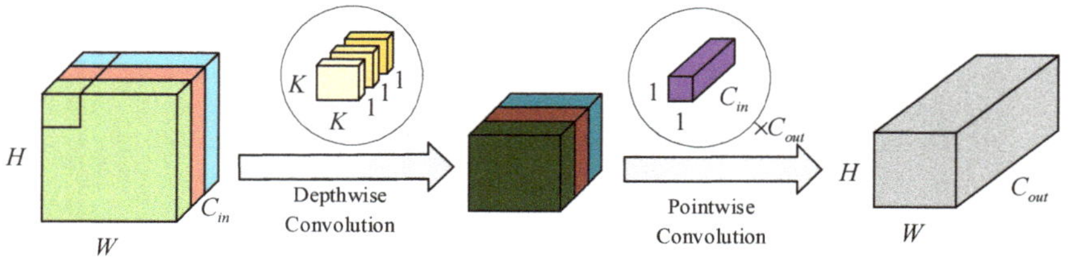

Figure 3. Schematic diagram of depthwise and pointwise convolutions.

3.2. Improved Conformer Encoder

This study introduces architectural improvements to the Conformer encoder for the quantitative state evaluation of relay protection equipment. The key modifications are as follows:

(1) Replacement of layer normalization with batch normalization;
(2) Repositioning the pre-normalization layer in each residual network to post-residual connections;

(3) Implementation of dynamic weighting coefficients to adaptively adjust contributions from front-end and rear-end feed-forward networks.

Vaswani et al. [34] proposed the original Transformer architecture for natural language processing (NLP) tasks, and its derivative Conformer model optimized for speech sequences utilizes layer normalization to handle variable-length inputs without padding alignment while preserving temporal dependencies. However, relay protection state data exhibit fixed-dimensional characteristics and lack intra-sample temporal relationships. Batch normalization is more effective in this context by mitigating gradient vanishing/explosion issues and enhancing the model training stability, thereby improving the prediction accuracy.

While pre-normalization facilitates gradient propagation through direct input pathways and reduces attenuation through non-linear transformations, it becomes redundant for shallow neural networks in our application. Repositioning normalization layers after residual connections stabilize the training convergence in such architectures.

Conventional implementations fix the front/rear feed-forward network weights at 0.5. Given that the model structure for relay protection state evaluation tasks does not require stacking multiple encoder layers to significantly increase complexity like NLP models, our innovation employs trainable coefficients (α) to dynamically balance their contributions, strengthening the model's non-linear fitting capacity.

The architecture of the improved Conformer encoder is shown in Figure 4.

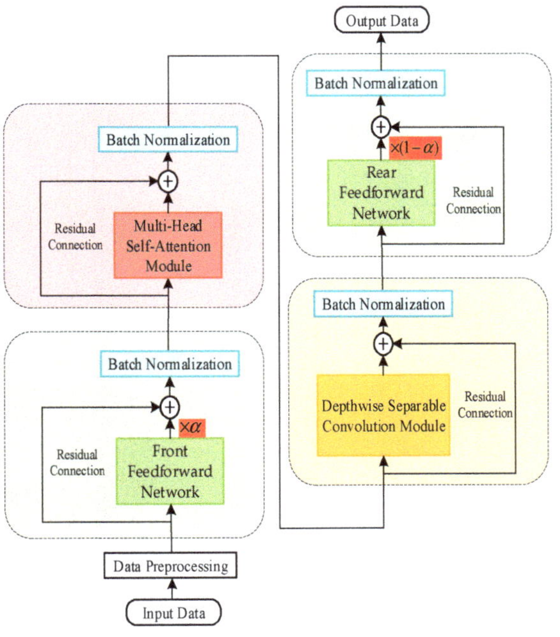

Figure 4. Improved Conformer encoder architecture.

3.3. APO Algorithm

This study employs a novel artificial protozoa optimizer (APO) [35] for the hyperparameter optimization of the model. Inspired by natural protozoan behaviors, the algorithm simulates their survival mechanisms through foraging, dormancy, and reproductive behaviors. The APO categorizes foraging behavior into autotrophic and heterotrophic modes. The foraging process facilitates local optimization, dormancy resets individuals, and reproduction induces minor variations.

Initially, a population of n protozoa X_i is randomly generated, where n denotes the population size (number of individuals) and $i \in [1, n]$ represents the protozoan index. The mathematical representation of X_i is given in Equation (5).

$$X_i = [x_i^1, x_i^2, \ldots, x_i^{dim}] \tag{5}$$

In the above equation, dim denotes the dimensionality of the vector. Each X_i must satisfy the constraint $X_{min} \leq X_i \leq X_{max}$, where the upper and lower bounds X_{max} and X_{min} are defined as follows:

$$X_{max} = [ub_1, ub_2, \ldots, ub_{dim}] \tag{6}$$

$$X_{min} = [lb_1, lb_2, \ldots, lb_{dim}] \tag{7}$$

where ub_i and lb_i denote the upper and lower bounds of the element in the i-th dimension, respectively.

Following the initialization of all protozoa individuals, the objective function $O_i = f(X_i)$ is utilized to calculate fitness values. This $f(\cdot)$ function can be designed for maximization (e.g., profit, throughput) or minimization (e.g., cost, error). A ranking function $Sort(\cdot)$ is defined under the minimization paradigm, which sorts individuals in ascending order of their objective function values such that superior solutions occupy leading positions in ascending order of their fitness values, as formalized in Equation (8).

$$X_i = Sort(X_i) \tag{8}$$

The APO employs three key parameters to differentiate the survival mode of each protozoan individual.

$$pf = pf_{max} \cdot rand \tag{9}$$

$$p_{ah} = \frac{1}{2} \cdot (1 + \cos(\frac{t}{T} \cdot \pi)) \tag{10}$$

$$p_{dr} = \frac{1}{2} \cdot (1 + \cos((1 - \frac{i}{n}) \cdot \pi)) \tag{11}$$

In Equation (9)–(11), pf represents the proportional fraction of protozoa engaged in dormancy and reproduction within the population, while 1-pf denotes the proportion allocated to foraging (i.e., autotrophic and heterotrophic). The parameter P_{ah} is compared against a system-generated random number *rand* to determine whether a protozoan adopts autotrophic or heterotrophic behavior. Similarly, P_{dr} is evaluated against *rand* to trigger dormancy or reproduction. Here, *rand* denotes a uniformly distributed random number within the interval [0, 1], t and T represent the current and maximum iteration counts, respectively, and i and n indicate the individual index and population size.

If $P_{ah} > rand$, the protozoa individuals are foraging in an autotrophic mode, otherwise in a heterotrophic mode. If $P_{dr} > rand$, the protozoa individuals are in dormancy, otherwise in reproduction. As shown in Equations (10) and (11), P_{ah} is the decreasing function of t, while P_{dr} is the increasing function of i. Therefore, P_{ah} decreases with the increase of t, so protozoa individuals tend to be more inclined toward heterotrophic mode as the iterative process proceeds. P_{dr} increases with the increase of i, so lower-ranked protozoa individuals tend to be more inclined toward dormancy mode. Finally, the higher-ranked individuals have significantly reduced errors after multiple rounds of iterative optimization. The following sections detail the four protozoan behaviors in the APO.

3.3.1. Autotrophic Behavior

For autotrophic behavior, individuals are updated using Equation (12):

$$X_i^{new} = X_i + f \cdot \left(X_j - X_i + \frac{1}{np} \cdot \sum_{k=1}^{np} w_a \cdot (X_{k-} - X_{k+}) \right) \odot M_f \tag{12}$$

$$f = rand \cdot (1 + \cos(\frac{t}{T} \cdot \pi)) \tag{13}$$

$$np_{max} = \left\lfloor \frac{n-1}{2} \right\rfloor \tag{14}$$

$$w_a = e^{-\left| \frac{f(X_{k-})}{f(X_{k+}) + eps} \right|} \tag{15}$$

$$M_f[di] = \begin{cases} 1, & \text{if } di \text{ is in } rand\text{perm}(\text{dim}, \left\lceil \text{dim} \cdot \frac{i}{n} \right\rceil) \\ 0, & \text{otherwise} \end{cases} \tag{16}$$

In Equation (12), X_i^{new} represents the updated individual while X_j denotes a randomly selected j-th protozoan. Equations (13)–(16) provide supplementary specifications to Equation (12), where f is the foraging factor, np indicates the number of neighbor pairs with a maximum limit np_{max}, X_{k-} refers to a protozoan randomly chosen from neighbor pairs with indices $k < i$, and X_{k+} corresponds to a protozoan selected from neighbor pairs with indices $k > i$. Boundary conditions dictate that if X_i is X_1, X_{k-} defaults to X_1, and if X_i is X_n, X_{k+} defaults to X_n. The weighting factor w_a governs autotrophic behavior, eps (2.2204×10^{-16}) is a minimal constant, and $f(\cdot)$ denotes the fitness function. The operator $\odot$ signifies the Hadamard product (element-wise multiplication), M_f is a binary search-oriented mapping vector of size $(1 \times dim)$ with elements 0 or 1, and di represents the dimension index within M_f.

3.3.2. Heterotrophic Behavior

For heterotrophic behavior, individuals are updated using Equation (17):

$$X_i^{new} = X_i + f \cdot (X_{near} - X_i + \frac{1}{np} \cdot \sum_{k=1}^{np} w_h \cdot (X_{i-k} - X_{i+k})) \odot M_f \tag{17}$$

$$X_{near} = (1 \pm Rand \cdot (1 - \frac{t}{T})) \odot X_i \tag{18}$$

$$w_h = e^{-\left| \frac{f(X_{i-k})}{f(X_{i+k}) + eps} \right|} \tag{19}$$

$$Rand = [rand_1, rand_2, \ldots, rand_{dim}] \tag{20}$$

Equations (18)–(20) provide the supplementary specifications for Equation (17). Here, X_{near} denotes a position in the vicinity of X_i, where the $\pm$ notation indicates that X_{near} can be sampled in different directions relative to the i-th protozoan. X_{i-k} represents the protozoan with index $i - k$ selected from the k-th neighbor pair, while X_{i+k} corresponds to the protozoan with index $i + k$. Boundary conditions ensure that if X_i is X_1, X_{i-k} defaults to X_1, and if X_i is X_n, X_{i+k} defaults to X_n. The weighting factor w_h governs heterotrophic behavior, and $Rand$ denotes a random vector with elements uniformly distributed in [0, 1].

3.3.3. Dormancy Behavior

For dormancy behavior, the model is defined as follows:

$$X_i^{new} = X_{min} + Rand \odot (X_{max} - X_{min}) \tag{21}$$

where X_{max} and X_{min} denote the upper and lower bound vectors, respectively.

3.3.4. Reproductive Behavior

For reproductive behavior, the model is defined using the following equation:

$$X_i^{new} = X_i \pm rand \cdot (X_{min} + Rand \odot (X_{max} - X_{min})) \odot M_r \tag{22}$$

$$M_r[di] = \begin{cases} 1, & \text{if } di \text{ is in } rand\text{perm}(\text{dim}, \lceil \text{dim} \cdot rand \rceil) \\ 0, & \text{otherwise} \end{cases} \tag{23}$$

where M_r is a mapping vector of size $(1 \times dim)$, with each element being either 0 or 1.

Autotrophic behavior and heterotrophic behavior are two different methods of local optimization. Dormancy behavior is equivalent to randomly resetting individuals and directly discarding the current solution in order to escape from local optima. Reproductive behavior is similar to mutation in genetic algorithms, causing significant changes in the current solution and also helping to escape from local optima.

The APO algorithm primarily involves parameters including population size (n), number of neighbor pairs for protozoa (np), maximum proportion fraction for dormancy and reproduction (pf_{max}), and maximum iteration count (T). To balance the model training accuracy and training time, the population size is set to 50 and the maximum iteration count to 50. The number of neighbor pairs for protozoa influences the randomness of individual updates during autotrophic and heterotrophic behaviors. For example, in autotrophic behavior, individuals are randomly selected from the current individual's neighboring pairs; a smaller number of neighbor pairs reduces randomness during selection, thereby enhancing the algorithm stability and accelerating convergence. The maximum proportion fraction for dormancy and reproduction determines the maximum proportion of individuals entering dormancy and reproductive states during iterations. A larger parameter improves the algorithm's global optimization capability but may weaken the local refinement searchability. Based on empirical evidence from the original references, we set the number of neighbor pairs to 1 and the maximum proportion fraction for dormancy and reproduction to 0.1.

3.4. Two-Stage APO Algorithm

The existing APO algorithm demonstrates strong exploration capabilities in the solution space, yet its probabilistic behavior selection mechanism risks the non-deterministic dormancy of high-quality individuals, thereby compromising exploitation performance. To address this imbalance, this study introduces an improved APO algorithm that integrates three strategic refinements. First, a good point set theory-driven initialization method is adopted to minimize the performance fluctuations caused by random initialization. Second, a two-stage optimization framework is established, explicitly separating global exploration during the initial iterations from intensified local exploitation in the convergence stages. Finally, an elite retention strategy is incorporated in the later optimization stages to refine the convergence precision. Collectively, these improvements aim to harmonize exploration-exploitation dynamics while mitigating initialization-induced stochastic disturbances. The technical specifications of these improvements are elaborated below:

(1) **Good point set initialization:** Based on the good point set theory proposed by Hua, L. K. et al. [36], the initial population is constructed within the s-dimensional unit cube Gs. Let $r \in G_s$ be a good point. When the discrepancy function satisfies

$\phi(n) = C(r,\epsilon)n - 1 + \epsilon$ (where $C(r,\epsilon)$ is a constant dependent only on r and an arbitrary positive ϵ), the good point set is defined as:

$$P_n(k) = \left\{ \left(\left\{ r_1^{(n)} \cdot k \right\}, \left\{ r_2^{(n)} \cdot k \right\}, \ldots, \left\{ r_s^{(n)} \cdot k \right\} \right), 1 \leqslant k \leqslant n \right\} \tag{24}$$

Specifically, the parameter r is selected as $r = \{2cos(2\pi k/p) \mid 1 \leq k \leq s\}$, where p is the smallest prime number satisfying $(p-3)/2 \geq s$. This construction method rigorously ensures a uniform spatial distribution of the initial population in the solution space. Figure 5 shows the comparison results between the good point set initialization and random initialization visualized using two-dimensional free variables. The figure reveals that the points generated by good point set initialization are uniformly distributed, while random initialization samples exhibit significant local clustering and sparse spatial regions. These results indicate that a good point set initialization strategy effectively mitigates the incomplete spatial coverage issue inherent in traditional random sampling.

(2) **Two-stage optimization:** The iterative process is partitioned into two stages with distinct selection strategies demarcated by the median iteration count as a threshold. During the second stage, an exponential probability distribution replaces the original uniform distribution to select individuals undergoing "dormancy or reproduction". The probability density function is defined as

$$P(X = k) = \frac{c-1}{c^T - 1} c^{k-1}, k = 1, 2, \ldots, T \tag{25}$$

where c is the progression factor (c > 1), X denotes a protozoan individual, k represents the individual's ranked index, and T is the population size. For moderate population sizes (e.g., $T = 50$), $c = 1.05$ is recommended to avoid exponential explosion risks. The selection probabilities are computed via this exponential distribution and executed using a roulette wheel strategy. When duplicate selections occur, priority is given to replacing duplicates with the suboptimal nearest-neighbor individual. If suboptimal candidates are exhausted, global reselection is triggered until the required number of individuals is selected.

(3) **Elitism preservation strategy:** During the second optimization stage, a fixed percentage of the fittest individuals is retained and exempted from resetting.

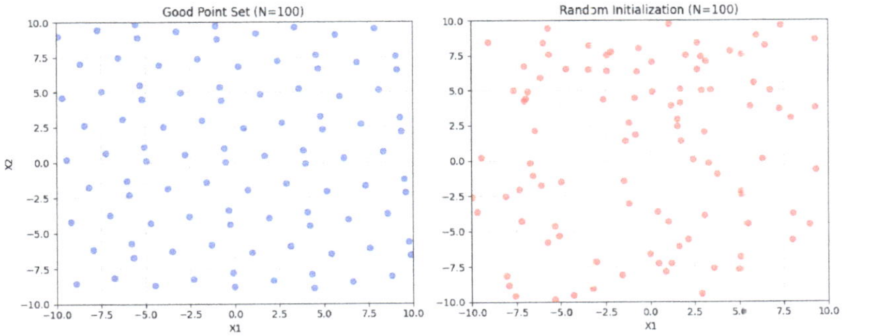

Figure 5. Comparison of results between good point set initialization and random initialization.

For a fair comparison, the parameters of the two-stage APO are consistent with the APO, and the selected parameters remain unchanged.

The optimization flowchart of the two-stage APO employed in this study is shown in Figure 6.

The following table shows the pseudocode for the two-stage APO algorithm (Algorithm 1).

Algorithm 1: Pseudo code of the proposed two-stage APO algorithm

```
Input: Initialize parameters n, dim, np, pf_max, and T
Output: The global optima X_best and f(X_best)
while t < T do
    Sort(X_i), i = 1, 2, ..., n;        // sort positions by fitness in ascending order
    pf = pf_max·rand;        // proportion fraction
    if t < 0.5T then
        Dr_index = randperm(n, ⌈n·pf⌉);
    else
        Dr_index is selected by Equation (25) and roulette wheel strategy
    end if
    for i = 1 : n do
        if i is in Dr_index then
            if P_dr > rand then
                Calculate X_i^new using Equation (21);        // dormancy
            else
                Calculate X_i^new using Equation (22);        // reproduction
            end if
        else
            if P_ah > rand then
                Calculate X_i^new using Equation (12);        // foraging in an autotroph
            else
                Calculate X_i^new using Equation (17);        // foraging in a heterotroph
        end if
        Get the current new position.
        if t < 0.5T then
            if f(X_i^new) < f(X_i) then
                X_i ← X_i^new;
            else
                X_i ← X_i
            end if
        else
            Select the top m individuals with the best fitness as elites;
            Merge them with the new population to form a candidate pool of size n + m;
            Choose the top n individuals from this pool as the next generation.
        end if
    end for
    t ← t + 1
end while
return X_best and f(X_best)
```

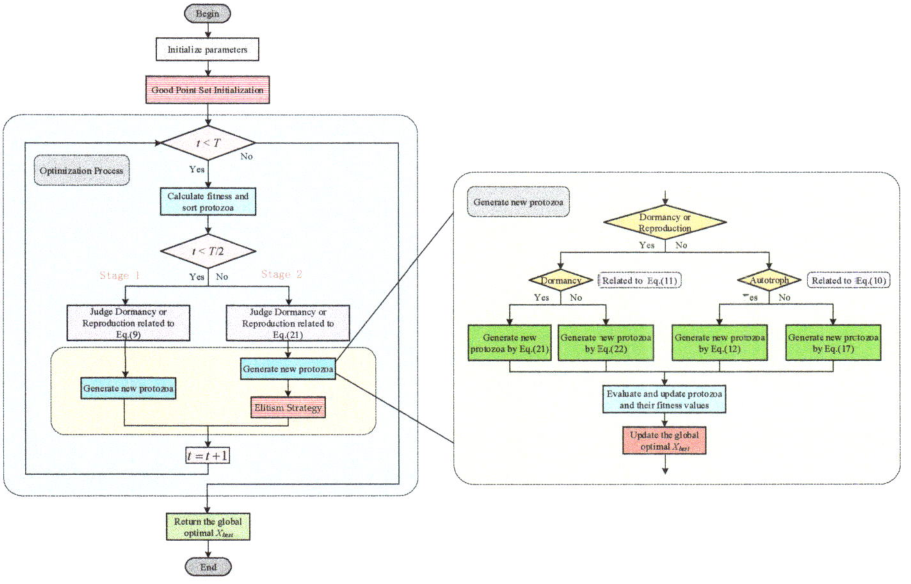

Figure 6. Flowchart of two-stage APO algorithm.

4. Model Training and Evaluation Metrics

4.1. Model Training

The training procedure for the relay protection equipment state quantification evaluation model consists of the following steps:

(1) **Dataset processing:** Given that the majority of data in the original relay protection equipment state evaluation dataset comprise high-score samples, synthetic low-score data were algorithmically generated to mitigate imbalance, while irrelevant feature columns with insufficient training samples were pruned. The processed dataset retained 42 feature columns and one device score column, partitioned into training, validation, and test sets in an 8:1:1 ratio.

(2) **Model architecture:** An improved Conformer model was constructed (detailed architecture in Figure 4), where the input data underwent feature projection via a fully connected layer followed by flattening. The model leverages multi-head attention mechanisms and depthwise separable convolutional modules within the Conformer blocks to capture cross-dimensional feature interactions, enhanced by two feed-forward networks for non-linear fitting capacity, culminating in a single-neuron FC layer for regression-based state score prediction.

(3) **Training model:** Training hyperparameters included a batch size of 128, 50 epochs, and an Adam optimizer with a dynamic learning rate (initial = 0.001; halved after five epochs of stagnant validation loss) using mean squared error (MSE) loss. The APO algorithm optimizes the Conformer hyperparameters via training and validation sets and then merges these sets for final model training, with the performance evaluated on the test set.

4.2. Model Evaluation Metrics

This study employs root mean square error (RMSE), mean absolute error (MAE), coefficient of determination (R^2), and adjusted R^2 to evaluate model performance. The mathematical formulations of these metrics are defined as follows:

$$RMSE = \sqrt{\frac{1}{n}\sum_{i=1}^{n}(y_i - \hat{y}_i)^2} \tag{26}$$

$$MAE = \frac{1}{n}\sum_{i=1}^{n}|y_i - \hat{y}_i| \tag{27}$$

$$R^2 = 1 - \frac{\sum_{i=1}^{n}(y_i - \hat{y}_i)^2}{\sum_{i=1}^{n}(y_i - \overline{y}_i)^2} \tag{28}$$

$$\text{Adjusted } R^2 = 1 - \frac{\sum_{i=1}^{n}(y_i - \hat{y}_i)^2/(n-k-1)}{\sum_{i=1}^{n}(y_i - \overline{y}_i)^2/(n-1)} \tag{29}$$

where n denotes the number of samples and k represents the number of features in the dataset.

The RMSE amplifies the difference between the predicted and true values through squaring, making it sensitive to large errors. The numerical units align with the original data, directly reflecting the absolute deviation level of the model predictions. MAE calculates the mean absolute value of prediction errors, demonstrating strong robustness to outliers. R^2 quantifies the model's ability to explain variance in the dependent variable while adjusted R^2 penalizes the number of independent variables to eliminate inflation effects caused by model complexity on R^2, making it more suitable for multiple regression analysis.

5. Case Study and Results Analysis

5.1. Data Sources and Preprocessing

This study utilizes the monitoring data from relay protection devices at substations in a southern Chinese region in the Year 2024, encompassing fundamental condition evaluation data, operational condition evaluation data, maintenance condition evaluation data, and ancillary factor evaluation data. Invalid feature columns with insufficient data were removed, and 42 valid features were retained to construct the dataset. To address the predominance of high-score samples in substation-provided data, synthetic low-score data were generated through rule-based guided corrections and expert knowledge integration. The final dataset contained 16,489 samples covering multiple device models within substations. In the complete dataset, there were 10,897 high-score samples accounting for 66.1%, corresponding to equipment normal status; 3198 attention-status samples accounting for 19.4%; abnormal-status and severe-abnormal sample numbers 1349 and 1276, accounting for 8.2% and 7.7%, respectively. Because substation monitoring data contain extremely few abnormal and severe-abnormal samples, small-sample generation algorithms are unsuitable for synthesis. Manual synthesis is employed to generate most of the data for sample construction. The entire dataset is randomly divided into training, validation, and test sets at an 8:1:1 ratio. All data are normalized to the range of 0–1 to eliminate differences in feature scales based on the features of the training set for model training. The divided validation and test sets each contained 1649 samples, maintaining a consistent data distribution across all three datasets. Each column of the data is normalized according to the characteristics of the corresponding column in the training set.

5.2. Parameter Optimization

The two-stage APO optimization algorithm is employed to optimize four hyperparameters in the improved Conformer encoder model: the number of neurons in the feed-forward network, the number of convolutional kernels in the convolutional layer, neuron dropout rate, and weight coefficient of the feed-forward network. Training hyperparameters include a batch size of 128 and 50 epochs and the Adam optimizer with a dynamic learning rate (initial = 0.001; halved after five epochs with stagnant validation loss). The APO configurations set the population size to 50, with 10 experimental trials conducted to select the optimal values. The optimized parameters are as follows: neurons: 351, convolutional kernels: 31, dropout rate: 0.1573, weight coefficient: 0.6903.

5.3. Method Validation and Result Analysis

The optimized model was retrained using the merged training set (combining the original training and validation sets) and evaluated on the test set. The test results are shown in Figures 7 and 8. Figure 7 presents the training and testing loss curves, while Figure 8 demonstrates the difference between the predicted and actual scores for 100 randomly selected test samples. During the initial training stages, both the training and testing losses decreased rapidly, converging to similar values as the iterations progressed, indicating effective model training. The observed error in training loss (occasionally exceeding testing loss) arises from dropout regularization during training, which is normal behavior.

Figure 7 shows the training and testing loss curves. The curves demonstrate that the two-stage APO method proposed in this paper achieves a rapid convergence speed and stable convergence behavior when optimizing the relay protection state quantification evaluation model. During the initial training stages, both the training and testing losses decreased rapidly, converging to similar values as the iterations progressed, indicating effective model training. The observed error in training loss (occasionally exceeding testing loss) arises from dropout regularization during training, which is normal behavior.

Figure 8 shows the difference between the predicted and actual scores for 100 randomly selected test samples. In the prediction results, due to differences in data distribution, the prediction scores for most devices are concentrated in the high-score range, with only a very small number of devices scoring below 60. The graph shows that high-score data exhibit smaller deviations between predicted and actual values, while low-score data (below 60) demonstrate certain prediction errors, indicating that the model is more accurate in identifying the status of normal equipment.

To validate the performance superiority of the proposed improved model, comparative experiments were conducted against normal deep learning models: CNN, MLP, CNN-BiGRU, CNN-BiLSTM, and a standard Transformer. The entire experiment was repeated ten times with the mean and best prediction metrics from each model selected for comparison. Figure 9 illustrates the mean prediction error across the machine learning approaches, and detailed performance parameter comparisons are summarized in Table 3. Table 4 shows the significance test results for the models in Table 3. Using a two-tailed Welch's t-test with a significance level of 0.05, pairwise comparisons are conducted between the model proposed in this paper and each model in Table 3.

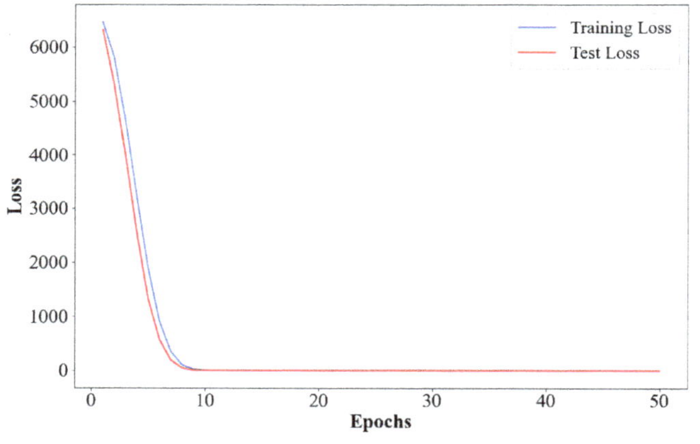

Figure 7. Training loss curves of the two-stage APO-IConf model.

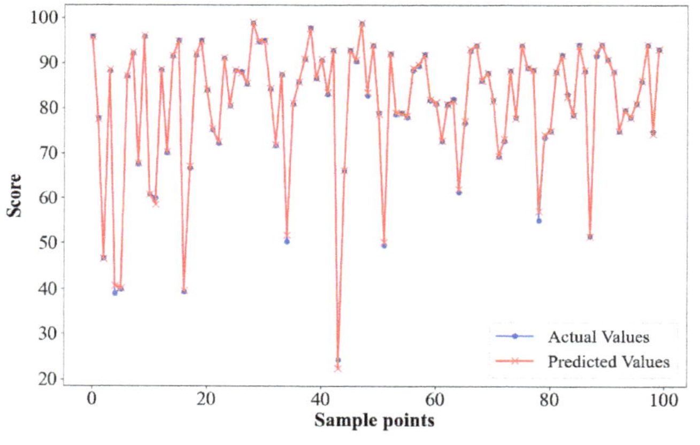

Figure 8. State quantification score prediction results.

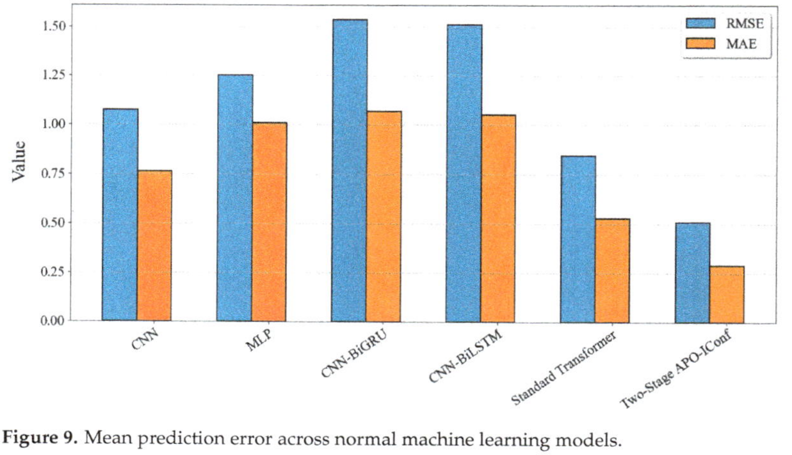

Figure 9. Mean prediction error across normal machine learning models.

Table 3. Performance comparison with normal machine learning models.

Model	Evaluation Index	RMSE	MAE	R^2	Adjusted R^2
CNN	Mean	1.0733	0.7627	0.9923	0.9920
	Best	1.0022	0.7154	0.9938	0.9937
MLP	Mean	1.2516	1.0091	0.9915	0.9913
	Best	1.2310	0.9395	0.9922	0.9920
CNN-BiGRU	Mean	1.5370	1.0681	0.9854	0.9849
	Best	1.3803	1.0352	0.9906	0.9903
CNN-BiLSTM	Mean	1.5116	1.0509	0.9878	0.9875
	Best	1.3674	0.9766	0.9908	0.9906
Standard Transformer	Mean	0.8455	0.5277	0.9958	0.9956
	Best	0.7987	0.4852	0.9961	0.9959
Two-Stage APO-IConf	Mean	**0.5064**	**0.2893**	**0.9985**	**0.9984**
	Best	**0.4857**	**0.2754**	**0.9988**	**0.9988**

Bold model indicates the optimal model, and bold values indicate the best value under that metric.

Table 4. Welch's *t*-test results with the proposed mode as the baseline.

Welch's *t*-Test		CNN	MLP	CNN-BiGRU	CNN-BiLSTM	Standard Transformer
RMSE	Test statistic	17.30	66.29	15.54	12.91	16.83
	p-value	2.75×10^{-9}	5.64×10^{-17}	4.72×10^{-8}	2.97×10^{-7}	6.47×10^{-9}
MAE	Test statistic	16.77	49.50	15.06	12.55	6.49
	p-value	6.62×10^{-9}	2.90×10^{-18}	6.51×10^{-8}	3.94×10^{-7}	7.61×10^{-5}

Welch's *t*-test is a statistical method for comparing whether there is a significant difference between the means of two independent samples. It does not assume equal variances between samples, making it more broadly applicable than the traditional *t*-test. The test yields a test statistic and *p*-value, which represents the ratio of the between-group mean difference to the sample variability, where larger absolute values indicate more pronounced group differences. The *p*-value denotes the probability of observing the current data or more extreme cases under the assumption of no difference between samples, with smaller values providing stronger evidence of statistically significant group differences.

As demonstrated by the experimental results, the two-stage APO-IConf model achieves significant advantages in the quantitative evaluation of the relay protection equipment state. Compared to the standard Transformer, it reduces the mean RMSE by 40.1% (from 0.8455 to 0.5064) and mean MAE by 45.2% (from 0.5277 to 0.2893) while attaining optimal performance in both R^2 (Best: 0.9988) and adjusted R^2 (Best: 0.9988) metrics. The model also shows substantial improvements over traditional methods, reducing the RMSE by 52.8% compared to CNN (Mean: 1.0733) and 59.5% compared to MLP (Mean: 1.2516). Both the mean and best values significantly outperform hybrid architectures like CNN-BiGRU and CNN-BiLSTM, demonstrating superior training efficacy across all evaluation metrics.

In Table 4, the proposed model has a very small *p*-value for all models, indicating statistically significant performance differences from the traditional machine learning baselines. This confirms that the performance improvement is attributable to model efficacy rather than to random factors.

To thoroughly validate the effectiveness of each component in the proposed method, experiments are conducted and compared with other Conformer models. Model 1 is the standard Conformer; Model 2 is the dynamic weighting coefficients Conformer, which adopts dynamic weighting coefficients on the standard Conformer architecture; Model 3 is post-BN-Conformer, which replaces the Pre-LN structure with a post-BN structure in the standard Conformer architecture; Model 4 is Improved Conformer; Model 5 is APO-Conformer; Model 6 is Two-Stage APO-Conformer; Model 7 is APO-IConformer; Model 8 is Two-Stage APO-IConformer. Standard Conformer serves as the baseline. This systematic evaluation quantifies the incremental contributions of architectural modifications to the final performance in the quantitative state evaluation of relay protection equipment. Figure 10 illustrates the prediction error across the other Conformer models, and detailed performance parameter comparisons are summarized in Table 5. Table 6 displays the significance test results for the models listed in Table 5. The testing process is the same as before, except that a standard Conformer (Model 1) is used as the baseline. Bold results indicate that the model shows no significant superiority over the benchmark.

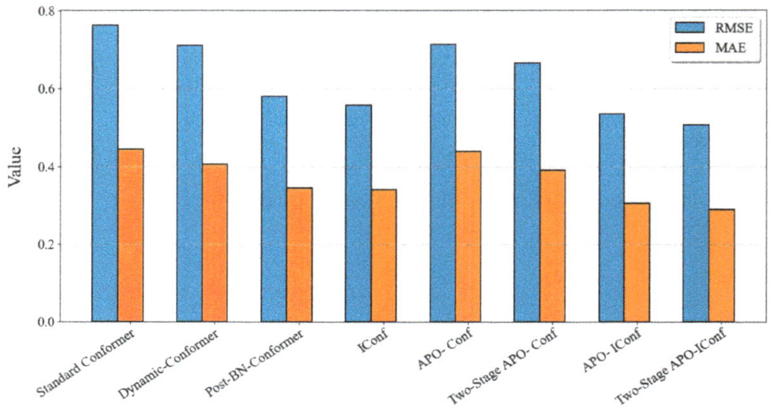

Figure 10. Predictive mean error across other Conformer models.

Table 5. Performance comparison with other Conformer models.

No.	Model	Evaluation Index	RMSE	MAE	R^2	Adjusted R^2
1	Standard Conformer	Mean	0.7626	0.4449	0.9965	0.9963
		Best	0.7260	0.4133	0.9969	0.9967
2	Dynamic-Conformer	Mean	0.7110	0.4069	0.9971	0.9970
		Best	0.6756	0.3845	0.9977	0.9976
3	Post-BN-Conformer	Mean	0.5797	0.3456	0.9980	0.9979
		Best	0.5604	0.3189	0.9981	0.9981
4	IConf	Mean	0.5572	0.3413	0.9982	0.9981
		Best	0.5321	0.3242	0.9983	0.9982
5	APO-Conf	Mean	0.7134	0.4156	0.9970	0.9968
		Best	0.6472	0.3967	0.9978	0.9976
6	Two-Stage APO- Conf	Mean	0.6654	0.3918	0.9976	0.9975
		Best	0.6320	0.3430	0.9978	0.9977
7	APO- IConf	Mean	0.5342	0.3063	0.9983	0.9982
		Best	0.5089	0.2821	0.9985	0.9984
8	**Two-Stage APO-IConf**	Mean	**0.5064**	**0.2893**	**0.9985**	**0.9984**
		Best	**0.4857**	**0.2754**	**0.9988**	**0.9988**

Bold model indicates the optimal model, and bold values indicate the best value under that metric.

This experiment demonstrates that the proposed two-stage APO-IConf (Model 8) achieves comprehensive superiority in the quantitative evaluation of relay protection equipment state, reducing mean RMSE by 33.6% (from 0.7626 to 0.5064) and mean MAE by 35.0% (from 0.4449 to 0.2893) compared to the baseline standard Conformer (Model 1) while improving the mean R^2 metric from 0.9965 to 0.9985.

Table 6. Welch's *t*-test results with standard Conformer (Model 1) as the baseline.

Welch's *t*-Test		Model 2	Model 3	Model 4	Model 5	Model 6	Model 7	Proposed Model
RMSE	Test statistic	−2.85	−18.00	−22.43	**−1.54**	−7.08	−15.81	−21.36
	p-value	1.43×10^{-2}	3.15×10^{-12}	1.90×10^{-14}	**1.53×10^{-1}**	6.27×10^{-6}	9.55×10^{-10}	4.37×10^{-13}
MAE	Test statistic	**−1.34**	−5.10	−5.38	**−1.96**	−2.48	−7.41	−10.54
	p-value	**1.95×10^{-1}**	7.48×10^{-5}	4.04×10^{-5}	**7.28×10^{-2}**	2.36×10^{-2}	7.22×10^{-7}	1.15×10^{-8}

Bold values mean the Welch's *t*-test results are not significant.

Models 1 through 4 form a controlled experimental group, where Model 2 and Model 3 represent single-aspect improvements over Model 1, and Model 4 combines both modifications. The experimental data demonstrate that both improvement methods (post-BN structure and dynamic weighting coefficient) enhance the model performance. Notably, the performance gain from implementing the post-BN structure was more substantial than that from adjusting the dynamic weighting coefficients. This indicates that post-BN modification serves as an effective enhancement approach, contributing more significantly to overall performance improvements than dynamic weighting coefficient optimization. While single-aspect modifications yield statistically insignificant improvements, the enhanced Conformer demonstrates significant gains.

Relative to the single-modification IConf (Model 4), the integration of the two-stage APO optimization strategy further reduces the mean RMSE by 9.1% (from 0.5572 to 0.5064) and mean MAE by 15.2% (from 0.3413 to 0.2893), validating the necessity of algorithm-architecture co-design. When compared to optimized variants, the proposed two-stage APO-IConf (Model 8) exhibits a 23.9% mean RMSE reduction (from 0.6654 to 0.5064) and 26.2% mean MAE reduction (from 0.3918 to 0.2893) over Two-Stage APO-Con (Model 6), as well as a 5.2% mean RMSE reduction (from 0.5342 to 0.5064) and 5.6% mean MAE reduction (from 0.3063 to 0.2893) against APO-IConf (Model 7), confirming that the synergistic combination of the improved Conformer architecture and two-stage APO strategy drives performance breakthroughs, as structural deficiencies in Model 6 and non-staged optimization in Model 7 result in significantly higher errors.

As shown in Table 6, the proposed model demonstrates the strongest statistical significance, followed by Model 4. Both exhibit the largest absolute values of the test statistics and smallest *p*-values, indicating robust significance. While Model 2 and Model 5 show improvements in both mean and best values, their statistical significance remains weak, potentially lacking statistical significance.

The two-stage APO-IConf achieves optimal RMSE (0.4857), MAE (0.2754), R^2 (0.9988), and adjusted R^2 (0.9988), providing quantitative state evaluation and guidance for developing maintenance strategies for substations.

6. Conclusions

To address the challenges of ambiguous status boundaries and hyperparameter selection in relay protection equipment evaluation, this study proposes a state quantification method based on a two-stage APO-optimized improved Conformer model. Experimental validation using real-world datasets confirms the effectiveness of the method. Key conclusions are:

(1) This paper proposes an improved Conformer architecture by replacing pre-layer normalization with post-batch normalization and introducing dynamic weight coefficients to regulate feed-forward network connectivity, which significantly improves the feature fusion capability and fitting accuracy for relay protection data.

(2) The two-stage APO algorithm integrates good point set initialization and elitism preservation strategies, achieving dynamic equilibrium between global exploration and local exploitation in the hyperparameter space of the Conformer, which effectively resolves traditional parameter selection difficulties.
(3) The two-stage APO-IConf model exhibits a 23.9% mean RMSE reduction (from 0.6654 to 0.5064) and 26.2% mean MAE reduction (from 0.3918 to 0.2893) over the Two-Stage APO-Con (Model 6), as well as a 5.2% mean RMSE reduction (from 0.5342 to 0.5064) and 5.6% mean MAE reduction (from 0.3063 to 0.2893) compared to APO-IConf (Model 7), confirming that the synergistic combination of the improved Conformer architecture and two-stage APO strategy drives performance breakthroughs.
(4) The two-stage APO-IConf model autonomously learns expert rule-based scoring patterns and field adjustment logic from data, enabling objective and rapid state evaluation without manual intervention in new inspection scenarios.

Existing device data already include dozens of substation protection devices, such as busbar and differential protection devices. In the future, testing can be conducted in actual substations to apply the research work to substations. For example, it can be combined with digital twin models of substations to perform online state evaluations, thereby improving the accuracy.

This paper still has the following limitations. First, the training data relies on expert scoring, which inherently embodies the subjective quantification of domain priors and may be influenced by experts' cognitive frameworks, knowledge blind spots, and evaluation criteria. Such implicit biases may be assimilated by the model, causing its scoring results to deviate from the objective truth condition. Second, although RMSE and MAE eliminate the state boundary problem in traditional classification methods and instead use continuous scores to evaluate the state of devices, these metrics are unable to distinguish the direction of error and are unable to reflect the error distribution. Therefore, the next stage of research will focus on other relevant metrics and whether we can train the corresponding models for devices with different score ranges.

Author Contributions: Y.L. led the conceptualization, methodology, and original draft preparation. Validation was carried out by Y.L., M.Z. and S.Z., while formal analysis was performed by Y.L. and M.Z.; S.Z. managed the investigation, and resources were provided by S.Z. and Y.Z.; Y.Z. was responsible for data curation. Writing—review and editing involved Y.L., M.Z. and Y.Z., with visualization by M.Z. and Y.Z. Supervision was provided by Y.L. and S.Z., project administration by Y.L. and Y.Z., and funding acquisition by Y.L. All authors have read and agreed to the published version of the manuscript.

Funding: This work was supported by the grant from China Southern Power Grid Co., Ltd. (Project No. 030100KC23110037).

Data Availability Statement: Data are contained in the article.

Conflicts of Interest: Authors Yanhong Li, Min Zhang, and Shaofan Zhang were employed by the Guangzhou Power Supply Bureau of Guangdong Power Grid Co., Ltd. Author Yifan Zhou was employed by the company NR Engineering Co., Ltd. The authors declare that this study received funding from China Southern Power Grid Co., Ltd. (Project No. 030100KC23110037). The funder was not involved in the study design, collection, analysis, interpretation of data, writing of this article, or the decision to submit it for publication.

Abbreviations

The following abbreviations are used in this manuscript:

APO	Artificial protozoa optimize
APO-Conf	Artificial protozoa optimize-Conformer
APO-IConf	Artificial protozoa optimize-improved Conformer
BNDT	Bayesian network decision trees
BP	Back propagation
CNN	Convolutional neural network
CNN-BiGRU	Convolutional neural network-bidirectional gated recurrent unit
CNN-BiLSTM	Convolutional neural network-bidirectional long short term memory network
CRC	Cyclic redundancy check
CT-N wire	Current transformer neutral wire
FAHP	Fuzzy analytical hierarchy process
FFN	Feed-forward network
GAN	Generative adversarial networks
GOOSE	Generic object oriented substation event
IConf	Improved Conformer
LSSVM	Least squares support vector machines
MHSA	Multi-head self-attention
MLP	Multilayer perceptron
NLP	Nature language processing
Post-BN	Post-batch normalization
Pre-LN	Pre-layer normalization
PT-N wire	Potential transformer neutral wire
Two-stage APO	Two-stage artificial protozoa optimizer
Two-stage APO-Conf	Two-stage artificial protozoa optimizer-Conformer
Two-stage APO-IConf	Two-stage artificial protozoa optimizer-improved Conformer

References

1. Kang, C.; Yao, L. Key Scientific Issues and Theoretical Research Framework for Power Systems with High Proportion of Renewable Energy. *Autom. Electr. Power Syst.* **2017**, *41*, 2–11.
2. Ou, K.; Gao, S.; Wang, Y.; Zhai, B.; Zhang, W. Assessment of the Renewable Energy Consumption Capacity of Power Systems Considering the Uncertainty of Renewables and Symmetry of Active Power. *Symmetry* **2024**, *16*, 1184. [CrossRef]
3. Yuan, X.; Zhang, M.; Chi, Y.; Ju, P. Basic Challenges of and Technical Roadmap to Power-electronized Power System Dynamics Issues. *Proc. CSEE* **2022**, *42*, 1904–1916.
4. Qiu, Y.; Lu, S.; Lu, H.; Luo, E.; Gu, W.; Zhuang, W. Flexibility of Integrated Energy System: Basic Connotation, Mathematical Model and Research Framework. *Autom. Electr. Power Syst.* **2022**, *46*, 16–43.
5. National Development and Reform Commission; National Energy Administration. The 14th Five-Year Plan for Building a Modern Energy System (2021–2025). Available online: https://www.gov.cn/zhengce/zhengceku/2022-03/23/content_5680759.htm (accessed on 26 April 2025).
6. Wang, L.; Zhang, J.; Zeng, Z.; Yu, H.; Zheng, X. Research and Application of the Random Failure Indicators in Status Evaluation of Relay Protection. *Electr. Power* **2013**, *46*, 87–90.
7. Wang, Y.; Liao, H.; Yuan, X.; Chen, J.; Xu, Z.; Luo, C. Development and application of relay protection condition evaluation system based on fault information processing system. *Power Syst. Prot. Control.* **2014**, *42*, 134–139.
8. Zhang, L.; Wang, G.; Cao, L.; Dai, Z.; Kou, B. Smart status evaluation and early warning approach for highly-reliable protection systems based on GAN model and random forest algorithm. *J. Electr. Power Sci. Technol.* **2021**, *36*, 104–112.
9. Wang, L.; Guo, P.; Kang, Y.; Yan, Z. Research on Relay Protection Equipment Maintenance Decision-Making Method Based on Risk Assessment. In Proceedings of the 2023 IEEE PES GTD International Conference and Exposition (GTD), Istanbul, Turkiye, 22–25 May 2023; pp. 231–235.

10. Zheng, S.; Yang, X.; Du, J.; Dong, P.; Li, Y.; Guo, P. Indexes and Methods of Multi-Dimensional Comprehensive Evaluation of Relay Protection. In Proceedings of the 2022 4th International Conference on System Reliability and Safety Engineering (SRSE), Guangzhou, China, 15–18 December 2022; pp. 307–311.
11. Dong, Z.; Li, H.; Tang, Y.; Yin, H.; Yin, J. Research and Application of Intelligent Diagnosis of Health Status of Relay Protection Equipment. In Proceedings of the 2022 International Conference on Intelligent Transportation, Big Data & Smart City (ICITBS), Hengyang, China, 26–27 March 2022; pp. 847–853.
12. Hu, H.; Kang, Y.; Zhang, Y.; Ye, X.; Guo, P.; Yan, Z. A Relay Protection State Evaluation Method with Multiple Influencing Factors. In Proceedings of the 2023 IEEE Sustainable Power and Energy Conference (iSPEC), Chongqing, China, 28–30 November 2023; pp. 1–5.
13. Jin, L.; Zhou, Z.; Zhan, R.; Yang, G.; Zhang, Y. Optimal Layout and Overheat Monitoring for Components of Highly Reliable Relay Protection Equipment. *IEEE Access* **2023**, *11*, 85615–85625. [CrossRef]
14. Fu, H.; Liu, Q.; Wang, Y.; Zhou, S.; Wang, K.; Wang, C.; Song, W.; Li, Y. Fuzzy Assessment of Hierarchical State of Relay Protection Equipment in Intelligent Substation Based on Entropy Weight DSmT. In Proceedings of the 2023 International Conference on Sensing, Measurement & Data Analytics in the era of Artificial Intelligence (ICSMD), Xi'an, China, 2–4 November 2023; pp. 1–6.
15. Sun, H.; Zhang, G.; Gao, B.; Wang, Y.; Li, Y.; Li, Y. Fuzzy Comprehensive Evaluation of Relay Protection Equipment Status in Intelligent Substations Based on Combination Weighting Method. *Electr. Meas. Instrum.* **2020**, *57*, 23–28.
16. Xu, C.; Wang, Y.; Zhao, L.; Gao, J.; Huang, L.; Ying, L. Fuzzy comprehensive evaluation of intelligent substation relay protection system state based on information trend prediction and combination weighting. *Electr. Power Autom. Equip.* **2018**, *38*, 162–168.
17. Zhang, J.; Xue, A.; Zhang, L.; Zhou, Z.; Yang, G.; Zhang, H.; Wang, W.; Wang, Z. Research on Health Status Evaluation of Relay Protection Based on Combinatorial Weighting Model. In Proceedings of the 2019 IEEE 3rd International Electrical and Energy Conference (CIEEC), Beijing, China, 7–9 September 2019; pp. 618–622.
18. Xu, C.; Ying, L.; Luo, X.; Zhao, L.; Gao, J.; Huang, L.; Tan, J. Method of weight updating for state evaluation of relay protection equipment based on Bayesian theory. *Eng. J. Wuhan Univ.* **2017**, *50*, 738–744.
19. Zhao, X.; Yu, Z.; Zhang, X. Fuzzy comprehensive state evaluation method of variable weight of relay protection based on variable membership degree. *Power Syst. Prot. Control.* **2017**, *45*, 22–29.
20. Zhou, Y.; Ou, R.; Li, D.; Liao, X.; Yang, Y. Health status assessment of secondary equipment based on interval PCA and fuzzy comprehensive evaluation. *Manuf. Autom.* **2023**, *45*, 104–109.
21. Alaerjan, A.; Jabeur, R.; Ben, H.; Karray, M.; Ksantini, M. Improvement of Smart Grid Stability Based on Artificial Intelligence with Fusion Methods. *Symmetry* **2024**, *16*, 459. [CrossRef]
22. Kezunovic, M.; Baembitov, R.; Mohamed, T. No Silver Bullet: Artificial Intelligence Is Not a Panacea, but It Works for Fault Analysis and Outage Management. *IEEE Power Energy Mag.* **2024**, *22*, 78–88. [CrossRef]
23. Ye, Y.; Wang, W.; Liu, H.; Wang, Q.; Wang, T.; Zhao, Z. Research on State Warning of Relay Protection Device Based on BP Neural Network. *Mod. Sci. Instrum.* **2022**, *39*, 195–201.
24. Jia, Y.; Ying, L.; Wang, D.; Zhang, J. Defect Prediction of Relay Protection Systems Based on LSSVM-BNDT. *IEEE Trans. Ind. Inform.* **2020**, *17*, 710–719. [CrossRef]
25. Zhou, D. Research on Digital Twin of Relay Protection Simulation Deduction and State Prediction Technology in the Intelligent Substation. Master's Thesis, Huazhong University of Science and Technology, Wuhan, China, 2022.
26. Liu, Y. Research on Reliability Evaluation Method of Protective Equipment and Software Design. Master's Thesis, North China Electric Power University, Beijing, China, 2020.
27. Dong, Y. State Evaluation and Fault Prediction of Protection System Based on Digital Twin. Master's Thesis, Shandong University, Jinan, China, 2023.
28. Zhu, Q.; Xu, Z.; Wang, L. Analysis and Comparison of Corporation Control Activities Assessment Based on Fuzzy Comprehensive Method and BP Neural Network Method. *Manag. Rev.* **2013**, *25*, 113–123.
29. Wang, F. A Comparative Study of Speed Control Schemes of PMSM Based on Fuzzy PID Control and BP Neural Network PID Control. *Micromotors* **2020**, *53*, 103–107.
30. Gulati, A.; Qin, J.; Chiu, C.C.; Parmar, N.; Zhang, Y.; Yu, J.; Han, W.; Wang, S.; Zhang, Z.; Wu, Y. Conformer: Convolution-Augmented Transformer for Speech Recognition. *arXiv* **2020**, arXiv:2005.08100.
31. Ramachandran, P.; Zoph, B.; Le, Q.V. Searching for Activation Functions. *arXiv* **2017**, arXiv:1710.05941.
32. Dai, Z.; Yang, Z.; Yang, Y.; Carbonell, J.; Le, Q.V.; Salakhutdinov, R. Transformer-Xl: Attentive Language Models beyond a Fixed-Length Context. *arXiv* **2019**, arXiv:1901.02860.
33. Shaw, P.; Uszkoreit, J.; Vaswani, A. Self-Attention with Relative Position Representations. *arXiv* **2018**, arXiv:1803.02155.
34. Vaswani, A.; Shazeer, N.; Parmar, N.; Uszkoreit, J.; Jones, L.; Gomez, A.N.; Kaiser, Ł.; Polosukhin, I. Attention Is All You Need. *Adv. Neural Inf. Process. Syst.* **2017**, *30*. Available online: https://proceedings.neurips.cc/paper_files/paper/2017/hash/3f5ee243547dee91fbd053c1c4a845aa-Abstract.html (accessed on 5 May 2025).

35. Wang, X.; Snášel, V.; Mirjalili, S.; Pan, J.S.; Kong, L.; Shehadeh, H.A. Artificial Protozoa Optimizer (APO): A Novel Bio-Inspired Metaheuristic Algorithm for Engineering Optimization. *Knowl.-Based systems* **2024**, *295*, 111737. [CrossRef]
36. Hua, L.K.; Wang, Y. *Applications of Number Theory to Numerical Analysis*; Springer Science & Business Media: Berlin/Heidelberg, Germany, 2012; ISBN 3-642-67829-7.

symmetry

MDPI

Article

Integrated I-ADALINE Neural Network and Selective Filtering Techniques for Improved Power Quality in Distorted Electrical Networks

Yap Hoon [1,*], Kuew Wai Chew [1,*] and Mohd Amran Mohd Radzi [2]

[1] Lee Kong Chian Faculty of Engineering and Science, Universiti Tunku Abdul Rahman, Kajang 43000, Selangor, Malaysia
[2] Department of Electrical and Electronic Engineering, Faculty of Engineering, Universiti Putra Malaysia, Serdang 43400, Selangor, Malaysia; amranmr@upm.edu.my
* Correspondence: hoonyap@utar.edu.my (Y.H.); chewkw@utar.edu.my (K.W.C.)

Abstract

Adaptive Linear Neuron (ADALINE) is a well-known neural network method that has been utilized for generating a reference current intended to regulate the operation of shunt-typed active harmonic filters (SAHFs). These filters are essential for improving power quality by mitigating harmonic disturbances and restoring current waveform symmetry in power systems. While the latest variant, Simplified ADALINE, offers notable advantages over its predecessors, such as a reduced complexity and faster learning speed, its performance has primarily been evaluated under stable grid conditions, leaving its performance under distorted environments largely unexplored. To address this gap, this work introduces two key modifications to the Simplified ADALINE framework: (1) the integration of a new phase-tracking algorithm based on the concept of orthogonality and selective filtering, and (2) transitioning from the direct current control (DCC) to an indirect current control (ICC) mechanism. Test environments featuring distorted grids and nonlinear rectifier loads are simulated in MATLAB/Simulink software to evaluate the performance of the proposed method against the existing Simplified ADALINE method. The key findings demonstrate that the proposed method effectively handled harmonic distortion and noise disturbance. As a result, the associated SAHF achieved an additional reduction in %THD (by 10.77–13.78%), a decrease in reactive power (by 58.3 VAR–67 VAR), and improved grid synchronization with a smaller phase shift (by 0.9–1.2°), while also maintaining proper waveform symmetry even in challenging grid conditions.

Keywords: active power filter; neural network; power quality; power system symmetry; synchronizer; two-phase orthogonal components

1. Introduction

The extensive utilization of nonlinear power electronics loads in power systems such as variable speed drives, uninterruptible power supplies, and energy-efficient lighting has made it increasingly difficult to manage harmonic disturbances [1,2]. These loads often disrupt the inherent symmetry of the power system, introducing unbalanced and distorted current waveforms. If left unmitigated, such asymmetry in current waveforms can cause a range of detrimental effects, such as increased heat losses, reduced system efficiency, voltage distortion, and potential damage to sensitive equipment [3–5].

The loss of waveform symmetry not only degrades the power quality but also negatively affects the reliability and stability of the power system operation. Therefore, preserving symmetry in both current and voltage waveforms is crucial to maintaining a balanced power flow and prevent equipment malfunction. Proactive mitigation strategies are necessary in order to address these disturbances and ensure the overall grid power quality. Recent studies highlight the role of shunt-type active harmonic filters (SAHFs), which utilize voltage source inverter topologies, as effective solutions for restoring waveform symmetry and mitigating harmonic distortion [3,6,7]. These filters help compensate for unbalanced and distorted currents, thereby improving both the quality and symmetry of the overall power system waveforms.

However, the effectiveness of an SAHF in mitigating power quality issues heavily depends on the accuracy and speed of its control strategies. The controller typically includes four core algorithms: harmonic currents extraction, DC-link voltage control, phase tracking, and current control algorithms [8–10]. Among these, the extraction of harmonic currents is considered the most critical. As the first algorithm executed within the controller, accurate extraction ensures that the harmonic components disrupting the symmetry of current waveforms are properly identified and targeted. Furthermore, the reference current must be precisely synchronized with the grid through the phase-tracking algorithm to maintain the symmetry and balance of the compensated current. Together, these processes enable the effective mitigation of waveform asymmetry and harmonic distortion, thereby preserving power quality [8,11,12].

Various studies have explored the techniques for harmonic currents extraction, as detailed in the literature [10,13]. Prominent techniques include the synchronous reference frame (SRF) [14,15] and instantaneous power theory [4,16] from the time-domain category, Fourier transform (FT) [17,18] from the frequency-domain category, and artificial neural network (ANN) [5,19,20] from the artificial intelligence (AI) category. Among these, the techniques based on the ANN concept are particularly noteworthy for their exceptional ability to provide rapid and accurate estimations of reference currents. These techniques are highly regarded for their self-adaptive capabilities, parallel processing power, and robust fault tolerance [19,21,22], making them a standout choice in dynamic and complex systems.

In the reference current generation, Adaptive Linear Neuron (ADALINE) [5,20,22,23] is a highly regarded architecture, recognized for its efficiency within ANN-based solutions. This approach operates by utilizing an appropriate weight adjustment mechanism, with the Widrow–Hoff (W–H) algorithm being particularly popular due to its simplicity and efficiency in minimizing the mean square error between the actual and predicted signals [24,25]. However, the conventional W–H algorithm faces challenges when attempting to learn the characteristics of multiple harmonic components, which necessitates more repetitive cycles and longer training times. To address these limitations, a modified W–H algorithm has been introduced. This modification directs the system to focus solely on learning the characteristics of the fundamental component, rather than the full spectrum of harmonic components, by fine-tuning the learning rate [21,25]. As a result, the modified algorithm significantly improves the speed and precision of the learning process, leading to fewer repetitive cycles and a considerable reduction in training time. However, both the conventional and modified W–H algorithms depend on conventional phase-tracking algorithms such as the zero-crossing detector (ZCD) [25] and phase-locked loop (PLL) [11,26] to align the phase of the generated reference current with the operating power system. This dependency has restricted their applicability, particularly to situations involving a stable supply voltage and steady-state operations.

A further investigation revealed that the modified W–H algorithm still contained redundant features and notable shortcomings, including a large mean square error and slow convergence, which negatively affected its overall performance. To address these issues, the algorithm was simplified to remove redundancy, resulting in the development of Simplified ADALINE (S-ADALINE) [21]. This revised version incorporates a new weight updating technique, called the fundamental active current (FAC), which helps to reduce the mean square error and further speed up the learning process, thereby enhancing the mitigation performance of the connected SAHF. Furthermore, a new phase-tracking algorithm, derived from the ADALINE concept, was integrated, effectively eliminating the dependence on ZCD and PLL [12,22]. With the implementation of the S-ADALINE algorithm, the associated SAHF now exhibits enhanced reliability, operating effectively under a stable supply voltage, and across both steady-state and dynamic conditions, even when subjected to varying load conditions.

Nevertheless, in practice, distortions in the utility grid are common and can compromise the reliability of the S-ADALINE algorithm, which was originally designed to operate under stable (sinusoidal and balanced) supply voltage conditions. As described in [12,22], the integrated ADALINE-based phase-tracking algorithm functions by converting the measured supply voltage directly into a unity form. Consequently, when the supply voltage is affected by distortions, the accuracy of the extracted phase information may degrade, resulting in improper phase synchronization and a failure to maintain waveform symmetry. Furthermore, an inherent drawback of the S-ADALINE algorithm is its tendency to generate a harmonic-based reference current, necessitating the gate pulses for the SAHF to be derived according to the direct current control (DCC) mechanism [27,28]. As reported in [8,29], the DCC mechanism lacks the ability to manage switching ripples caused by the operation of the SAHF, which further undermines the mitigation of asymmetrical and distorted waveforms. In contrast, the indirect current control (ICC) mechanism, which derives the gate pulses based on the evaluation of the supply current against a fundamental-based reference current [27,30], has been reported to address these shortcomings of the DCC mechanism by offering an improved ripple management and better preservation of waveform symmetry. Despite these advantages, no studies have yet explored integrating the S-ADALINE algorithm with the ICC mechanism to enhance harmonic mitigation and waveform symmetry.

With the increasing severity and unpredictability of harmonic disturbances, there is a pressing need to enhance the reliability and adaptability of the SAHF through adaptive AI approaches, in line with the recent advancements in artificial intelligence. Particularly, the development of improved reference current generation methods that maintain symmetry and effectively track fundamental components, even in highly distorted and dynamic grid conditions, is critical. In this work, the existing S-ADALINE algorithm is further developed by harnessing the strengths of the ICC mechanism. Additionally, the concept of orthogonality and selective filtering are integrated to enhance the overall reliability and performance of the system under such challenging grid environments. With these modifications, the resulting algorithm, termed Indirect ADALINE (I-ADALINE), offers the benefits of enhanced harmonic mitigation and reactive power compensation, along with reduced phase shift and improved accuracy in tracking the fundamental active current, thereby preserving the symmetry of current waveforms. The proposed I-ADALINE method is thoroughly tested under both steady-state and dynamic conditions, using a single-phase low-voltage distribution setup operating at 230 V and 50 Hz, with a distorted supply voltage and nonlinear rectifier loads to ensure a comprehensive performance evaluation.

The paper is organized as follows: Section 2 outlines the configuration of the power circuit and the controller for the single-phase SAHF system implemented in this work. In Section 3, the underlying principles of the proposed I-ADALINE algorithm are explained in detail. Section 4 provides a thorough analysis of the data, validating the design approach and demonstrating the effectiveness of the proposed algorithm. Finally, Section 5 summarizes the key outcomes and contributions of this work.

2. Single-Phase SAHF and Associated Control Algorithms

The architecture of the single-phase SAHF system along with the details of its controller is illustrated in Figure 1. In this work, a traditional voltage source inverter (VSI) is configured to function as an SAHF, while the harmonics are generated from rectifier-based circuits that supply either capacitive or inductive loads. The nonlinear behavior of the rectifier-based circuits will create harmonics in the load current i_L; i.e., the load current will contain both fundamental i_F and harmonic i_H components, thereby contaminating the supply current i_S. Mathematically, their relationship can be expressed as follows:

$$i_s = i_L = i_F + i_H \quad (1)$$

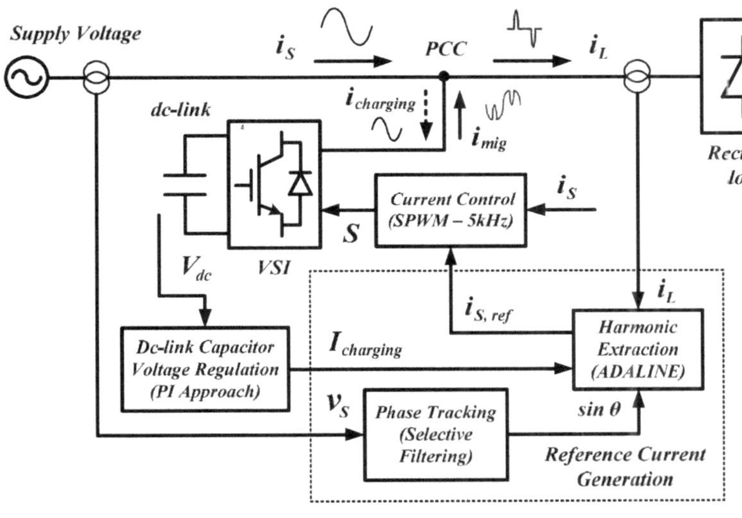

Figure 1. Single-phase SAHF system along with the details of its controller.

To resolve this concern, the SAHF is strategically positioned between the supply voltage v_S and the load, linked at a common point (PCC). The controller then manages the SAHF to deliver mitigation current i_{mig}, which neutralizes all the harmonic components according to Expression (2). At the same time, for the proper operation, the SAHF is also controlled to draw a small amount of charging current $i_{charging}$ from the system to charge the DC-link capacitor. As shown, when the mitigation current i_{mig} is adjusted to match the harmonic components i_H, the resulting supply current i_S becomes purely the fundamental current i_F . This effectively restores the system to a clean, harmonic-free power quality.

$$i_s = i_F + \left(i_H - i_{mig}\right) + i_{charging} \quad (2)$$

For the effective operation of the SAHF, the applied controller is structured with three primary modules, each responsible for generating reference current $i_{S,ref}$, regulating DC-link voltage V_{dc} and delivering the gate pulses S, respectively. This work, however, primarily focuses on the generation of the reference current, introducing a method called Indirect ADALINE (I-ADALINE). Detailed information about this approach will be provided in the following section. Meanwhile, to regulate the DC-link voltage, a proportional-integral (PI) controller [16,26,31] is applied to estimate the amount of charging current I_{dc} needed. For the generation of gate pulses, a method based on sinusoidal pulse-width modulation (SPWM) is adapted [8,28,32].

3. Reference Current Generation Based on I-ADALINE Concepts

This section describes the development of the proposed Indirect ADALINE (I-ADALINE) algorithm, starting with an in-depth analysis of the existing Simplified ADALINE (S-ADALINE) algorithm. Establishing this baseline provides a foundation for detailing the modifications performed, ultimately leading to the development of the proposed I-ADALINE algorithm. Lastly, the workflow is included to illustrate the structured organization and sequential execution of the algorithms implemented in this work, thereby facilitating a clearer understanding of the overall control strategy and the interactions between its components.

In general, the ADALINE algorithm with a conceptual model shown in Figure 2 is an approach to extract fundamental components of a periodic function. According to the Fourier series, a periodic function f can be represented as the sum of the sine and cosine functions. This representation allows any periodic signal to be approximated by a combination of these basis functions (sine and cosine waves) at different frequencies:

$$f(\theta) = W_0 + \sum_{n=1,2,3,\ldots}^{N} \left(W_{n(sin)} sin(n\theta) + W_{n(cos)} cos(n\theta) \right) \quad (3)$$

where W_0 is the DC offset term, typically zero for AC signals, and $W_{n(sin)}$ and $W_{n(cos)}$ are the weight factors (amplitude) of the sine and cosine terms for each harmonic n, to a maximum number N. Meanwhile, variable θ is defined as $\theta = \omega t + \varnothing$ with angular frequency ω, time t, and phase offset $\varnothing$.

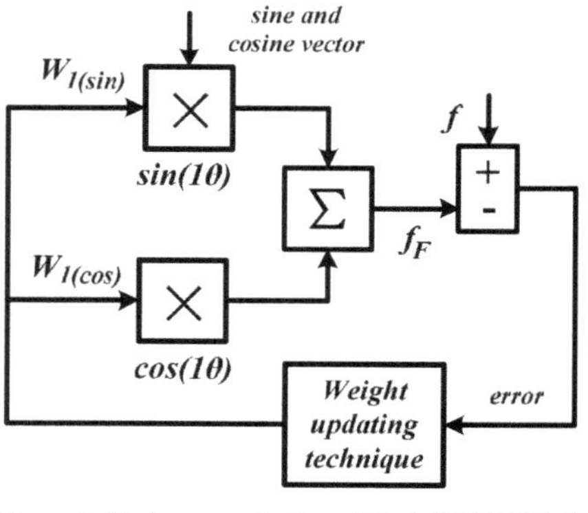

Figure 2. Basic conceptual model of ADALINE algorithm.

However, instead of trying to learn the whole periodic function f, a standard ADALINE algorithm focuses only on the fundamental frequency component f_F (the first harmonic, $n = 1$):

$$f_F = W_{1(sin)} sin(1\theta) + W_{1(cos)} cos(1\theta); n = 1 \quad (4)$$

The goal of the ADALINE algorithm is to find the weights $W_{1(sin)}$ and $W_{1(cos)}$ that best represent this fundamental component using a weight-updating technique. During the learning process, in each iteration, the ADALINE algorithm adjusts these weights based on the error (the difference between the estimated signal and the actual signal). This process uses a learning rate to control how much the weights are updated at each step. By the end of the learning process, the ADALINE algorithm gives the correct weights for the fundamental frequency component, which can then be used to recreate the main part of the original signal.

3.1. Characteristics of the Simplified ADALINE (S-ADALINE) Algorithm

Figure 3 illustrates a block diagram representing the operational concept of the S-ADALINE algorithm. As has already been clearly described in [21], the algorithm incorporates a weight-updating technique, called the fundamental active current (FAC). This approach dynamically adjusts the weight factor of the sine term based on the magnitude of the fundamental active current within the power system. Unlike the standard ADALINE algorithm that updates both the sine and cosine terms, the S-ADALINE algorithm only updates the sine term because it is assumed that the fundamental active current has only a sine component. The overall learning process performed by the FAC updating technique can be simplified as follows:

$$I_{F(updated)} = I_{F(present)} + (i_L - i_{Fest})\, \gamma_i \sin\theta \tag{5}$$

where I_F is the magnitude of the fundamental active current, i_L is the measured load current, i_{Fest} is the estimated fundamental active current, γ_i is the learning rate, and $sin\theta$ is the synchronization phase delivered by the phase-tracking algorithm.

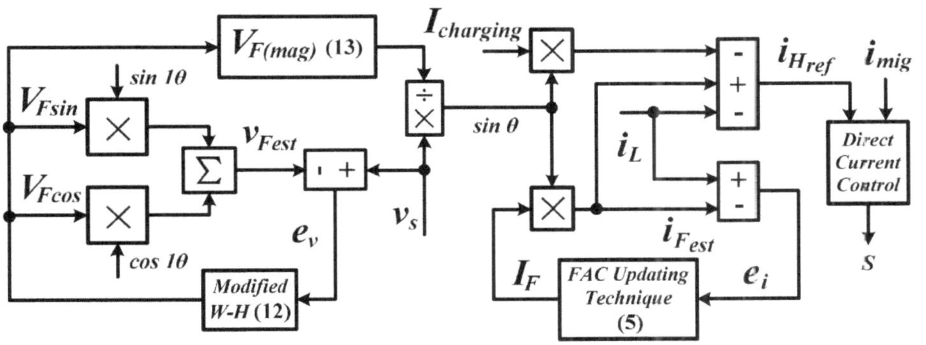

Figure 3. Reference current generation based on the existing S-ADALINE concept.

In each iteration, i_L will be compared with i_{Fest} and the resulted error e_i will be processed to update magnitude I_F which will then be applied in the next loop. However, the measured load current i_L contains harmonic components which cannot be fully described by a single sine component; thus, the error generated will be large. To handle this issue, a learning rate γ_i of 0.0001 is required [21]. Over successive iterations, the estimated i_{Fest} progressively aligns with the profile of the fundamental active current. Using the updated I_F, along with the synchronization phase, the charging current $I_{charging}$, and the measured load current i_L, a harmonic-based reference current i_{Href} can be derived as shown in Expression (6). This harmonic-based reference current necessitates the gate pulses S to be derived based on the direct current control (DCC) mechanism, which requires measuring the actual mitigation current i_{mig}.

$$i_{Href} = \left(I_F - I_{charging}\right)\sin\theta - i_L \tag{6}$$

On the other hand, to support the generation of the reference current, a phase-tracking algorithm based on the ADALINE concept is adopted. The complete operating principle of this algorithm has already been clearly described in [12,22]. In brief, the ADALINE-based phase-tracking algorithm utilizes a modified online W–H algorithm with a learning rate γ_v of 0.01 to incrementally update two weight factors $V_{F_{sin}}$ (sine term) and $V_{F_{cos}}$ (cosine term), which represent the magnitude factor V_F of the supply voltage v_s. The updating process can be summarized as follows:

$$V_{F(updated)} = V_{F(present)} + \gamma_v e_v \left[\frac{P}{P^T P}\right] \tag{7}$$

$$V_F = \begin{bmatrix} V_{F_{sin}} \\ V_{F_{cos}} \end{bmatrix} \tag{8}$$

$$P = \begin{bmatrix} sin1\theta \\ cos1\theta \end{bmatrix} \tag{9}$$

$$P^T P = \begin{bmatrix} sin1\theta & cos1\theta \end{bmatrix} \begin{bmatrix} sin1\theta \\ cos1\theta \end{bmatrix} = 1 \tag{10}$$

$$e_v = v_S - v_{Fest} \tag{11}$$

where e_v is the resulting error, P is the phase factor containing the fundamental sine ($sin1\theta$) and cosine ($cos1\theta$) terms of the supply voltage, and v_{Fest} is the estimated fundamental voltage. Together, Expression (7) can be finalized as follows:

$$V_{F(updated)} = V_{F(present)} + 0.01\,(v_S - v_{Fest}) \begin{bmatrix} sin1\theta \\ cos1\theta \end{bmatrix} \tag{12}$$

Eventually, the effective magnitude of the fundamental voltage can be obtained using Expression (13). Finally, the synchronization phase $sin\theta$ of the operating system is extracted by directly taking the unity representation of the supply voltage, according to Expression (14).

$$V_{F(mag)} = \sqrt{{V_{F_{sin}}}^2 + {V_{F_{cos}}}^2} \tag{13}$$

$$\sin\theta = \frac{v_S}{V_{F(mag)}} \tag{14}$$

As can be observed from Expression (14), the measured supply voltage v_s is directly applied to establish the synchronization phase without pre-filtering to remove any potential distortions in the utility grid. Hence, any distortions in the grid can disrupt the accurate determination of the synchronization phase, leading to a misalignment of the mitigating current injected by the SAHF. Resolving this misalignment is crucial to ensuring a stable and reliable system operation.

3.2. Proposed Modifications

This work proposes two key modifications to the existing S-ADALINE algorithm, resulting in the development of the I-ADALINE algorithm as presented in Figure 4. First, a new phase-tracking algorithm which utilizes orthogonality and the selective filtering concept is integrated, replacing the previous dependency of the ADALINE-based phase-tracking algorithm. This approach leverages the orthogonal properties of the signal components to enable the effective pre-filtering of unwanted disturbances, which results in a more reliable phase estimation under challenging environments. In operation, assuming the supply voltage measured from the utility grid is represented in the time-domain as $v_s(t) = V\ sin\ (\omega t)$, two orthogonal periodic signals can be generated by delaying the measured $v_s(t)$ by a quarter of its period T to create a phase-shifted version of the signal. Mathematically, the following relationship holds:

$$\begin{bmatrix} v_s \angle \theta \\ v_s \angle (\theta - 90^\circ) \end{bmatrix} = \begin{bmatrix} v_s(t) \\ v_s(t - \frac{T}{4}) \end{bmatrix} = \begin{bmatrix} V\ sin\ (\omega t) \\ V\ sin\ (\omega t - \frac{\pi}{2}) \end{bmatrix} = \begin{bmatrix} V\ sin\ (\omega t) \\ -\ V\ cos\ (\omega t) \end{bmatrix} \quad (15)$$

where $v_s \angle \theta$ and $v_s \angle (\theta - 90^\circ)$ represent two orthogonal periodic signals of the measured supply voltage.

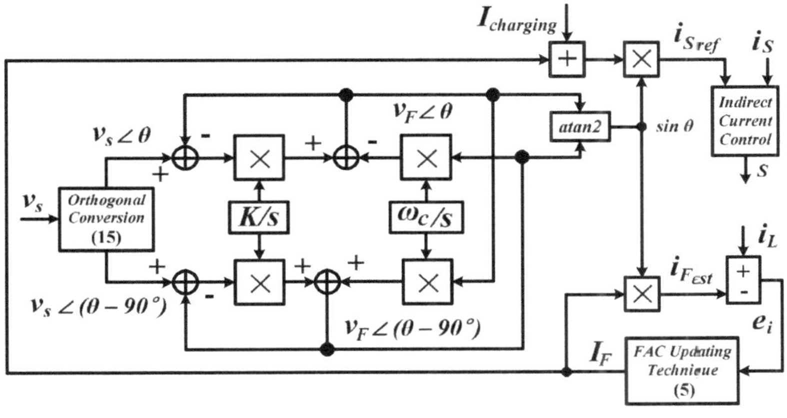

Figure 4. Reference current generation based on the proposed I-ADALINE concept.

A selective filter is then applied to extract the fundamental components from these orthogonal signals. To obtain the transfer function of the selective filter, the integration process of the input orthogonal periodic signals needs to be studied as follows:

$$C_{\alpha\beta}(t) = e^{j\omega_c t} \int e^{-j\omega_c t} R_{\alpha\beta}(t) dt. \quad (16)$$

where $R_{\alpha\beta}(t)$ and $C_{\alpha\beta}(t)$ are used to represent the input and output orthogonal signals of the integration process with angular frequency ω_c. Next, by applying the Laplace transformation to convert the expression from the time-domain to the s-domain, and by presenting them as a ratio of the output to input, the resulting transfer function $H(s)$ is obtained.

$$H(s) = \frac{C_{\alpha\beta}(s)}{R_{\alpha\beta}(s)} = \frac{C_\alpha(s) + jC_\beta(s)}{R_\alpha(s) + jR_\beta(s)} = \frac{s + j\omega_c}{s^2 + {\omega_c}^2} \quad (17)$$

To enable the adjustment of the selective filter's performance, a parameter known as selectivity factor K is introduced to the transfer function, which is reformulated as Expressions (18) and (19). It is important to note that a smaller value of K enhances the filtering selectivity but reduces the convergence speed, whereas a larger value improves the convergence speed at the expense of selectivity.

$$\frac{C_\alpha(s) + jC_\beta(s)}{R_\alpha(s) + jR_\beta(s)} = K\frac{(s + K) + j\omega_c}{(s + K)^2 + {\omega_c}^2} \quad (18)$$

$$C_\alpha(s) + jC_\beta(s) = K\left(\frac{(s+K)}{(s+K)^2 + \omega_c^2} + \frac{j\omega_c}{(s+K)^2 + \omega_c^2}\right)(R_\alpha(s) + jR_\beta(s)) \quad (19)$$

Next, by equating the real and imaginary sections, the following expressions can be obtained:

$$C_\alpha(s) = \frac{K(s+K)}{(s+K)^2 + \omega_c^2} R_\alpha(s) - \frac{K\omega_c}{(s+K)^2 + \omega_c^2} R_\beta(s) \quad (20)$$

$$C_\beta(s) = \frac{K(s+K)}{(s+K)^2 + \omega_c^2} R_\beta(s) + \frac{K\omega_c}{(s+K)^2 + \omega_c^2} R_\alpha(s) \quad (21)$$

Finally, by solving Expressions (20) and (21) simultaneously, the transfer function of the selective filter can be simplified as follows.

$$C_\alpha(s) = \frac{K}{s}(R_\alpha(s) - C_\alpha(s)) - \frac{\omega_c}{s} C_\beta(s) \quad (22)$$

$$C_\beta(s) = \frac{K}{s}(R_\beta(s) - C_\beta(s)) + \frac{\omega_c}{s} C_\alpha(s) \quad (23)$$

Here, to suit the model applied in this work for extracting the fundamental components from the supply voltage, the output orthogonal signal $C_\alpha(s)$ is replaced with $v_F \angle \theta$ and $C_\beta(s)$ is replaced with $v_F \angle (\theta - 90^\circ)$, while the input orthogonal signal $R_\alpha(s)$ is replaced with $v_s \angle \theta$ and $R_\beta(s)$ is replaced with $v_s \angle (\theta - 90^\circ)$. As a result, the final transfer function of the selective filter applied in this work can be summarized as follows:

$$v_F \angle \theta = \frac{K}{s}(v_s \angle \theta - v_F \angle \theta) - \frac{\omega_c \, v_F \angle (\theta - 90^\circ)}{s} \quad (24)$$

$$v_F \angle (\theta - 90^\circ) = \frac{K}{s}[v_s \angle (\theta - 90^\circ) - v_F \angle (\theta - 90^\circ)] + \frac{\omega_c \, v_F \angle \theta}{s} \quad (25)$$

where $v_F \angle \theta$ and $v_F \angle (\theta - 90^\circ)$ represent the two orthogonal periodic signals of the extracted fundamental components of the supply voltage with angular velocity ω_c = 100 π rad/s, and K = 20 is the selectivity factor applied in this work. Finally, the synchronization phase $sin\theta$ is determined using the trigonometry function below:

$$sin\theta = sin[atan2\,(v_F \angle \theta, v_F \angle (\theta - 90^\circ))] \quad (26)$$

In addition, to further enhance the mitigation performance via the minimization of switching ripples caused by the switching activities of the SAHF, the proposal involves transitioning from the DCC to the indirect current control (ICC) mechanism for the control of the SAHF. This transition requires adapting the existing S-ADALINE framework to align with the requirements of the ICC mechanism. Specifically, instead of generating a harmonic-based reference current, the modified algorithm will now produce a fundamental-based reference current i_{Sref}, as outlined in Expression (27).

$$i_{Sref} = \left(I_F + I_{charging}\right) \sin\theta \quad (27)$$

This fundamental-based reference current enables the generation of gate pulses S, which are derived from the measurement of the supply current i_S rather than the mitigation current i_{mig}. Unlike the DCC mechanism, the ICC mechanism operates by obtaining precise information about the switching ripples present in the supply current. Hence, any ripples contained in the supply current will be directly evaluated in the control system loop and eventually be minimized. In other words, this shift in approach improves the mitigation performance of the SAHF by directly addressing the limitations of the DCC mechanism.

3.3. Workflow of the Control Strategy

To facilitate a clear and structured understanding of how the various algorithms interact and operate sequentially in this work, an overview of the workflow is presented in Figure 5.

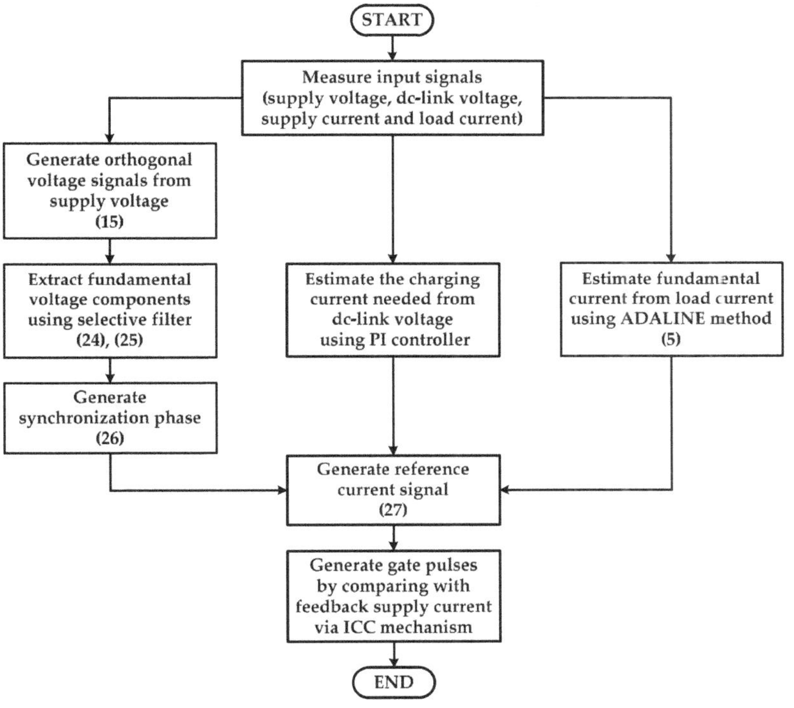

Figure 5. Overall workflow of the control algorithms applied.

As a summary, the control algorithms operate according to the following sequences:

(i) Measure the input signals (supply voltage v_s, DC-link voltage V_{dc}, supply current i_S, and load current i_L);

(ii) Generate orthogonal voltage signals $v_s \angle \theta$ and $v_s \angle (\theta - 90^\circ)$ from the measured supply voltage (Expression (15));

(iii) Extract fundamental voltage components $v_F \angle \theta$ and $v_F \angle (\theta - 90^\circ)$ using a selective filter (Expressions (24) and (25));

(iv) Generate synchronization phase $sin\theta$ (Expression (26));

(v) Estimate fundamental current i_F from the measured load current using the ADALINE method (Expression (5));

(vi) Estimate the charging current $I_{charging}$ from the measured DC-link voltage using a PI controller;

(vii) Generate reference current i_{Sref} (Expression (27));

(viii) Generate gate pulses S by comparing between the reference current and the measured supply current (feedback) via the ICC mechanism.

4. Results and Discussion

The single-phase SAHF system, along with its control algorithms and the newly proposed I-ADALINE method, is developed and simulated within the MATLAB/Simulink (R2012a) environment. The complete simulation model, which integrates the power circuit and control components, is illustrated in Figure 6. Meanwhile, the key simulation parameters used in the study are summarized in Table 1 for clarity.

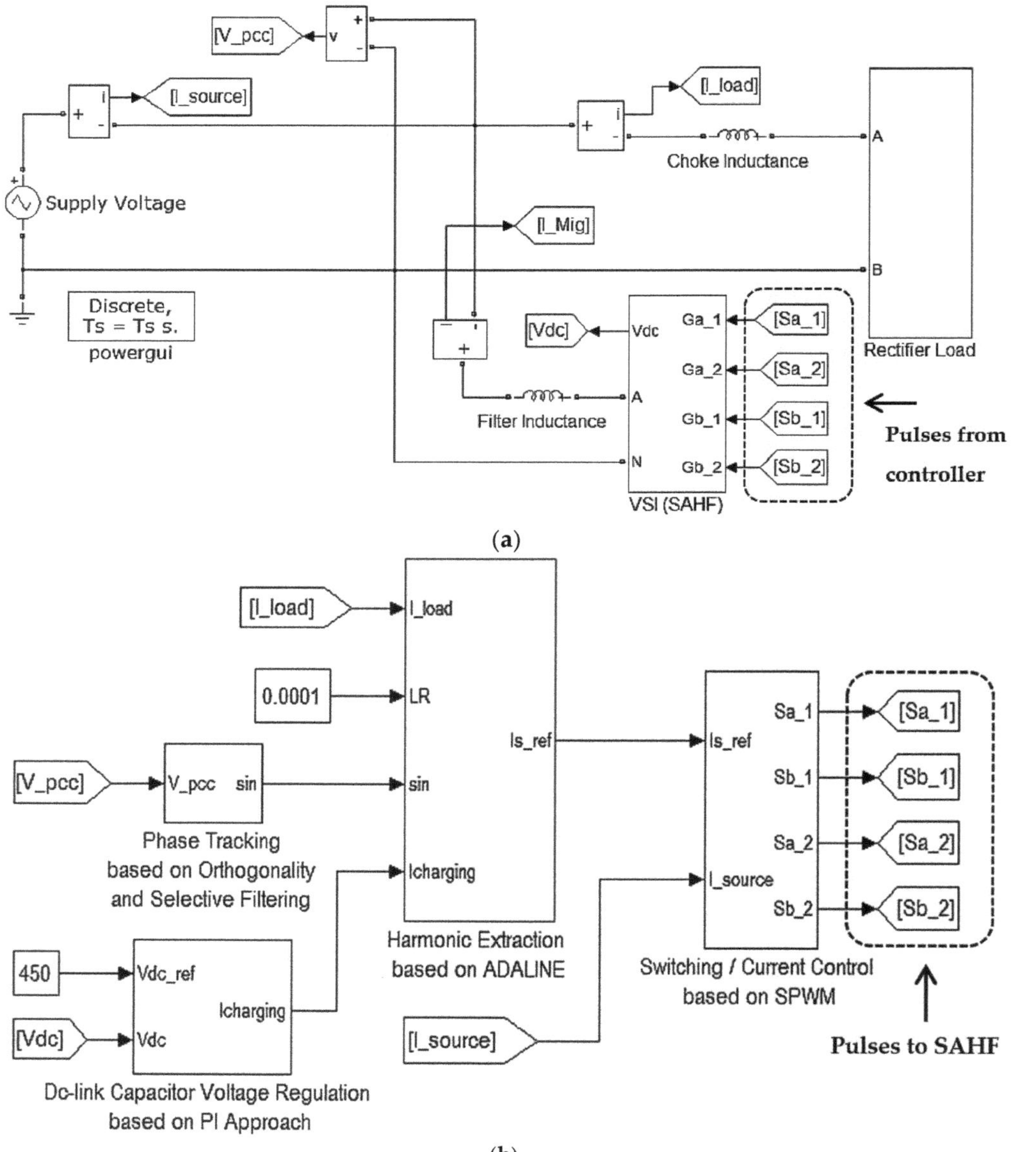

(b)

Figure 6. Simulation model of single-phase SAHF system showing arrangement of (**a**) power circuit and (**b**) control algorithms applied.

Table 1. Fundamental parameters used in the simulation of this study.

Parameters	Details
Sinusoidal Supply Voltage	Fundamental = 230 V (rms), 50 Hz
Distorted Supply Voltage (THD = 17.53%)	3rd harmonic = 13.04% 5th harmonic = 8.70% 7th harmonic = 6.52% 9th harmonic = 4.35%
Noise Disturbance (SNR = 20.89 dB)	3304 Hz component = 4.35% 4004 Hz component = 2.17% 5722 Hz component = 3.91% 7302 Hz component = 6.52%
Choke Inductance	3 mH
RC load: (Uncontrolled bridge rectifier supplying resistor-capacitor load)	R = 30 Ω, C = 470 μF
RL load: (Uncontrolled bridge rectifier supplying resistor-inductor load)	R = 10 Ω, L = 160 mH
Dc-link Capacitor/Desired Voltage	3300 μF/450 V
Filter Inductance	5 mH
Switching Frequency	5 kHz

In this setup, three types of supply voltages are applied, namely, sinusoidal, distorted, and distorted, with noise disturbance supply voltages. The sinusoidal supply voltage is set at a fundamental magnitude of 230 Vrms, operating at a frequency of 50 Hz. Meanwhile, the distorted supply voltage is characterized by a total harmonic distortion (THD) of 17.53%, indicating the presence of significant harmonic components. Finally, noise disturbance with a signal-to-noise ratio (SNR) of 20.89 dB is added to the distorted supply voltage, creating a challenging scenario that combines both harmonic distortion and noise interference. Two rectifier-based nonlinear loads are then connected to create a harmonic-rich environment. The first load, consisting of an uncontrolled bridge rectifier, a resistor and a capacitor, is named as the RC load. Meanwhile, the second load, which features a similar rectifier but with a resistor and an inductor, is named as the RL load.

To ensure a comprehensive evaluation, the proposed method is assessed under six scenarios, namely, Scenario A1 (sinusoidal supply and RC load), Scenario A2 (sinusoidal supply and RL load), Scenario B1 (distorted supply and RC load), Scenario B2 (distorted supply and RL load), Scenario C1 (distorted supply with noise and RC load), and Scenario C2 (distorted supply with noise and RL load). For benchmarking purposes, the existing S-ADALINE method is also re-simulated and tested under identical conditions. This allows for a direct comparison, highlighting the advantages of the proposed I-ADALINE method. In addition to these cases, a separate dynamic operation test is also conducted considering the load variation from RC to RL. Furthermore, a sensitivity analysis is performed to investigate the effect of the varying supply voltage total harmonic distortion (THD) and signal-to-noise ratio (SNR) on the performance of the proposed algorithm. These analyses aim to evaluate the robustness and adaptability of the I-ADALINE method under more challenging and realistic operating conditions, ensuring its practical applicability in dynamic power system environments.

First, the phase-tracking performance of each method is evaluated under four distinct conditions: the presence of harmonic distortion, the presence of a 20.89 dB noise disturbance, the presence of both harmonics and noise disturbance, and a 40° phase jump. Each method is tested for its ability to accurately track the desired phase under these challenging conditions, offering valuable insights into their respective strengths and limitations when subjected to complex signal disturbances. The resulting phase information extracted by both methods is shown in Figure 7, with a detailed comparison highlighted in Table 2. As presented in Figure 7, both methods are revealed to be effective when dealing with the sinusoidal supply voltage. However, when the supply voltage is affected by harmonic distortion and noise (starting from time = 1 s), the phase information extracted by the existing S-ADALINE method can be observed to display undesired ripples and noise components, thereby undermining its reliability.

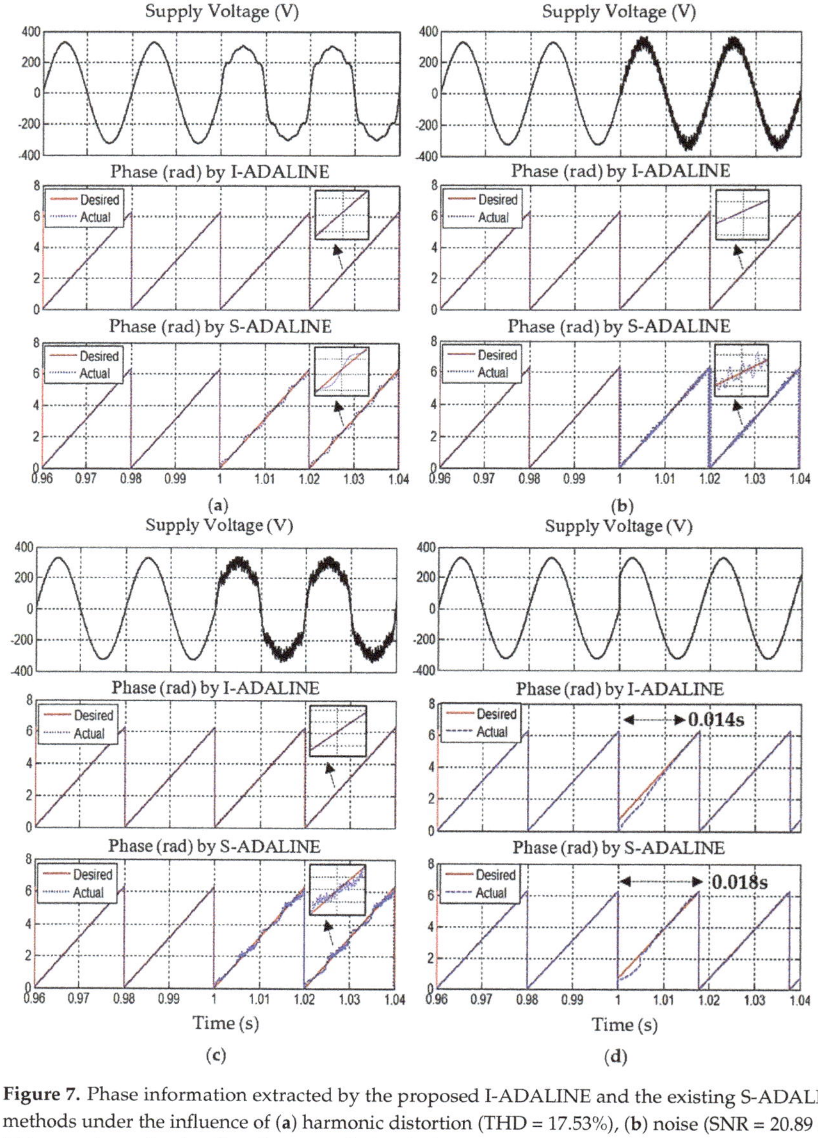

Figure 7. Phase information extracted by the proposed I-ADALINE and the existing S-ADALINE methods under the influence of (**a**) harmonic distortion (THD = 17.53%), (**b**) noise (SNR = 20.89 dB), (**c**) harmonics and noise disturbance, and (**d**) 40° phase jump.

Table 2. Comparison of phase tracking performance between the proposed I-ADALINE and the existing S-ADALINE methods.

Performance Aspect	I-ADALINE	S-ADALINE
Effectiveness in pure condition (Sinusoidal Input)	Performs effectively under sinusoidal input.	Performs effectively under sinusoidal input.
Response to harmonic distortion (THD = 17.53%)	Performs effectively despite the presence of harmonic distortion.	Fails to suppress harmonic distortion (evident by visible ripples).
Response to noise (SNR = 20.89 dB)	Performs effectively despite the presence of noise.	Fails to suppress noise (evident by visible noise).
Response to combine effect of harmonic distortion and noise	Performs effectively despite the presence of combined harmonics and noise.	Fails to suppress the combined effect of harmonics and noise (visible ripples and noise).
Response to phase jump (40° Phase Jump)	Recovery within 0.014 s.	Recovery within 0.018 s.
Adaptability	Adjust effectively to changes like harmonic distortion and noise (High Adaptability).	Adjust poorly to conditions like harmonic distortion and noise (Limited Adaptability).

In contrast, despite the presence of harmonics and noise, the proposed I-ADALINE method is found to have effectively preserved the correct phase information, remaining free from any ripples or noise components. Moreover, when the supply voltage experiences a sudden phase jump of 40°, the proposed I-ADALINE method is observed to demonstrate superior performance by achieving a recovery time of 0.014 s. In comparison, the existing S-ADALINE method takes 0.018 s, making the proposed I-ADALINE method 0.004 s faster.

Based on this evaluation, it can be concluded that the proposed I-ADALINE method outperforms the existing S-ADALINE method in both phase-tracking accuracy and recovery speed. It offers high adaptability by maintaining accurate phase tracking even when the supply voltage is influenced by harmonic distortion and/or noise, while also providing a faster recovery time. This makes the proposed I-ADALINE method highly reliable in real-world applications, where disturbances are common. In contrast, the existing S-ADALINE method is constrained to operate only in disturbance-free environments and exhibits a slower performance with a longer recovery time when subjected to a phase shift.

Next, the proposed I-ADALINE method is evaluated across the six specified scenarios to determine its effectiveness in improving the mitigation performance of the SAHF, with outcomes directly compared to those of the existing S-ADALINE method. Figures 8–10 present the waveforms obtained from this evaluation, while Table 3 provides a detailed comparison to facilitate a comprehensive interpretation of the findings.

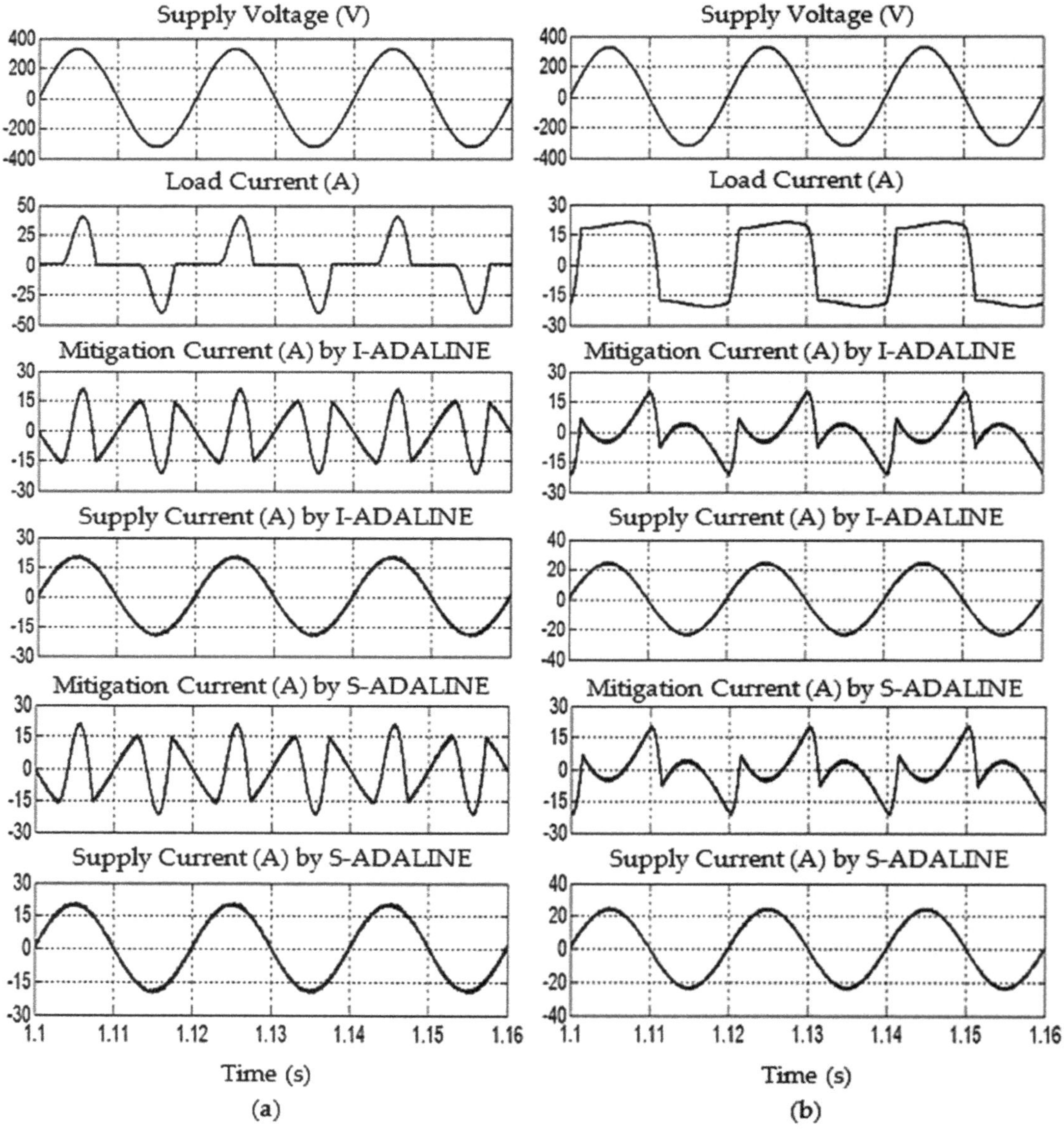

Figure 8. Steady-state simulation waveforms obtained by operation of SAHF integrated with the proposed I-ADALINE and SAHF integrated with the existing S-ADALINE methods under (**a**) Scenario A1 and (**b**) Scenario A2.

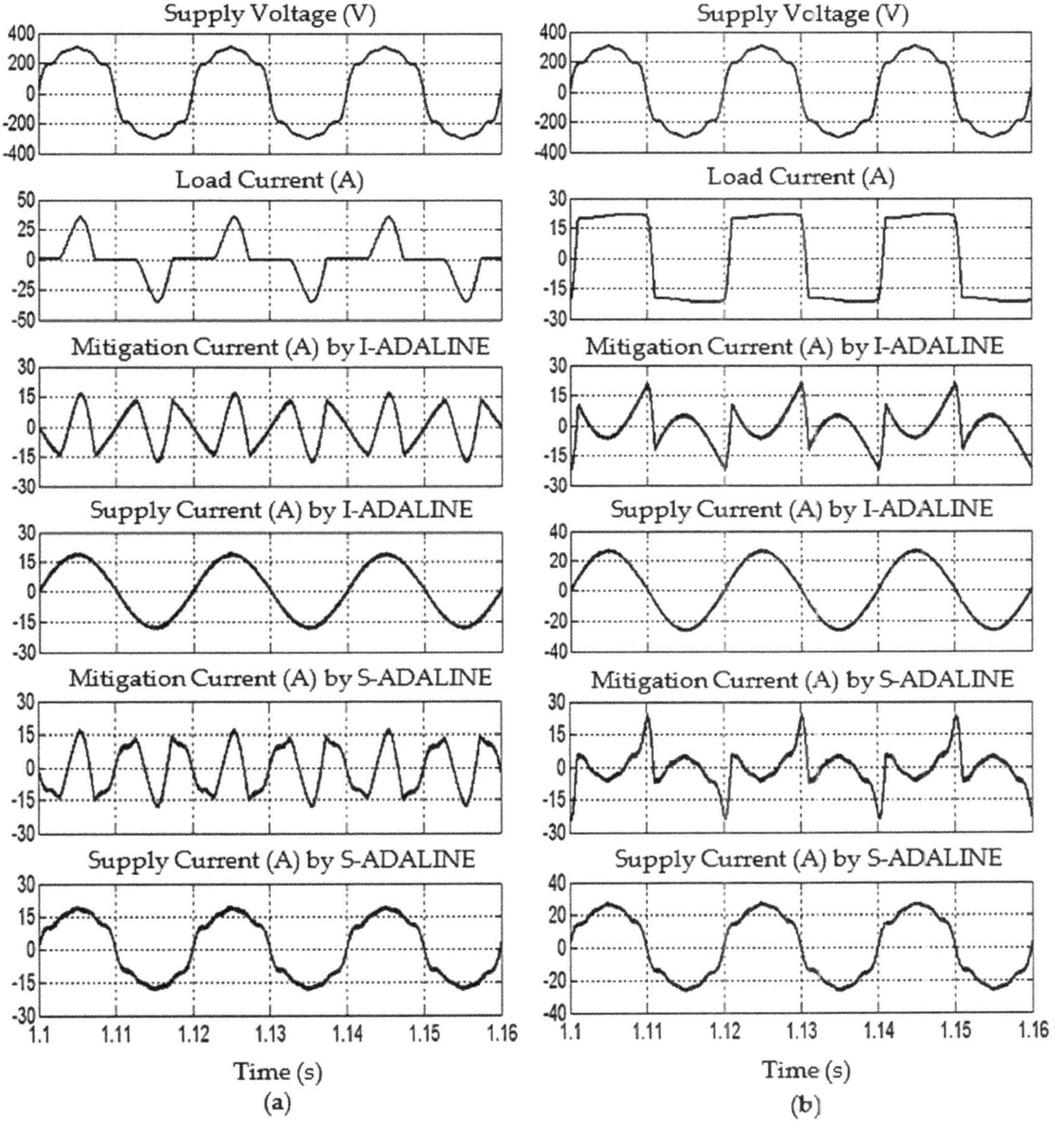

Figure 9. Steady-state simulation waveforms obtained by operation of SAHF integrated with the proposed I-ADALINE and SAHF integrated with the existing S-ADALINE methods under (**a**) Scenario B1 and (**b**) Scenario B2.

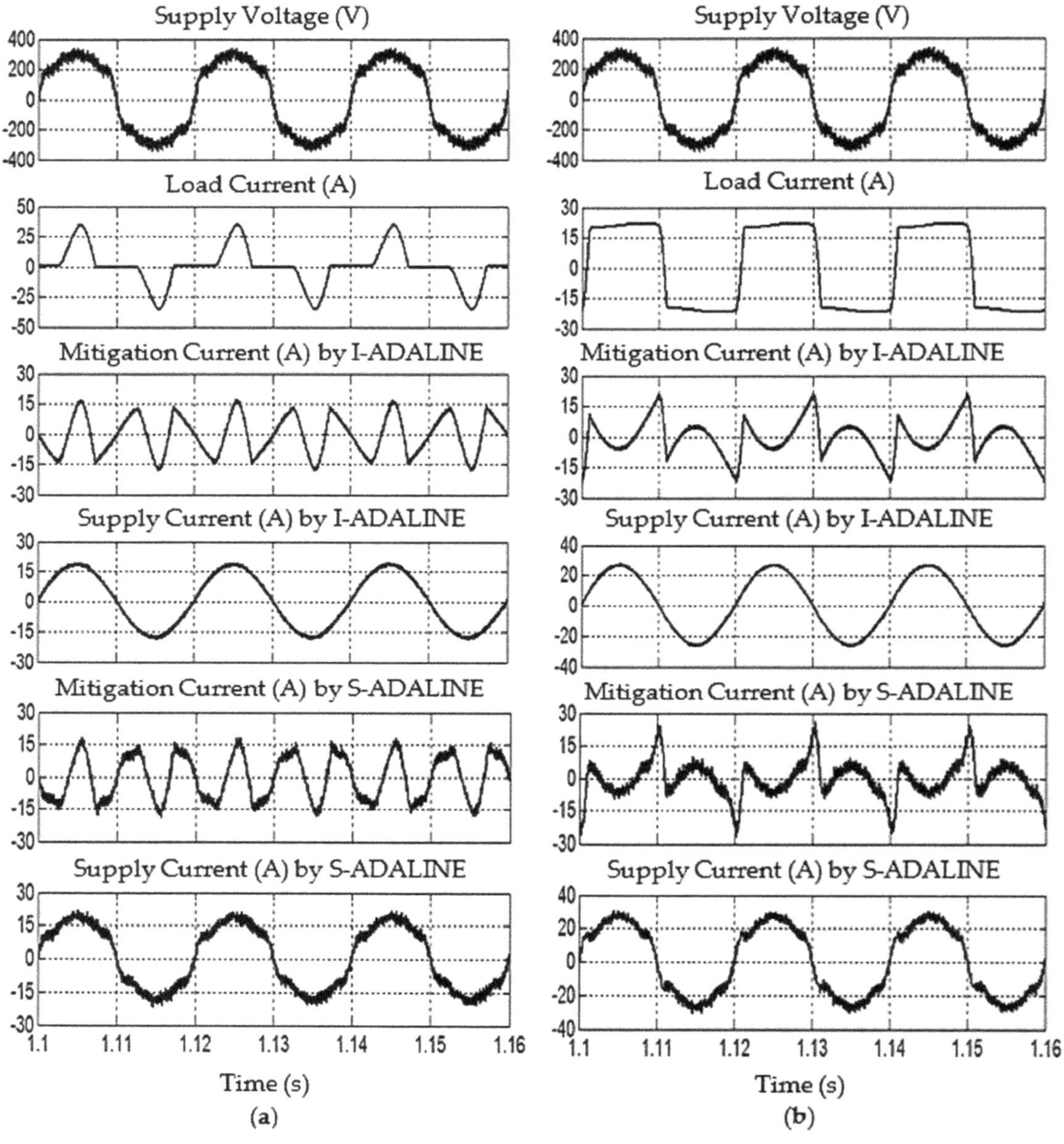

Figure 10. Steady-state simulation waveforms obtained by operation of SAHF integrated with the proposed I-ADALINE and SAHF integrated with the existing S-ADALINE methods under (**a**) Scenario C1 and (**b**) Scenario C2.

Table 3. Comparison of mitigation performance between the proposed I-ADALINE and the existing S-ADALINE methods.

Test Condition	THD (%)	Phase Shift (°)	Reactive Power (VAR)	Power Factor	Fundamental Current (A)	Accuracy of Fundamental Current (%)	Current Waveform
Before activation of SAHF							
Scenario A1	80.38	7.5	417.0	0.772	13.80	–	Distorted
Scenario A2	35.34	20.3	1248.3	0.884	17.83	–	Distorted
Scenario B1	73.06	4.3	220.7	0.805	12.77	–	Distorted
Scenario B2	39.57	14.0	1055.6	0.902	19.02	–	Distorted
Scenario C1	72.72	4.1	207.3	0.806	12.78	–	Distorted
Scenario C2	39.51	13.9	1050.1	0.903	19.03	–	Distorted
SAHF integrated with I-ADALINE method							
Scenario A1	3.19	0.2	12.5	0.999	13.95	98.92	Sinusoidal
Scenario A2	2.58	0.1	3.7	0.999	16.97	95.17	Sinusoidal
Scenario B1	3.64	0.2	10.4	0.999	12.98	98.38	Sinusoidal
Scenario B2	2.81	0.1	11.1	0.999	18.76	98.63	Sinusoidal
Scenario C1	3.71	0.2	10.2	0.999	12.99	98.38	Sinusoidal
Scenario C2	2.77	0.1	11.5	0.999	18.77	98.63	Sinusoidal
SAHF integrated with S-ADALINE method							
Scenario A1	3.25	0.2	12.5	0.999	13.95	98.92	Sinusoidal
Scenario A2	2.61	0.1	3.7	0.999	16.97	95.17	Sinusoidal
Scenario B1	14.41	1.4	73.6	0.989	13.38	95.44	Distorted
Scenario B2	14.45	1.0	76.7	0.989	19.32	98.44	Distorted
Scenario C1	15.77	1.2	68.5	0.987	13.81	92.54	Distorted
Scenario C2	16.55	1.0	78.5	0.986	20.10	94.67	Distorted

From Figure 8, the sinusoidal waveform of all supply currents and their in-phase alignment with the supply voltage clearly demonstrate that both methods successfully directed the connected SAHF to restore waveform symmetry through effective harmonic mitigation and reactive power compensation under Scenarios A1 and A2. As recorded in Table 3, both methods reduced the supply current THD value to below 5%, thereby complying with the IEEE 519 standard [33,34], and ensured a near-unity power factor by minimizing the reactive power. This contributes to maintaining a symmetrical power flow between the supply and the load. Meanwhile, in terms of fundamental current tracking, both methods demonstrated comparable accuracy, achieving 98.92% and 95.17% in Scenarios A1 and A2, respectively. However, due to the improved switching ripple reduction in the proposed I-ADALINE method enabled by the ICC mechanism, it achieved slightly better waveform symmetry, with an additional THD reduction of 0.06% and 0.03% in Scenarios A1 and A2, respectively. These results highlight the ability of the proposed method to enhance signal symmetry under sinusoidal operating conditions.

On the other hand, Figure 9 clearly shows that the SAHF integrated with the existing S-ADALINE method failed to restore the sinusoidal symmetrical shape of the supply current in both Scenarios B1 and B2. As reported in Table 3, the THD values obtained are 14.41% and 14.45%, respectively, both exceeding the 5% limit specified in the IEEE 519 standard [33,34]. This elevated distortion reflects a persistent asymmetry in the current waveform. Moreover, the existing method is also found to cause a larger phase shift (1.4° in Scenario B1 and 1.0° in Scenario B2) and increased reactive power (73.6 VAR in Scenario B1 and 76.7 VAR in Scenario B2), which collectively degrade the power factor to 0.989. Next, the fundamental current tracking accuracy is also observed to be lower, with values of 95.44% and 98.44% in Scenarios B1 and B2, respectively.

In contrast, integrating the proposed I-ADALINE method enabled the connected SAHF to consistently preserve the sinusoidal and symmetrical shape of the supply current under the same distorted conditions. Specifically, the proposed method significantly reduces the THD values to 3.64% and 2.81% in Scenarios B1 and B2, respectively, restoring waveform symmetry and improving the power quality. Moreover, the proposed method also achieves substantially smaller phase shifts (0.2° in Scenario B1 and 0.1° in Scenario B2) and reactive power (10.4 VAR in Scenario B1 and 11.1 VAR in Scenario B2), thereby enhancing the power factor to 0.999 and promoting a more symmetrical power exchange. The fundamental current tracking accuracy has also improved, reaching 98.38% and 98.63% for Scenarios B1 and B2, respectively, which further confirms the effectiveness of the proposed method in maintaining waveform symmetry under distorted supply conditions.

Finally, under the additional influence of noise, as can be observed from Figures 10 and 11, the SAHF integrated with the existing S-ADALINE method again failed to restore the sinusoidal and symmetrical shape of the supply current in both Scenarios C1 and C2. This resulted in higher THD values of 15.77% and 16.55%, respectively, indicating persistent waveform asymmetry. Additionally, a larger phase shift (1.2° in Scenario C1 and 1.0° in Scenario C2) and increased reactive power (68.5 VAR in Scenario C1 and 78.5 VAR in Scenario C2) are also observed from the existing method, causing the power factor to degrade further to 0.987 and 0.986, respectively. More importantly, the added impact of noise has led to the lowest fundamental current tracking accuracy of 92.54% and 94.67% for Scenarios C1 and C2, respectively.

In contrast, the proposed I-ADALINE method enabled the connected SAHF to consistently maintain the sinusoidal and symmetrical shape of the supply current despite the presence of noise. The recorded THD values are significantly reduced to 3.71% and 2.77% for Scenarios C1 and C2, respectively, effectively restoring waveform symmetry and improving the power quality. Furthermore, the proposed method consistently achieved smaller phase shifts (0.2° in Scenario C1 and 0.1° in Scenario C2) and a lower reactive power (10.2 VAR in Scenario C1 and 11.5 VAR in Scenario C2), once again resulting in an improved power factor of 0.999. The fundamental current tracking accuracy is also substantially enhanced, reaching 98.38% and 98.63% for Scenarios C1 and C2, further demonstrating the robustness of the I-ADALINE method in preserving waveform symmetry under noisy and distorted conditions.

In addition to steady-state performance, a dynamic load test is carried out to evaluate the system behavior under varying load conditions, specifically involving a transition from the RC to RL load and a distorted supply with noise disturbance, as illustrated in Figure 12. As observed, following the load change at time = 1 s, both the I-ADALINE and S-ADALINE methods exhibit comparable dynamic responses, with recovery times within 0.06 s. This similarity is anticipated, as both methods utilize the same FAC updating technique and learning rate within their ADALINE frameworks. However, due to the incorporation of selective filtering and the ICC mechanism in the proposed I-ADALINE method, the quality of the mitigated supply current is significantly enhanced, exhibiting a continuous, sinusoidal, and symmetrical waveform with minimal distortion. In contrast, the output from the S-ADALINE method remains distorted and shows a significant level of noise. This demonstrates that the proposed enhancements in the I-ADALINE method effectively improve the output quality without compromising dynamic performance.

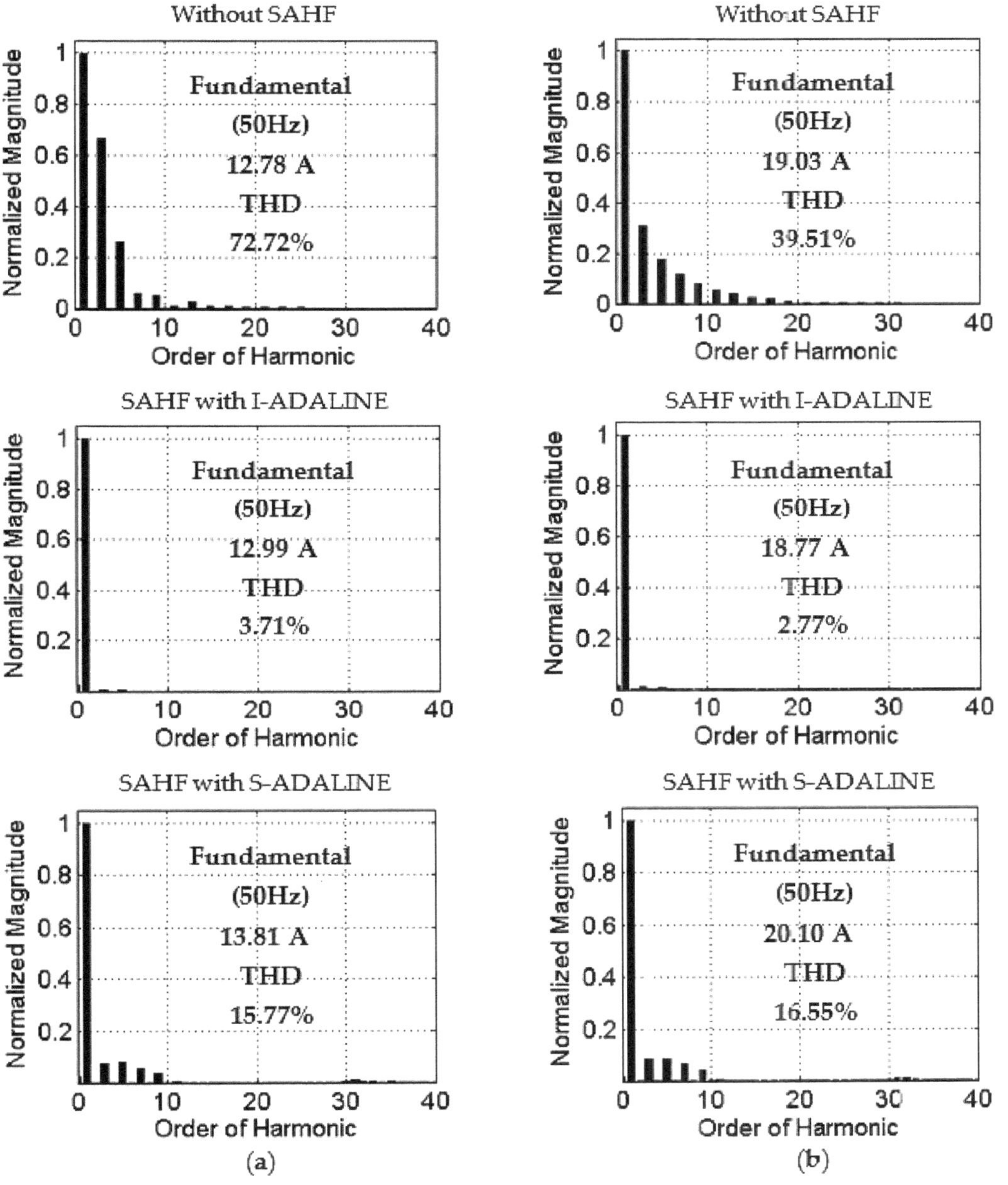

Figure 11. Harmonic spectrums of supply current obtained under Scenarios (**a**) C1 and (**b**) C2.

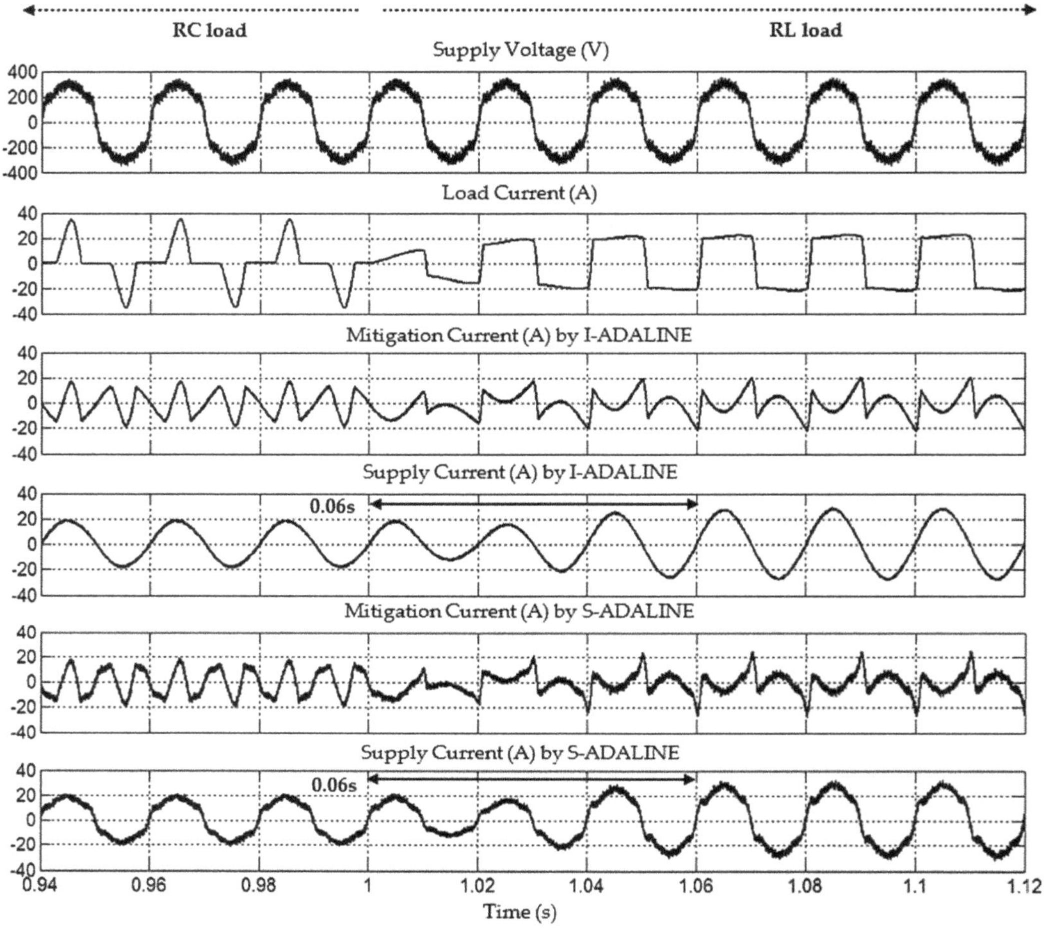

Figure 12. Dynamic-state simulation waveforms of SAHF integrated with the proposed I-ADALINE and SAHF integrated with the existing S-ADALINE methods, considering load variation from RC to RL and distorted supply condition with noise disturbance.

From the comparative analysis performed, the proposed I-ADALINE method demonstrates clear advantages over the existing S-ADALINE method. The key improvements, as highlighted in Table 4, include a significant reduction in %THD (by 10.77–13.78%), a decrease in phase shift (by 0.9–1.2°), and a substantial drop in reactive power (by 58.3 VAR–67 VAR). These enhancements contribute to an increase in power factor values (by 0.010 to 0.013). Furthermore, the proposed method also achieves better fundamental current tracking which improved the accuracy (by 0.19% to 5.84%). Importantly, the I-ADALINE method consistently restores and maintains the sinusoidal and symmetrical shape of the supply current waveform, effectively mitigating the harmonic distortion and reactive power caused by various nonlinear loads, even when the supply is affected by severe harmonic distortion and noise disturbances. This results in a superior overall power quality and more symmetrical power flow between the supply and load, demonstrating the robustness and effectiveness of the proposed method in challenging operating environments.

Table 4. Highlights of benefits offered by the proposed I-ADALINE in comparison to the existing S-ADALINE methods.

Performance Parameter	Scenario B1	Scenario B2	Scenario C1	Scenario C2
Reduction in THD (%)	10.77	11.64	12.06	13.78
Reduction in phase shift (°)	1.2	0.9	1.0	0.9
Reduction in reactive power (VAR)	63.2	65.6	58.3	67.0
Improvement in power factor	0.010	0.010	0.012	0.013
Accuracy of fundamental current	Improved by 2.94%	Improved by 0.19%	Improved by 5.84%	Improved by 3.96%
Quality of current waveform	Improved	Improved	Improved	Improved

To further support the findings from the initial comparative analysis, a sensitivity study is added to examine how variations in the supply voltage total harmonic distortion (THD) and signal-to-noise ratio (SNR) influence the performance of the proposed I-ADALINE method. In this study, the supply voltage THD is adjusted between 5.16% and 25.16%, while the SNR is adjusted between 10.24 dB and 30.24 dB. The performance metrics evaluated include the THD of the supply current, the phase shift, and the power factor of the system. The results summarized in Tables 5 and 6 provide strong evidence of the superior performance of the proposed I-ADALINE method compared to the existing S-ADALINE approach, under a range of operating conditions.

Table 5. Comparison of mitigation performance between the proposed I-ADALINE and the existing S-ADALINE methods under varying levels of supply voltage distortion and noise disturbance, with an applied RC load.

Supply Voltage Condition	THD (%)		Phase Shift (°)		Power Factor	
	I-ADALINE	S-ADALINE	I-ADALINE	S-ADALINE	I-ADALINE	S-ADALINE
Distorted Supply Voltage—%THD Variation						
THD = 5.16%	3.26	5.53	0.2	0.6	0.999	0.998
THD = 10.16%	3.40	9.99	0.2	0.6	0.999	0.995
THD = 15.16%	3.61	11.94	0.2	1.3	0.999	0.992
THD = 20.16%	3.85	15.16	0.2	1.7	0.999	0.988
THD = 25.16%	4.03	19.17	0.3	2.0	0.999	0.981
Distorted Supply Voltage (THD = 17.53%) added with noise disturbance—SNR Variation						
SNR = 30.24 dB	3.66	14.68	0.2	1.3	0.999	0.989
SNR = 25.24 dB	3.69	14.98	0.2	1.3	0.999	0.988
SNR = 20.24 dB	3.78	16.26	0.2	1.3	0.999	0.986
SNR = 15.24 dB	3.94	19.72	0.1	1.2	0.999	0.980
SNR = 10.24 dB	4.20	22.43	0.1	1.3	0.999	0.975

Table 6. Comparison of mitigation performance between the proposed I-ADALINE and the existing S-ADALINE methods under varying levels of supply voltage distortion and noise disturbance, with an applied RL load.

Supply Voltage Condition	THD (%)		Phase Shift (°)		Power Factor	
	I-ADALINE	S-ADALINE	I-ADALINE	S-ADALINE	I-ADALINE	S-ADALINE
Distorted Supply Voltage—%THD Variation						
THD = 5.16%	2.46	5.11	0.1	0.3	0.999	0.998
THD = 10.16%	2.55	9.90	0.1	0.3	0.999	0.995
THD = 15.16%	2.63	11.91	0.1	1.0	0.999	0.992
THD = 20.16%	2.90	15.15	0.1	1.3	0.999	0.988
THD = 25.16%	3.31	19.71	0.2	1.4	0.999	0.980
Distorted Supply Voltage (THD = 17.53%) added with noise disturbance—SNR Variation						
SNR = 30.24 dB	2.59	14.84	0.1	1.0	0.999	0.989
SNR = 25.24 dB	2.61	15.44	0.1	0.9	0.999	0.988
SNR = 20.24 dB	2.81	16.86	0.1	1.0	0.999	0.985
SNR = 15.24 dB	2.93	19.23	0.2	1.1	0.999	0.981
SNR = 10.24 dB	3.25	20.74	0.3	1.6	0.999	0.978

One of the key observations from the study is that, as the THD in the supply voltage increases, there is a corresponding increase in the THD of the supply current. This trend is observed for both the RC and RL load conditions and is consistent with the behavior expected in power systems where harmonic-rich voltage sources tend to propagate distortion into the load current. Despite this natural increase in current distortion, the I-ADALINE method demonstrates a remarkable ability to contain and mitigate its effects. For example, in the case of the RC load, the supply current THD with the I-ADALINE method increases only slightly from 3.26% to 4.03% when the supply voltage THD increases from 5.16% to 25.16%. In contrast, the S-ADALINE method shows a much steeper rise in the current THD, from 5.53% to 19.17% over the same range. A similar pattern is evident with the RL load, where the I-ADALINE method keeps the current THD below 3.31% even at the highest distortion level, while the current THD escalates to 19.71% for the S-ADALINE method. These results clearly demonstrate that, while current distortion does increase with worsening voltage conditions, the proposed I-ADALINE method is significantly more effective at suppressing this distortion compared to the existing S-ADALINE method.

In addition to the improved harmonic mitigation, the proposed I-ADALINE method also shows superior performance in controlling the phase shift between the voltage and current, which is critical for maintaining synchronization and minimizing reactive power flow. Across all test conditions, the I-ADALINE method maintains the phase shift within a narrow range, typically below 0.3°, while the S-ADALINE method experiences larger deviations, especially under high distortion or noise conditions. This phase accuracy contributes to the near-unity power factor achieved by the I-ADALINE method, which remains at 0.999 in all cases. In contrast, the power factor associated with the S-ADALINE method degrades with increasing distortion, dropping to as low as 0.981 and 0.980 in the RC and RL load conditions, respectively.

Furthermore, the robustness of the I-ADALINE method under noise disturbances is evident from its stable performance across a wide range of SNRs, from 30.24 dB down to 10.24 dB. Even in the presence of significant noise, the supply current THD, phase shift, and power factor remain largely unaffected with the I-ADALINE method, indicating a strong adaptability and lower sensitivity to disturbances. On the other hand, the S-ADALINE method exhibits an obvious decline in performance under the same noise levels, indicating a greater sensitivity to disturbances.

As a summary, by utilizing the advantages of selective filtering and the ICC mechanism, the proposed I-ADALINE method has demonstrated clear and consistent superiority over the existing S-ADALINE method in mitigating the harmonic distortion, minimizing the phase shift, and maintaining the near-unity power factor under a wide range of supply voltage distortions and noise disturbances. Consequently, the proposed method reliably preserves the sinusoidal and symmetrical supply current waveforms, even under severe distortion and challenging operating conditions. This results in an enhanced power quality and more efficient energy transfer between the supply and load, making the approach a robust and dependable solution for modern power systems facing nonlinear and variable conditions.

Future work could explore the integration of the I-ADALINE method with other advanced machine-learning methods such as deep learning or reinforced learning to further enhance its performance in dynamic and complex systems. These AI-driven approaches offer promising capabilities for self-adaptation, real-time learning, and increased resilience against unpredictable disturbances, which are particularly valuable in dynamic and complex power systems. While this study has focused on enhancing the ADALINE-based approach, future work could include a comparative analysis between I-ADALINE and other neural network models such as artificial neural networks (ANNs), recurrent neural networks (RNNs), long short-term memory (LSTM) networks, convolutional neural networks (CNNs), or hybrid models. Each of these models exhibits unique strengths in handling nonlinearities and dynamic behavior. Benchmarking them against I-ADALINE would provide deeper insight into the relative advantages and help identify the most suitable architecture for specific power quality applications.

Additionally, the application of the proposed method could be further validated under real-world conditions involving a broader range of power quality disturbances. This includes scenarios with significant voltage distortion, supply frequency fluctuations, dynamic load behaviors, and the intermittent nature of renewable energy sources. Evaluating the method under such diverse conditions would enhance its practical relevance and deployment readiness. The sensitivity analysis also represents an important area for future study. Investigating how variations in key system parameters and initial conditions affect the performance of the I-ADALINE method would provide critical insights into its robustness and scalability for real-world implementation.

Another promising research direction involves extending the I-ADALINE framework with diagnostic capabilities, such as the automatic classification and source identification of power quality disturbances. As modern power systems become increasingly populated with power electronic converters, electric vehicle chargers, and nonlinear industrial equipment, the ability to identify which specific consumer or load is responsible for the power quality degradation becomes essential. Such features would enable more targeted and effective mitigation strategies, allowing grid operators to proactively manage and maintain regulatory compliance in increasingly complex electrical networks.

Building on these directions, the most immediate and practical continuation of this work is the development of a laboratory-scale prototype that implements the I-ADALINE method within a physical power system environment. Such an experimental setup would facilitate the validation of the proposed approach under real-world operating conditions, accounting for factors such as electrical noise, hardware limitations, and unpredictable disturbances, which are often challenging to replicate accurately in simulation environments.

In summary, future research should focus on enhancing the I-ADALINE framework through advanced machine-learning integration, a comparative analysis with other neural models, sensitivity evaluation, and diagnostic capabilities such as disturbance classification. The immediate step is developing a laboratory-scale prototype to validate real-time performance under realistic conditions. These efforts will help bridge the gap between simulation and deployment, supporting the adoption of intelligent, adaptive solutions for modern power quality management.

5. Conclusions

This study proposes an approach called I-ADALINE, which aims to enhance synchronization and improve the mitigation performance of SAHF under challenging grid environments. It is basically achieved by incorporating an alternative phase-tracking approach that builds upon the concept of orthogonality and selective filtering, while also utilizing the benefits of the ICC mechanism. Together, these elements enable the method to address harmonic distortion, accurately generate the reference current, and preserve current waveform symmetry, which are crucial for maintaining the overall power quality. Extensive simulation tests, which include scenarios of highly nonlinear loads, voltage distortions, and noise disturbances, were conducted to validate the operational concept and assess the performance of the proposed method.

A comparative analysis against the existing S-ADALINE method demonstrates that the proposed I-ADALINE method offers notable improvements in both the phase-tracking accuracy and adaptability. These advantages are particularly evident under adverse conditions such as harmonic distortion, noise disturbance, and phase jumps. More importantly, the I-ADALINE method is also revealed to deliver superior mitigation performance for hte SAHF, as evidenced by an additional reduction in %THD (by 10.77–13.78%), a further decrease in reactive power (by 58.3 VAR–67 VAR), improved grid synchronization with a smaller phase shift (by 0.9–1.2°), and an improvement in the tracking accuracy of the fundamental current (by 0.19–5.84%). These findings confirm the effectiveness of I-ADALINE in maintaining power quality and waveform symmetry under challenging grid environments.

However, it is important to note a few limitations of the proposed method. While it demonstrates reliable performance under distorted but frequency-stable conditions, its ability to maintain precise phase angle tracking under time-varying and highly dynamic grid frequency conditions remains limited. This highlights the need for further enhancements to improve its robustness against frequency fluctuations. Another limitation is that the performance can be affected by changes in system conditions and settings. For example, different loads or disturbances may require parameter adjustments to maintain optimal results, which was not fully studied in this work. The proposed method is particularly suited for industrial and commercial applications involving nonlinear loads such as variable speed drives, rectifiers, and renewable energy interfacing, where the power quality is critical. It also shows promise for residential environments with the growing penetration of power electronics. Future work will focus on extending the I-ADALINE framework by incorporating advanced machine-learning techniques, developing diagnostic capabilities for disturbance classification and source identification, conducting sensitivity analyses, and progressing toward laboratory-scale prototyping. These efforts aim to bridge the gap between simulation and practical deployment, enhancing the adaptability and resilience in increasingly complex power systems.

Author Contributions: Conceptualization, Y.H.; methodology, Y.H. and K.W.C.; software, Y.H. and K.W.C.; validation, Y.H., K.W.C. and M.A.M.R.; formal analysis, Y.H.; investigation, Y.H. and K.W.C.;

resources, Y.H., K.W.C. and M.A.M.R.; writing—original draft preparation, Y.H.; writing—review and editing, Y.H., K.W.C. and M.A.M.R.; supervision, M.A.M.R.; project administration, Y.H. and K.W.C.; funding acquisition, Y.H. and K.W.C. All authors have read and agreed to the published version of the manuscript.

Funding: This research was supported by the Ministry of Higher Education, Malaysia (MoHE), through the Fundamental Research Grant Scheme (FRGS/1/2024/TK08/UTAR/02/4).

Data Availability Statement: The data is contained within the article.

Conflicts of Interest: The authors declare no conflicts of interest.

Abbreviations

The following abbreviations are used in this manuscript:

ADALINE	Adaptive linear neuron
ANN	Artificial neural network
DCC	Direct current control
FAC	Fundamental active current
ICC	Indirect current control
PCC	Point of common coupling
SAHF	Shunt-typed active harmonic filter
SNR	Signal-to-noise ratio
THD	Total harmonic distortion
VSI	Voltage source inverter

References

1. Lumbreras, D.; Gálvez, E.; Collado, A.; Zaragoza, J. Trends in power quality, harmonic mitigation and standards for light and heavy industries: A review. *Energies* **2020**, *13*, 5792. [CrossRef]
2. Smadi, I.A.; Albatran, S.; Alsyouf, M.A. Power quality improvement of a class of reduced device count inverter. *Simul. Model. Pract. Theory* **2019**, *96*, 101939. [CrossRef]
3. Li, D.; Wang, T.; Pan, W.; Ding, X.; Gong, J. A comprehensive review of improving power quality using active power filters. *Electr. Power Syst. Res.* **2021**, *199*, 107389. [CrossRef]
4. Imam, A.A.; Kumar, R.S.; Al-Turki, Y.A. Modeling and simulation of a pi controlled shunt active power filter for power quality enhancement based on p-q theory. *Electronics* **2020**, *9*, 637. [CrossRef]
5. Asadi, Y.; Eskandari, M.; Mansouri, M.; Chaharmahali, S.; Moradi, M.H.; Tahriri, M.S. Adaptive neural network for a stabilizing shunt active power filter in distorted weak grids. *Appl. Sci.* **2022**, *12*, 8060. [CrossRef]
6. Hasan, K.; Othman, M.M.; Meraj, S.T.; Mekhilef, S.; Abidin, A.F.B. Shunt active power filter based on savitzky-golay filter: Pragmatic modelling and performance validation. *IEEE Trans. Power Electron.* **2023**, *38*, 8838–8850. [CrossRef]
7. Nour, A.M.M.; Helal, A.A. A voltage sensorless technique for a shunt active power filter adopting band pass filter under abnormal supply conditions. *Electr. Power Syst. Res.* **2025**, *238*, 111133. [CrossRef]
8. Hoon, Y.; Radzi, M.A.M.; Hassan, M.K.; Mailah, N.F. Control algorithms of shunt active power filter for harmonics mitigation: A review. *Energies* **2017**, *10*, 2038. [CrossRef]
9. Liang, X.; Andalib-Bin-Karim, C. Harmonics and mitigation techniques through advanced control in grid-connected renewable energy sources: A review. *IEEE Trans. Ind. Appl.* **2018**, *54*, 3100–3111. [CrossRef]
10. Green, T.C.; Marks, J.H. Control techniques for active power filters. *IEE Proc. Electr. Power Appl.* **2005**, *152*, 369–381. [CrossRef]
11. Qasim, M.; Kanjiya, P.; Khadkikar, V. Artificial-neural-network-based phase-locking scheme for active power filters. *IEEE Trans. Ind. Electron.* **2014**, *61*, 3857–3866. [CrossRef]
12. Hoon, Y.; Radzi, M.A.M.; Zainuri, M.A.A.M.; Zawawi, M.A.M. Shunt active power filter: A review on phase synchronization control techniques. *Electronics* **2019**, *8*, 791. [CrossRef]
13. Bhattacharya, A.; Chakraborty, C.; Bhattacharya, S. Shunt compensation. *IEEE Ind. Electron. Mag.* **2009**, *3*, 38–49. [CrossRef]
14. Campos-Gaona, D.; Peña-Alzola, R.; Monroy-Morales, J.L.; Ordonez, M.; Anaya-Lara, O.; Leithead, W.E. Fast selective harmonic mitigation in multifunctional inverters using internal model controllers and synchronous reference frames. *IEEE Trans. Ind. Electron.* **2017**, *64*, 6338–6349. [CrossRef]

15. Kumar, R.; Bansal, H.O.; Gautam, A.R.; Mahela, O.P.; Khan, B. Experimental investigations on particle swarm optimization based control algorithm for shunt active power filter to enhance electric power quality. *IEEE Access* **2022**, *10*, 54878–54890. [CrossRef]
16. Ouchen, S.; Betka, A.; Gaubert, J.-P.; Abdeddaim, S. Simulation and real time implementation of predictive direct power control for three phase shunt active power filter using robust phase-locked loop. *Simul. Model. Pract. Theory* **2017**, *78*, 1–17. [CrossRef]
17. Terriche, Y.; Guerrero, J.M.; Vasquez, J.C. Performance improvement of shunt active power filter based on non-linear least-square approach. *Electr. Power Syst. Res.* **2018**, *160*, 44–55. [CrossRef]
18. Sozanski, K.; Szczesniak, P. Advanced control algorithm for three-phase shunt active power filter using sliding dft. *Energies* **2023**, *16*, 1453. [CrossRef]
19. Iqbal, M.; Jawad, M.; Jaffery, M.H.; Akhtar, S.; Rafiq, M.N.; Qureshi, M.B. Neural networks based shunt hybrid active power filter for harmonic elimination. *IEEE Access* **2021**, *9*, 69913–69925. [CrossRef]
20. Janpong, S.; Areerak, K.; Areerak, K. Harmonic detection for shunt active power filter using adaline neural network. *Energies* **2021**, *14*, 4351. [CrossRef]
21. Zainuri, M.A.A.M.; Radzi, M.A.M.; Soh, A.C.; Mariun, N.; Rahim, N.A. Simplified adaptive linear neuron harmonics extraction algorithm for dynamic performance of shunt active power filter. *Int. Rev. Model. Simul.* **2016**, *9*, 144–154.
22. Rahman, N.F.A.; Radzi, M.A.M.; Soh, A.C.; Mariun, N.; Rahim, N.A. Dual function of unified adaptive linear neurons based fundamental component extraction algorithm for shunt active power filter operation. *Int. Rev. Electr. Eng.* **2015**, *10*, 544–552.
23. Karthikeyan, M.; Sharmilee, K.; Balasubramaniam, P.M.; Prakash, N.B.; Babu, M.R.; Subramaniyaswamy, V.; Sudhakar, S. Design and implementation of ann-based sapf approach for current harmonics mitigation in industrial power systems. *Microprocess. Microsyst.* **2020**, *77*, 103194. [CrossRef]
24. Merabet, L.; Saad, S.; Abdeslam, D.O.; Merckle, J. Direct neural method for harmonic currents estimation using adaptive linear element. *Electr. Power Syst. Res.* **2017**, *152*, 61–70. [CrossRef]
25. Radzi, M.A.M.; Rahim, N.A. Neural network and bandless hysteresis approach to control switched capacitor active power filter for reduction of harmonics. *IEEE Trans. Ind. Electron.* **2009**, *56*, 1477–1484. [CrossRef]
26. Demirdelen, T.; Kayaalp, R.İ.; Tumay, M. Simulation modelling and analysis of modular cascaded multilevel converter based shunt hybrid active power filter for large scale photovoltaic system interconnection. *Simul. Model. Pract. Theory* **2017**, *71*, 27–44. [CrossRef]
27. Adel, M.; Kandil, T. Assessment of direct and indirect current control techniques applied to active power filters. *Recent Adv. Electr. Electron. Eng.* **2020**, *13*, 1256–1265. [CrossRef]
28. Sao, J.K.; Patidar, R.D.; Swain, S.D. Analysis of direct and indirect current control techniques in p-q theory based shunt active power filter. In Proceedings of the 2024 IEEE International Conference on Smart Power Control and Renewable Energy (ICSPCRE), Rourkela, India, 19–21 July 2024; pp. 1–6.
29. Adel, M.; Zaid, S.; Mahgoub, O. Improved active power filter performance based on an indirect current control technique. *J. Power Electron.* **2011**, *11*, 931–937. [CrossRef]
30. Gao, C.; He, S.; Fang, X.; Davari, P.; Leung, K.N.; Loh, P.C.; Blaabjerg, F. Current-limiting control strategy for indirect-current-controlled active power filter. *IEEE Trans. Power Deliv.* **2024**, *39*, 3551–3554. [CrossRef]
31. Li, G.; Luo, A.; He, Z.; Ma, F.; Chen, Y.; Wu, W.; Zhu, Z.; Guerrero, J.M. A dc hybrid active power filter and its nonlinear unified controller using feedback linearization. *IEEE Trans. Ind. Electron.* **2021**, *68*, 5788–5798. [CrossRef]
32. Bushra, E.; Zeb, K.; Ahmad, I.; Khalid, M. A comprehensive review on recent trends and future prospects of pwm techniques for harmonic suppression in renewable energies based power converters. *Results Eng.* **2024**, *22*, 102213. [CrossRef]
33. Taghvaie, A.; Warnakulasuriya, T.; Kumar, D.; Zare, F.; Sharma, R.; Vilathgamuwa, D.M. A comprehensive review of harmonic issues and estimation techniques in power system networks based on traditional and artificial intelligence/machine learning. *IEEE Access* **2023**, *11*, 31417–31442. [CrossRef]
34. *IEEE Std 519-2022*; IEEE Standard for Harmonic Control in Electric Power Systems. Institute of Electrical and Electronics Engineers: New York, NY, USA, 2022.

symmetry

MDPI

Article

Distributed Active Support from Photovoltaics via State–Disturbance Observation and Dynamic Surface Consensus for Dynamic Frequency Stability Under Source–Load Asymmetry

Yichen Zhou [1,*], Yihe Gao [1], Yujia Tang [2,3,4], Yifei Liu [1], Liang Tu [2,3,4], Yifei Zhang [1], Yuyan Liu [2,3,4], Xiaoqin Zhang [1], Jiawei Yu [2,3,4] and Rui Cao [1]

[1] Key Laboratory of Distributed Energy Storage and Microgrid of Hebei Province (North China Electric Power University), Baoding 071003, China; gaoyihe2025@163.com (Y.G.)

[2] State Key Laboratory of HVDC, Electric Power Research Institute, China Southern Power Grid, Guangzhou 510663, China; yujw4@csg.cn (J.Y.)

[3] National Energy Power Grid Technology R&D Centre, Guangzhou 510663, China

[4] Guangdong Provincial Key Laboratory of Intelligent Operation and Control for New Energy Power System, Guangzhou 510663, China

* Correspondence: hbdldx_zyc@ncepu.edu.cn

Abstract

The power system's dynamic frequency stability is affected by common-mode ultra-low-frequency oscillation and differential-mode low-frequency oscillation. Traditional frequency control based on generators is facing the problem of capacity reduction. It is urgent to explore new regulation resources such as photovoltaics. To address this issue, this paper proposes a distributed active support method based on photovoltaic systems via state–disturbance observation and dynamic surface consensus control. A three-layer distributed control framework is constructed to suppress low-frequency oscillations and ultra-low-frequency oscillations. To solve the high-order problem of the regional grid model and to obtain its unmeasurable variables, a regional observer estimating both system states and external disturbances is designed. Furthermore, a distributed dynamic frequency stability control method is proposed for wide-area photovoltaic clusters based on the dynamic surface control theory. In addition, the stability of the proposed distributed active support method has been proven. Moreover, a parameter tuning algorithm is proposed based on improved chaos game theory. Finally, simulation results demonstrate that, even under a 0–2.5 s time-varying communication delay, the proposed method can restrict the frequency deviation and the inter-area frequency difference index to 0.17 Hz and 0.014, respectively. Moreover, under weak communication conditions, the controller can also maintain dynamic frequency stability. Compared with centralized control and decentralized control, the proposed method reduces the frequency deviation by 26.1% and 17.1%, respectively, and shortens the settling time by 76.3% and 42.9%, respectively. The proposed method can effectively maintain dynamic frequency stability using photovoltaics, demonstrating excellent application potential in renewable-rich power systems.

Keywords: distributed control; photovoltaic power generation; frequency oscillation suppression; dynamic surface control; disturbance observer; chaos game optimization

1. Introduction

In renewable-rich power systems, dynamic frequency stability is increasingly threatened by common-mode ultra-low-frequency oscillation (ULFO) and differential-mode low-frequency oscillations (LFOs) under source–load asymmetry disturbances [1]. In 2005, inter-area LFOs accompanied by a ULFO (0.015 Hz) occurred in the European power grid, leading to the loss of synchronization of multiple synchronous units, large-scale power outages and economic losses. In 2020, a 0.5 Hz LFO with a 0.02 Hz ULFO were observed in the UK power grid, leading to grid frequency deviation that exceeded ±0.2 Hz.

Currently, LFO control primarily relies on additional damping control, such as power system stabilizers (PSSs). However, their effective bandwidth spans only 0.1–2.5 Hz, and ULFOs remain beyond their reach [2,3]. ULFOs are typically mitigated via long-term power regulation (e.g., GPSS). However, GPSS's response speed is too slow to match the rapid dynamics of LFOs [4,5]. Moreover, related studies suggest that ULFO and LFO are in fact coupled [6]. Ultra-low-frequency power swings can excite LFO modes, whose fast dynamics in turn disturb ULFO control [7]. In 2021, though ULFOs at the Hami PV plant in Xinjiang were successfully controlled, the 0.8 Hz LFO persisted and drove the frequency deviation beyond ±0.2 Hz. Therefore, coordinated control of both ULFO and LFO is essential, and single-band methods alone cannot solve the problem.

In power systems increasingly dominated by new energy, the decreasing proportion of synchronous machines has led to a rapid deterioration in system regulation capabilities [8,9]. Now photovoltaic (PV) systems, as the leading new energy source, offer millisecond-level power regulation (<100 ms). Thus, PV systems can rapidly inject or absorb power during grid disturbances, providing advantages in suppressing ULFO and LFO. In addition, the China National Energy Administration's Notice [10] requires centralized PV systems to feature enhanced frequency regulation and real-time operational data reporting. Therefore, a critical technical challenge is how to leverage the active and rapid power support capabilities of PV clusters for dynamic frequency stability.

Currently, there are three main methods for dynamic frequency stability control: centralized, decentralized, and distributed control. Centralized control centralizes decision-making in a single controller to achieve global optimization. Ref. [11] proposed a VSC-HVDC frequency control method by centrally regulating the output of VSC-HVDC. Ref. [12] designed a centralized H∞ controller to suppress ULFO and LFO in hydro–solar complementary systems. Ref. [13] proposes a centralized model predictive controller for frequency stability in microgrids. However, the inherent defects of centralized control are particularly prominent in high-proportion PV power systems: ① A central controller failure will cripple the network's oscillation suppression, and the large-scale PV plants will further amplify the fault impact. ② Centralized control relies on wide-area communication networks. Since PV plants are widely distributed and numerous, any large communication delay can significantly reduce the control performance. ③ The need for centralized controllers to process real-time status from all grid PV units creates a computational burden that grows exponentially with installed capacity (from GW to TW). This makes real-time control infeasible.

Decentralized control operates by dividing system into multiple locally communicating subsystems, thereby removing reliance on global communication. Ref. [14] proposes decentralized event-triggered load frequency control to address excessive communication burden associated with centralized control. Ref. [15] proposes a decentralized load frequency control based on the differential evolution algorithm for multi-region systems. Despite this advantage, decentralized control still struggles to utilize PV systems to suppress ULFO/LFO: ① Relying solely on local data, photovoltaic controllers struggle to identify and suppress inter-area LFOs. ② Without effective communication and coordination, controllers may conflict and amplify oscillations.

Distributed control can significantly enhance system flexibility and adaptability by assigning tasks to multiple computing nodes that communicate. Addressing those challenges, ref. [16] introduces a leader–follower distributed control strategy for faster voltage control response. Ref. [17] proposes a sliding mode control-based adaptive algorithm for PV primary frequency regulation. However, distributed frequency control still has shortcomings: (1) It focuses on suppressing a single oscillation type without addressing ULFO and LFO concurrently. (2) It requires high-quality communication. (3) It lacks specific applications designed for large-scale PV clusters to provide effective active frequency support.

To address these challenges, this paper proposes a distributed active frequency support method based on photovoltaic systems. A three-layer distributed control framework is constructed to suppress LFOs and ULFOs. To solve the high-order problem of regional grid and to obtain its unmeasurable variables, a regional observer estimating both system states and external disturbances is designed. Furthermore, a distributed dynamic frequency stability control method is proposed for wide-area PV clusters based on the dynamic surface control theory. In addition, the stability of the proposed distributed active support method has been proven. Moreover, a parameter tuning algorithm is proposed based on improved chaos game theory. Finally, simulation results demonstrate that the proposed method can effectively maintain dynamic frequency stability using photovoltaics, demonstrating excellent application potential in renewable-rich power systems.

Main contributions of this paper include the following: ① This paper proposes a regional observer with dual observation for state and disturbance variables. This approach effectively tackles key challenges in wide-area power system with large-scale PVs, including high model order, multiplicity of elements, and the presence of unmeasurable variables.

② To date, the PV-based approach has not yet been established for dynamic frequency stability control in multi-area power systems. We propose a distributed dynamic frequency stability control method for wide-area photovoltaic clusters, which effectively leverages photovoltaics to maintain system frequency stability with high performance.

③ Existing distributed frequency control strategies struggle to simultaneously suppress LFO and ULFO and rely heavily on communication. This paper proposes a new distributed PV frequency control approach by integrating the multi-agent consensus algorithm (MAC) and dynamic surface control (DSC), which can effectively suppress both oscillations even with weak communications and large delays.

④ In contrast to existing control methods for which parameter tuning often relies on ad hoc trial and error, we not only prove the stability of the proposed method but also propose a parameter tuning algorithm based on chaos game optimization (CGO).

2. Cloud Control Architecture for Dynamic Frequency Active Support of Wide-Area Photovoltaic Clusters

2.1. Cloud-Based Control Architecture for Dynamic Frequency Active Support of Wide-Area Photovoltaic Clusters

(1) Photovoltaic Participation in Frequency Regulation Structure and Dynamic Frequency Stability Issues

The structure of photovoltaic participation in power system frequency regulation is shown in Figure 1. When a power deficit occurs in the power system, the frequency of the center of inertia (CoI) of region i, calculated using synchronous phasor measurement units (PMUs), will experience a deviation. Based on the WAMS system, the photovoltaic cluster detects Δf_i and rapidly adjusts the photovoltaic output power through active frequency droop control at each power station, achieving a rapid response to the frequency deviation and thereby accelerating the power system frequency recovery process.

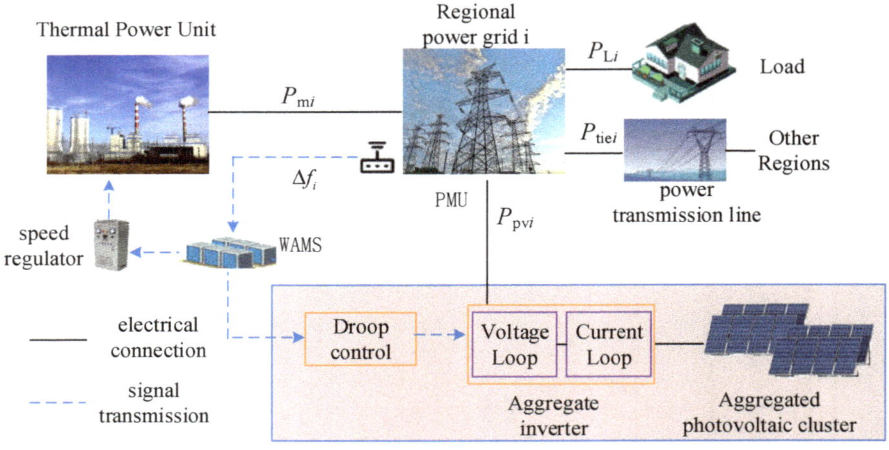

Figure 1. Primary frequency regulation process of PV in *i*-th regional power system.

The above process primarily addresses single-frequency tuning issues and does not consider frequency oscillation issues. Taking a typical two-region interconnected system as an example, the simulation results are shown in Figure 2. When power disturbances occur in the region, the system's frequency response exhibits significant common-mode ultra-low-frequency and differential-mode low-frequency oscillations, threatening the system's frequency stability. Therefore, this paper focuses on the issue of dynamic frequency stability and, based on a photovoltaic-participating system frequency regulation structure, further investigates a distributed dynamic frequency stability active support control method for a wide-area photovoltaic cluster.

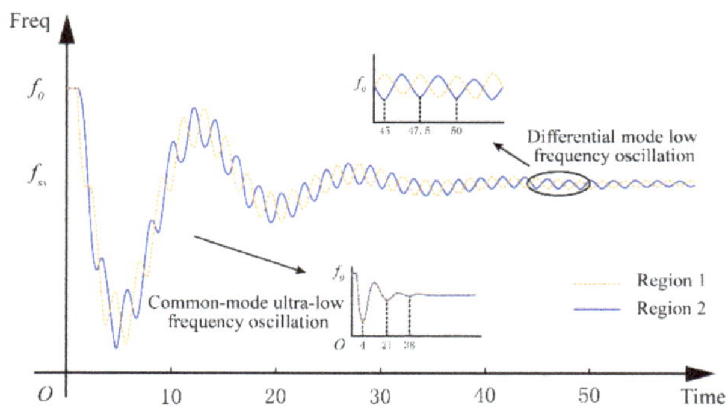

Figure 2. Common-mode ULFO and differential-mode LFO in power systems.

(2) Cloud control architecture actively supported by a wide-area photovoltaic cluster

Based on the similarity of frequency changes at various nodes in the power system and the geographical interconnection structure, large power grids can typically be divided into multiple interconnected regional systems. The dynamic frequency active support cloud control architecture for PV clusters in wide-area power systems consists of three layers:

terminal, edge, and cooperative control layers, as shown in Figure 3. The relationships among the three layers in each area are as follows.

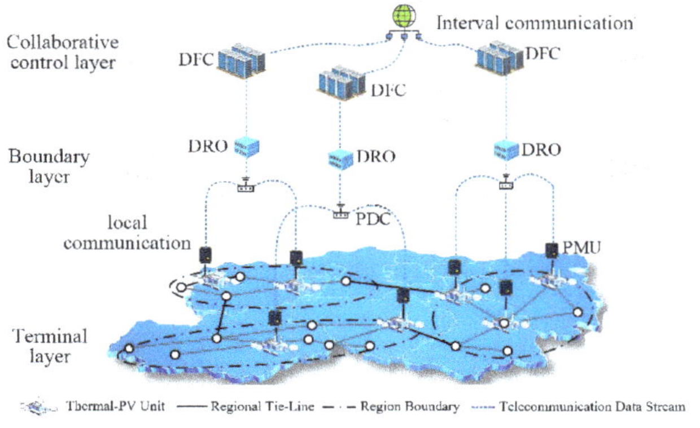

Figure 3. Three-layer framework for distributed frequency stability control.

In terms of the control hierarchy architecture, each area consists of a terminal layer, an edge layer, and a collaborative control layer. The control relationships among the three layers in each area are as follows:

At the terminal layer, each region uses PMUs to measure regional frequency deviations and tie-line power deviations, which are transmitted to the regional phasor data concentrator (PDC) at the edge layer.

At the edge layer, the DRO utilizes the data collected by the PDC from various measurement points to calculate the CoI frequency of the region as input, and outputs the state variables and disturbance variables of the frequency response model for photovoltaic participation in frequency regulation. The observation results are then transmitted to the regional controller.

At the collaborative control layer, the distributed frequency controller (DFC) exchanges frequency information with other regional DFCs via WAMS, and comprehensively considers the state variables and disturbance variables provided by the DROs of each region to formulate control signals through multi-region collaboration. The control signals are then transmitted in reverse through the edge layer to the inverter control chips of photovoltaic power plants at the terminal layer, dynamically adjusting photovoltaic output to suppress frequency oscillations and enhance system frequency dynamic stability.

2.2. *Fundamental Model for Dynamic Frequency Active Support*

(1) Physical system model

This subsection focuses on PV clusters and establishes their system frequency response (SFR) model for participation in power system frequency regulation. The voltage control and current control modules of the PV cluster are aggregated and modeled as first-order inertial elements, represented by the time constants T_{Vi} and T_{Ii} [18], respectively. Considering the time delay in acquiring the aggregated CoI frequency for region i, this delay is modeled as a first-order inertial element with an equivalent time constant T [19]. Given that the mechanical power change ΔP_{mi} of thermal power units is relatively slow and of small magnitude, it can be collectively treated as an external disturbance along with tie-line power deviation ΔP_{tiei} and load power deviation ΔP_{Li}. M_{i} denotes the equivalent inertia of region i, and D_{i} represents its equivalent damping. The equivalent inertia M_{i} synthesizes the virtual inertia from the rapid power regulation of PV units and the mechanical inertia of conventional thermal units. The equivalent damping D_i integrates the active damping provided by the PV droop control and the natural damping characteristics of thermal units. The overall inertial response and damping effect of the area against frequency disturbances are ultimately characterized by the transfer function $1/M_is + D_i$. Based on this modeling framework, the SFR model for region i is illustrated in Figure 4.

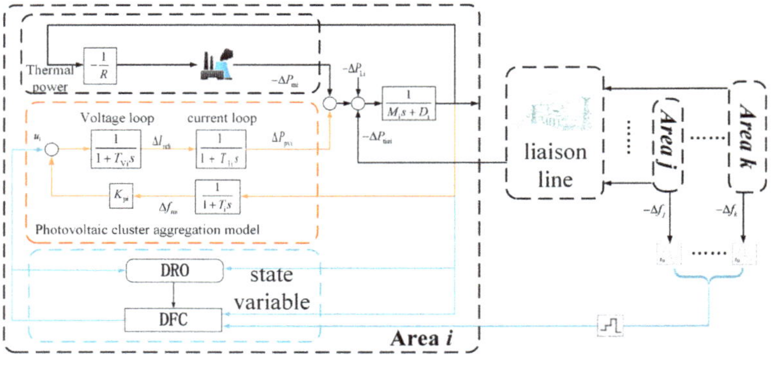

Figure 4. Frequency response model of PV systems participating in frequency regulation in regional power systems.

In Figure 4, the SFR model of photovoltaic clusters participating in power system frequency regulation can be established as the state space equation shown in Equation (1):

$$\begin{cases} \Delta\dot{f}_{\text{mi}} = \frac{1}{T_{\text{i}}}(-\Delta f_{\text{mi}} + K_{\text{pi}}\Delta f_{\text{i}}) \\ \Delta\dot{f}_{\text{i}} = \frac{1}{M_{\text{i}}}(-D_{\text{i}}\Delta f_{\text{i}} + \Delta P_{\text{pvi}} - \Delta P_{\text{tie,i}} - \Delta P_{\text{Li}} - \Delta P_{\text{mi}}) \\ \Delta\dot{P}_{\text{pvi}} = \frac{1}{T_{\text{Ii}}}(-\Delta P_{\text{pvi}} + \Delta I_{\text{refi}}) \\ \Delta\dot{I}_{\text{refi}} = \frac{1}{T_{\text{Vi}}}(-\Delta I_{\text{refi}} + \Delta f_{\text{mi}} + u_{\text{i}}) \\ y_{\text{i}} = \Delta f_{\text{i}} \\ u_{\text{i}} = f\left(\Delta f_{\text{i}}, \Delta f_{\text{dj}}\right) \end{cases} \tag{1}$$

where Δf_{mi} is the aggregated frequency deviation accounting for the CoI frequency. Δf_{dj} is the aggregated frequency deviation that accounts for the inter-area communication delay. Δf_{i}, ΔP_{pvi}, and ΔI_{refi} are the aggregated CoI frequency deviation, the aggregated change in PV output active power, and the aggregated current reference value for region *i*, respectively. K_{pi} is the equivalent droop coefficient of the PV cluster in region *i*. $u_{\text{i}} = \sum_{j=1}^{n} (S_{\text{i}}/S_{\text{ij}})u_{\text{ij}}/S_i$ is the aggregated value of the control inputs u_{cij} from all PV plants within region i, where S_{ij} and S_{i} are the capacities of PV plant j and region i, respectively.

The dependency relationships among the state variables in Equation (1) are complex, which is inconvenient for control design. Therefore, a linear transformation operator, Equation (2), is employed to convert Equation (1) into the strict-feedback system described by Equation (3). It can be observed that Equation (3) exhibits a cascaded form, where the dynamics of each state variable depend only on the state variables themselves and external inputs, thereby facilitating control design.

$$\begin{gathered} x_{\text{i},1} = \Delta f_{\text{mi}}, x_{\text{i},2} = \frac{K_{\text{pi}}}{T_{\text{i}}}\Delta f_{\text{i}} \\ x_{\text{i},3} = \frac{K_{\text{pi}}}{M_{\text{i}}T_{\text{i}}}\Delta P_{\text{pvi}}, x_{\text{i},4} = \frac{K_{\text{pi}}}{M_{\text{i}}T_{\text{i}}T_{\text{Ii}}}\Delta I_{\text{ref}} \end{gathered} \tag{2}$$

$$\begin{cases} \dot{x}_{i,1} = x_{i,2} + g_{i,1} \\ \dot{x}_{i,2} = x_{i,3} + g_{i,2} - (\Delta P_{tie,i} + \Delta P_{Li} + \Delta P_{mi})/M_i T_i \\ \dot{x}_{i,3} = x_{i,4} + g_{i,3} \\ \dot{x}_{i,4} = g_{i,4} + g_{i,5} + u_i \\ y_i = x_{i,2} \end{cases} \quad (3)$$

where $u_i = u_i K_{pi}/M_i T_i T_{Ii} T_{Vi}$, $g_{i,1} = -x_{i,1}/T_i$, $g_{i,2} = -x_{i,2} D_i/M_i$, $g_{i,3} = -x_{i,3}/T_{Ii}$, $g_{i,4} = -x_{i,4}/T_{Vi}$, $g_{i,5} = x_{i,1} K_{pi}/M_i T_i T_{Ii} T_{Vi}$.

The error in the dynamic equation of the state variable $x_{i,2}$ is further modeled. Compared to the actual frequency, the regional frequency deviation $x_{i,2}$ (i.e., Δf) inevitably exhibits an error $d_{0,i}$. The disturbance variable, incorporating this error, is defined as: $d_{i,1} = (-\Delta P_{tie,i} - \Delta P_{Li} - \Delta P_{mi})/M_i T_i + d_{0,i}$ Consequently, Equation (3) can be modified as:

$$\begin{cases} \dot{x}_{i,1} = x_{i,2} + g_{i,1} \\ \dot{x}_{i,2} = x_{i,3} + g_{i,2} + d_{i,1} \\ \dot{x}_{i,3} = x_{i,4} + g_{i,3} \\ \dot{x}_{i,4} = g_{i,4} + g_{i,5} + u_i \\ y_i = x_{i,2} \end{cases} \quad (4)$$

(2) Information Communication Model

In the control framework of this paper, there are mainly two types of time delays in the information system: regional delay and inter-region delay.

Regional delay corresponds to the delay in frequency measurement by PMUi in region *i* and its transmission to DROi and DFCi, which includes PMU measurement, PMU-to-PDC signal transmission, DRO model calculation, and DRO-to-DFC signal transmission.

The regional delay encompasses the time delay in frequency measurement by PMUi in region *i* and its subsequent transmission to PDCi, DROi and DFCi, and the calculation time of DRO. Since this delay occurs between PMUs and DFCs within the same region, it is typically small and is therefore ignored in this paper.

The inter-region delay refers to the delay produced during the transmission of the frequency signal Δfj from PMUj to DFCj in Region j and then onward to DFCi in Region *i*, as illustrated in Figure 5. It shows the inter-region delay is primarily influenced by the longer communication delays between DFCs, making it a significant and non-negligible factor.

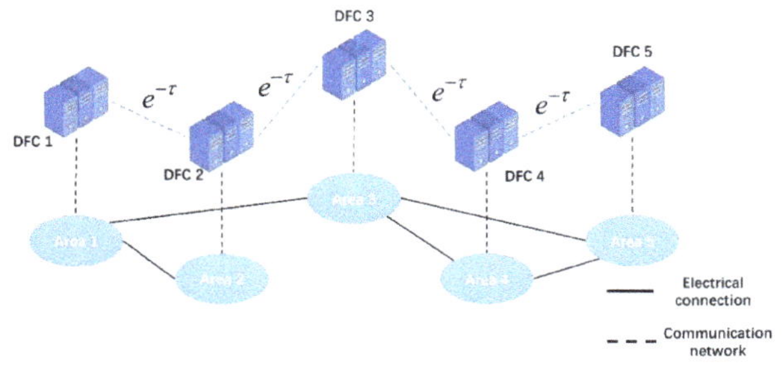

Figure 5. Communication Structure of distributed frequency stability control system.

The inter-region delay is modeled as follows. Differences in signal delays across different regions can make it difficult for the DFC to perform unified calculations. Therefore, a signal holder is used to retain the signal at time k, ensuring that all necessary regional signals are received before commencing calculations, so that inter-region delay $\tau > 0$ can be modeled as an interval delay with an upper bound, i.e., $\tau \in (0, \tau_{max}]$.

Figure 6 takes DFC4 as an example and specifically demonstrates the synchronization mechanism of the signal holder. At time k, DFC4 caches the input signals from each region via the signal holder until all associated signals arrive. The final delay is determined by the transmission time of the latest arriving signal, thereby eliminating the impact of delay differences.

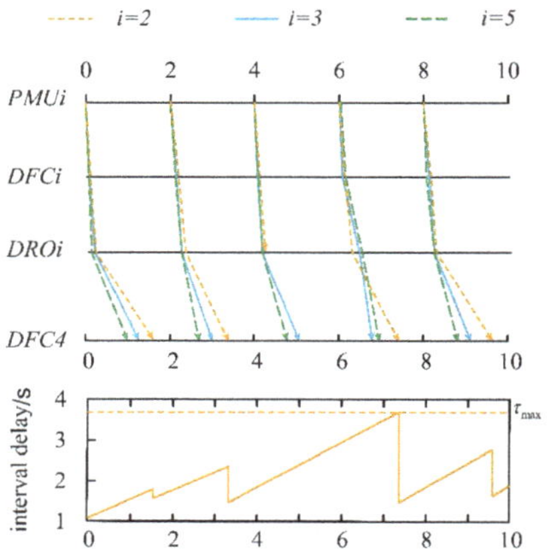

Figure 6. Transmission process of the inter-region signal and the corresponding inter-region delay.

3. Dual Observer for State and Disturbance

The problem of photovoltaic participation in power system frequency control involves the coupled dynamics of photovoltaic and thermal power units. Direct controller design suffers from issues such as multiple elements, high order, and unmeasurable variables. Therefore, this paper focuses on the photovoltaic entities participating in frequency control and uses an observer to construct a frequency response model for photovoltaic participation in frequency control.

Regarding the variables in Equation (1), the regional frequency deviation signal Δf_i obtained from WAMS calculations is known, and the PV output power $\Delta P_{\text{pv}i}$ can be obtained from its own power measurement devices. However, the aggregated values Δf_{mi}, ΔI_{refi}, and $d_{0,\text{i}}$ from each region are difficult to measure, resulting in the inability to directly measure $x_{i,1}$, $x_{i,4}$, $g_{i,1}$, $g_{i,4}$, $g_{i,5}$, and $d_{i,1}$ in Equation (4). Therefore, this section proposes a distributed regional observer (DRO) with dual observation functions for state and disturbance based on the input decoupling approach [20].

The state-space model corresponding to Equation (4) is:

$$\begin{aligned} \dot{x}_i(t) &= A_i x_i(t) + B_i u_i(t) + E d_i(t) \\ y_i(t) &= C_i x_i(t) \end{aligned} \tag{5}$$

where $x_i = \begin{bmatrix} x_{i,1} \\ x_{i,2} \\ x_{i,3} \\ x_{i,4} \end{bmatrix}$, $A_i = \begin{bmatrix} -\frac{1}{\tau_i} & 1 & 0 & 0 \\ 0 & -\frac{D_i}{M_i} & 1 & 0 \\ 0 & 0 & -\frac{1}{T_{Ii}} & 1 \\ \frac{K_{pi}}{M_i \tau_i T_{Ii} T_{Vi}} & 0 & 0 & -\frac{1}{T_{Vi}} \end{bmatrix}$, $E = \begin{bmatrix} 0 \\ 1 \\ 0 \\ 0 \end{bmatrix}$, $B_i = \begin{bmatrix} 0 \\ 0 \\ 0 \\ 1 \end{bmatrix}$, $C_i = \begin{bmatrix} 0 & 1 & 0 & 0 \end{bmatrix}$, $d_i = d_{i,1}$.

The dynamic equation of the DRO for the system in (5) is:

$$\begin{bmatrix} \dot{z}_i(t) \\ \dot{q}_i(t) \end{bmatrix} = \begin{bmatrix} F_i & 0 \\ -K_{\text{d}i}(K_{\text{d}i} + A_i) & -K_{\text{d}i} \end{bmatrix} \begin{bmatrix} z_i(t) \\ q_i(t) \end{bmatrix} + \begin{bmatrix} T_i B_{,i} \\ -K_{\text{d}i} B_i \end{bmatrix} u_i(t) + \begin{bmatrix} K_i \\ -K_{\text{d}i}(K_{\text{d}i} + A_i) H_i \end{bmatrix} y_i(t) \tag{6}$$

where $z_i \in \Re^{4\times1}$ and $q_i \in \Re^{4\times1}$ are auxiliary variables for the state observation $\hat{x}_i$ and disturbance observation $E\hat{d}_i$, respectively. $\hat{x}_i$ denotes the observed value of state x_i. $E\hat{d}_i$ represents the observed value of disturbance Ed_i. Matrices F_i, T_i, K_i, and H_i are parameters to be designed for the state observer. K_{di} is the parameter to be designed for the disturbance observer.

The observed state $\hat{x}_i$ and observed disturbance $E\hat{d}_i$ in the actual system can be obtained based on the dynamic model of these auxiliary variables. In the designed DRO, they exhibit the following relationship:

$$\begin{cases} \hat{x}_i = z_i + H_i y_i \\ E\hat{d}_i = q_i + K_{di} z_i + K_{di} H_i y_i \end{cases} \tag{7}$$

Based on Equations (6) and (7), the structure of the DRO designed in this paper can be obtained, as shown in Figure 7. The observer uses the input of the original system, i.e., the control signal u_i, and the output of the original system y_i, i.e., the regional frequency deviation signal Δf_i, as the input of the observer, and outputs the unknown state variables and disturbance quantity observation values $\hat{x}_i$ and $E\hat{d}_i$.

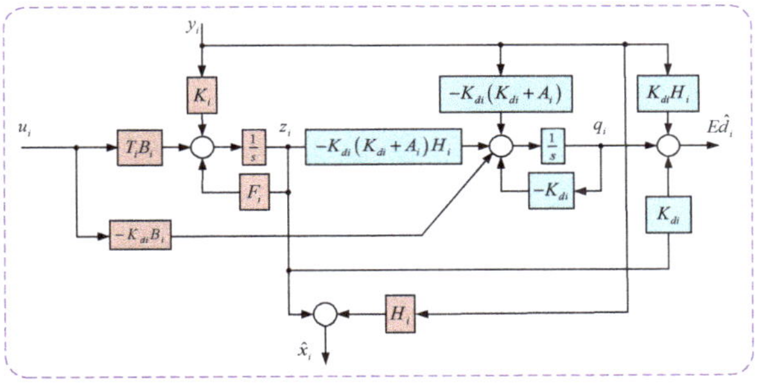

Figure 7. DRO block diagram structure.

Substituting the observed results from Equation (7) into Equation (4) yields the observed values $\hat{x}_{i,1}$, $\hat{x}_{i,4}$, $\hat{g}_{i,1}$, $\hat{g}_{i,4}$, $\hat{g}_{i,5}$ and $\hat{d}_{i,1}$, and the observed error is:

$$\begin{cases} \tilde{x}_{i,k} = \hat{x}_{i,k} - x_{i,k} \\ \tilde{g}_{i,m} = \hat{g}_{i,m} - g_{i,m} \\ \tilde{d}_{i,1} = \hat{d}_{i,1} - d_{i,1} \end{cases} \tag{8}$$

Here $\tilde{x}_{i,k}$, $\tilde{g}_{i,m}$ and $\tilde{d}_{i,1}$ are the observation errors of $x_{i,k}$, $g_{i,k}$ and $d_{i,1}$, respectively, $k \in \{1,4\}$ and $m \in \{1,4,5\}$.

Observation errors are significantly affected by relevant parameters, and parameters must be reasonably designed to eliminate the influence of unknown inputs and ensure stable convergence of observations. Therefore, Theorem 1 will be used to ensure the stability of this observer, and a parameter design method will be proposed.

Theorem 1. *By reasonably designing the parameter matrices F_i, T_i, K_i, and H_i to satisfy Equation (9), the state observation error can be guaranteed to converge asymptotically. Additionally, by designing K_{di}, the disturbance observation error can be guaranteed to be a bounded quantity and arbitrarily small within the closed set $\Omega_{\tilde{d}}$. The larger the value of $\lambda_{min}(K_i')$, the smaller the disturbance observation error.*

$$\begin{aligned} F_i &= A_i - H_i C_i A_i - K_{i1} C_i \\ K_{i2} &= F_i H_i \\ T_i &= I_i - H_i C_i \\ T_i E_i &= 0 \end{aligned} \tag{9}$$

Proof. The proof is given in Appendix A. □

For the state observer, according to Theorem 1, when the parameters satisfy Equation (9), the state observation part and the input part can be decoupled. Through observability decomposition and the pole placement method, the gain matrix K_{i} can be set to satisfy the necessary and sufficient conditions for state observation, thereby achieving the asymptotic stability of the state observation error.

For the disturbance observer, it can be seen from Equations (9) and (A7) that the observation error is not only related to the state observer but also mainly related to $K_{\mathrm{d}i}$. Therefore, the coefficient $K_{\mathrm{d}i}$ can be designed as a diagonal matrix with all diagonal elements being positive values, and the larger $\lambda_{\min}(K_{\mathrm{d}i})$ is, the smaller the disturbance observation error will be.

Based on the above analysis, we propose the following Algorithm 1:

Algorithm 1. Parameter calculation of DRO

Input: System matrix A, C, E
Output: Observer gain F, T, K, H
if rank(CE) $\neq$ rank(E)
  Output "UIO does not exist"
end if
$H \leftarrow E * inv(CE' * CE) * CE'$
$T \leftarrow I - H * C$
$A_1 \leftarrow T * A$
if (C, A_1) is observable
  Calculate K_1 using pole placement and calculate F and K according to Equations (9)–(11)
end if
Wo ← *obsv(C,* A_1*)*
n1 ← *rank(Wo)*
while *True*
P1 ← *the first* n_1 *rows of Wo*
P2 ← Randomly generate an $(n - n_1) \times n$ matrix
$P \leftarrow [P_1; P_2]$
If det(P) $\neq 0$
  return P
PAP_inv ← $P * A *$ inv(P)
CP_inv ← $C *$ inv(P)
A_{11} ←*The upper-left* $n_1 \times n_1$ *submatrix of PAP_inv*
A_{12} ←*The lower-left* $n_1 \times (n - n_1)$ *submatrix of PAP_inv*
A_{22} ←*The lower-right* $(n - n_1) \times (n - n_1)$ *submatrix of PAP_inv*
Kp ← place(A_{11}, C(1:n_1, :), expected poles)
$K_1 \leftarrow$ inv(P) * K_p
$K_2 \leftarrow$ randomly generate an $(n - n_1) * m$ matrix
$K \leftarrow K_1 + K_2$
$F \leftarrow A_1 - K_1 * C$
return F, K, T, H

4. Photovoltaic-Based Distributed Frequency Controller

In this section, the distributed DFC is designed based on the strict feedback system observation model provided by DRO for the cooperative control layer for PV. The MAC is employed to construct error surfaces to drive the system to its desired equilibrium. Subsequently, the DSC is utilized to construct the energy function of the error surface in a step-by-step manner, followed by the derivation of the virtual control law. Specifically, an initial error surface $e_{i,1}$ is constructed from the control objective, and the associated energy function $V_{i,1}$ is defined for $e_{i,1}$. To stabilize the derivative of $V_{i,1}$, a virtual control law $\alpha_{i,1}$ is designed. Then a second error surface $e_{i,2}$ and a virtual law $\alpha_{i,2}$ is derived to stabilize $\dot{V}_{i,2}$. This iterative procedure continues until the derivative of $V_{i,3}$ is obtained. At this stage, there is no need to define a new error surface. The final control law u_i is derived directly by stabilizing $\dot{V}_{i,3}$ thereby completing the design.

4.1. Error Surface Design Based on MAC

The error surface is designed using the MAC method. The design of the error surface $e_{i,1}$ will simultaneously consider the suppression of low-frequency oscillations and ultra-low-frequency oscillations. Considering that the regional aggregated frequency is observable for ultra-low-frequency oscillations, while the inter-regional aggregated frequency difference $\Delta f_i - \Delta f_j$ is observable for inter-area low-frequency oscillations, $e_{i,1}$ is designed as follows.

$$e_{i,1} = \sum_{j\in N_i} a_{i,j}(y_i - y_j(t-\tau)) + a_{i,0}y_i \tag{10}$$

where $t_\tau = t - \tau$, $a_{i,j}$ and $a_{i,0}$ are the adjacency weights and reference signal weights, respectively, which must satisfy the condition that the weights of mutually communicating nodes are not zero.

According to the definition of node out-degree in algebraic graph theory: $b_i = \sum_{j\in N_i} a_{i,j}$, Equation (10) can be further rearranged as:

$$e_{i,1} = c_{\mathrm{ab}i}y_i - \sum_{j\in N_i} a_{ij}y_j(t_\tau) \tag{11}$$

where $c_{\mathrm{ab}i} = b_i + a_{i,0}$.

In addition, the expected dynamics of state variables $x_{i,3}$ and $x_{i,4}$ are virtual control laws $\alpha_{i,1}$ and $\alpha_{i,2}$, respectively. In backstepping control, their deviations are typically defined as error surfaces $\bar{e}_{i,2}$ and $\bar{e}_{i,3}$:

$$\bar{e}_{i,k-1} = x_{i,k} - \alpha_{i,k-2} \tag{12}$$

where $k = 3, 4$.

However, defining errors $\bar{e}_{i,2}$ and $\bar{e}_{i,3}$ using Equation (12) would introduce high-order differentiation in the control law, leading to the "differential explosion" problem. Therefore, the DSC method is adopted, where $\alpha_{i,1}$ and $\alpha_{i,2}$ are input into the following first-order low-pass filter to obtain $\lambda_{i,1}$ and $\lambda_{i,2}$:

$$T_{\mathrm{f}i,k-2}\dot{\lambda}_{i,k-2} + \lambda_{i,k-2} = \alpha_{i,k-2} \tag{13}$$

where $T_{\mathrm{f}i,k-2}$ is the filter time constant, $\lambda_{i,k-2}(0) = \alpha_{i,k-2}(0)$

By replacing $\alpha_{i,1}$ and $\alpha_{i,2}$ with $\lambda_{i,1}$ and $\lambda_{i,2}$, the differential operations in the control law design are transformed into algebraic operations. Consequently, $e_{i,2}$ and $e_{i,3}$ are defined as:

$$e_{i,k-1} = x_{i,k} - \lambda_{i,k-2} \tag{14}$$

Comparing Equations (12) and (14), it follows that the error surface $\bar{e}_{i,k-1}$ equals the error surface $e_{i,k-1}$ minus the filtering error $\eta_{\mathrm{i,k-2}} = \lambda_{\mathrm{i,k-2}} - \alpha_{\mathrm{i,k-2}}$:

$$e_{i,k-1} = \bar{e}_{i,k-1} - \eta_{i,k-2} = x_{i,k} - \alpha_{i,k-2} - \eta_{i,k-2} \tag{15}$$

If we denote $\psi_{i,k-2} = \dot{\alpha}_{i,k-2}$, the dynamics of the filter output error can be expressed as: $\dot{\eta}_{i,k-2} = -\eta_{i,k-2}/T_{\mathrm{fi},k-2} - \psi_{i,k-2}$

Since $x_{\mathrm{i,k}}$ in Equation (15) must be obtained via the DRO, the observed value of the error surface is:

$$\hat{e}_{\mathrm{i,k-1}} = \hat{x}_{\mathrm{i,k}} - \lambda_{\mathrm{i,k-2}} \tag{16}$$

Substituting Equation (8) into Equation (16) yields the relationship:

$$\hat{e}_{\mathrm{i,k-1}} = e_{\mathrm{i,k-1}} + \tilde{x}_{\mathrm{i,k}} \tag{17}$$

4.2. Control Law Design Based on DSC

Control Law Design

This section designs control laws to ensure that the system state moves according to each virtual control law, guaranteeing stable convergence of the defined error surface and thereby suppressing low-frequency and ultra-low-frequency oscillations. The control laws are designed by constructing the energy function of the error surface and compensating for interference step by step.

(1) Design virtual control rate $\alpha_{\mathrm{i},1}$ for $x_{i,3}$.

For error $e_{i,1}$ in Equation (11), define the Lyapunov function:

$$V_{i,1} = \frac{1}{2}e_{i,1}^2 \tag{18}$$

By taking the derivative of Equation (18) and incorporating Equations (4), (10) and (14), we can obtain:

$$\dot{V}_{i,1} = e_{i,1}(c_{\mathrm{ab}i}(e_{i,2} + \eta_{i,1} + \alpha_{i,1} + g_{i,2} + d_{i,1}) - \sum_{j\in N_i} a_{ij}(x_{j,3}(t_\tau) + g_{j,2}(t_\tau) + d_{j,1}(t_\tau)) \tag{19}$$

To stabilize Equation (19), the virtual control law $\alpha_{i,1}$ is designed as follows:

$$\alpha_{i,1} = -g_{i,2} - \hat{d}_{i,1} - (c_{\mathrm{ab}i} + c_{i,1}/c_{\mathrm{ab}i})e_{i,1} + \frac{1}{c_{\mathrm{ab}i}}\sum_{j\in N_i} a_{ij}\left[x_{j,3}(t_\tau) + g_{j,2}(t_\tau) + \hat{d}_{j,1}(t_\tau) - e_{i,1}\right] \tag{20}$$

where $c_{i,1}$ is a parameter to be designed.

Substituting $\alpha_{i,1}$ into Equation (19) and considering Young's inequality, we obtain:

$$\dot{V}_{i,1} \le -c_{i,1}e_{i,1}^2 + \frac{1}{2}\eta_{i,1}^2 + \frac{1}{2}\tilde{d}_{i,1}^2 + \sum_{j\in N_i}\frac{a_{ij}}{2}\tilde{g}_{j,2}^2(t_\tau) + \sum_{j\in N_i}\frac{a_{ij}}{2}\tilde{d}_{j,1}^2(t_\tau) + c_{\mathrm{ab}i}e_{i,1}e_{i,2} \tag{21}$$

Equation (21) indicates that while $-c_{i,1}e_{i,1}^2 < 0$ contributes to stability, convergence of $e_{i,1}$ is coupled with error surface $e_{i,2}$ and filtering error $\eta_{i,1}$, necessitating further energy function construction and design of virtual control rate $\alpha_{i,2}$.

(2) Design virtual control rate $\alpha_{\mathrm{i},2}$ for $x_{i,4}$.

Define the Lyapunov function for the first term in Equation (14):

$$V_{i,2} = V_{i,1} + \frac{1}{2}e_{i,2}^2 + \frac{1}{2}\eta_{i,1}^2 \tag{22}$$

Taking the derivative of Equation (22) and combining Equations (4), (14) and (15) yields:

$$\dot{V}_{i,2} = \dot{V}_{i,1} + e_{i,2}(e_{i,3} + \eta_{i,2} + \alpha_{i,2} + g_{i,3} - \dot{\lambda}_{i,1}) + \eta_{i,1}(-\eta_{i,1}/T_{\mathrm{fi},1} - \psi_{i,1}) \tag{23}$$

To stabilize Equation (23), design the virtual control law $\alpha_{i,2}$ as follows:

$$\alpha_{i,2} = -g_{i,3} + \dot{\lambda}_{i,1} - c_{\mathrm{ab}i}e_{i,1} - (c_{i,2} + 1.5)e_{i,2} \tag{24}$$

where $c_{i,2}$ is a parameter to be designed.

Substituting $\alpha_{i,2}$ into Equation (23) and considering Young's inequality, we obtain:

$$\dot{V}_{i,2} \le \dot{V}_{i,1} - c_{\mathrm{abi}}e_{i,1}e_{i,2} - c_{i,2}e_{i,2}^2 + \frac{1}{2}\eta_{i,2}^2 + \frac{1}{2}\psi_{i,1}^2 + \left(\frac{1}{2} - \frac{1}{T_{\mathrm{fi},1}}\right)\eta_{i,1}^2 + e_{i,2}e_{i,3} \tag{25}$$

Equation (25) indicates that the stability of $e_{i,1}$, $e_{i,2}$, and $\eta_{i,1}$ is influenced by error surfaces $e_{i,3}$, $\eta_{i,2}$, and other factors, necessitating further construction of an energy function and design of control law u_i.

(3) Design Control Law u_i

Define the Lyapunov function for the second term in Equation (14):

$$V_{i,3} = V_{i,2} + \frac{1}{2}e_{i,3}^2 + \frac{1}{2}\eta_{i,2}^2 \tag{26}$$

Taking the derivative of Equation (26) and combining Equations (4) and (14) yields:

$$\dot{V}_{i,3} = \dot{V}_{i,2} + e_{i,3}(u_i + \hat{g}_{i,4} + g_{i,5} - \dot{\lambda}_{i,2}) - \eta_{i,2}\psi_{i,2} - \eta_{i,2}^2/T_{\mathrm{fi},2} \tag{27}$$

To stabilize Equation (27), design the control law u_i:

$$u_i = \dot{\lambda}_{i,2} - \hat{g}_{i,4} - g_{i,5} - e_{i,2} - (c_{i,3} + 2)\hat{e}_{i,3} \tag{28}$$

where $c_{i,3}$ is a parameter to be designed.

Substituting u_i into Equation (27) and considering Young's inequality, we obtain:

$$\dot{V}_{i,3} \le -c_{i,1}e_{i,1}^2 - c_{i,2}e_{i,2}^2 - c_{i,3}e_{i,3}^2 + \sum_{v=1}^{2}(1/T_{\mathrm{fi},v} - 1)\eta_{i,v}^2 + \frac{1}{2}\sum_{j\in N_i} a_{ij}\tilde{d}_{j,1}^2(t_\tau) + \frac{1}{2}(c_{i,3}+2)^2\tilde{x}_{i,4}^2 + \frac{1}{2}\left(\psi_{i,1}^2 + \psi_{i,2}^2 + \tilde{g}_{i,4}^2 + \tilde{d}_{i,1}^2\right) \tag{29}$$

The control law design is now complete. Using error surfaces (12), (14) and control laws (20), (24), (28), the photovoltaic-based distributed cooperative frequency controller is realized.

The flowchart of the proposed control method is shown in Figure 8. The overall view of the design and application process are shown in Figure 8a and the detailed information transmission process is given in Figure 8b. And the final control structure of DFC is depicted in Figure 9. In the PV-based distributed cooperative frequency control system, the measurement devices and DROs provide necessary information for DFCs to calculate its outputs. By integrating regional with neighborhood information, control signals for ULFO and LFO can be computed. This approach fully leverages the potential of PV plants to enhance dynamic frequency stability in wide-area power systems.

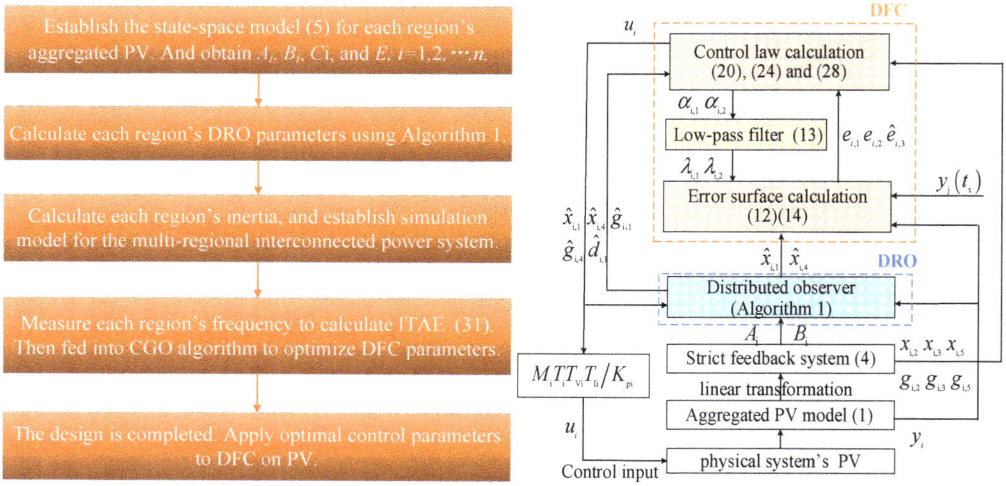

(a) The parameter design and application flowchart (b) The detailed information transmission process

Figure 8. The flowchart of the proposed control method.

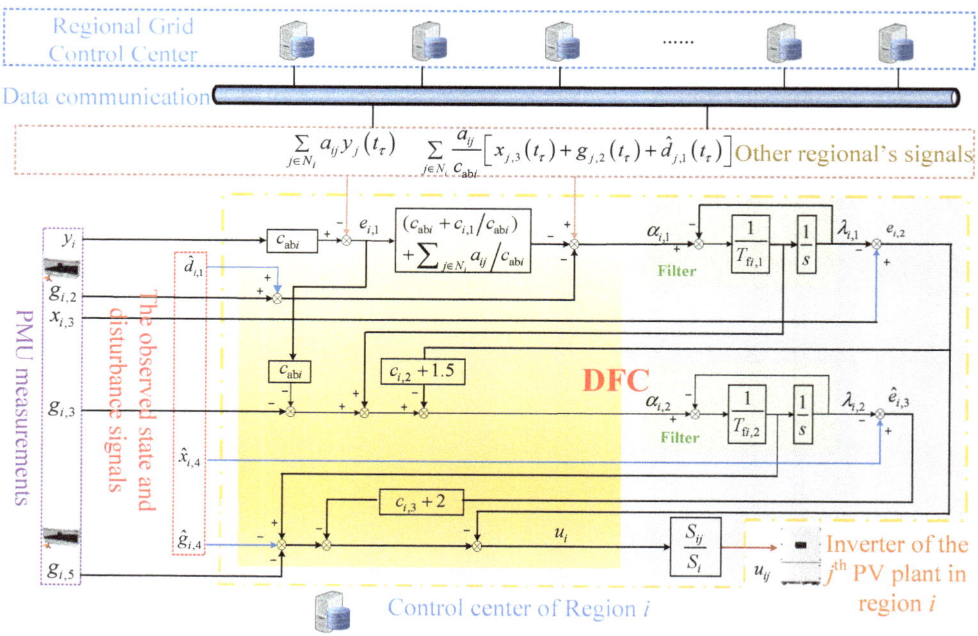

Figure 9. The detailed structure of DFC.

4.3. Control Law Parameter Design

As evident from the expressions of designed control law, the closed-loop stability performance of the entire system is critically dependent on parameter design. Therefore, this section first derives Theorem 2 on the stability of the distributed photovoltaic DFC, then proposes a parameter design method for the control law based on Theorem 2.

Theorem 2. *For each regional controlled system (1) under the action of DRO-based DFC, if the design parameters* $c_{i,k}$ *and* $T_{fi,v}$ *satisfy Equation (30), then for any* $P > 0$ *there exists a bounded compact set* Ω_i *such that when* $\mu \geq \kappa/P$*, all signals in the closed-loop system are semi-globally uniformly ultimately bounded, where* $\mu = min[2c_{i,k}, 2/T_{fi,v} - 2]$*,* $\kappa = 1/2 \cdot \sum_{i=1}^{N}\left(\sum_{v=1}^{2} A_{i,v}^2 + D_{SF,i}^2 + D_{DCO,i}^2\right)$.

$$c_{i,k} > 0, 1/T_{fi,v} - 1 > 0, i = 1, \ldots, N, v = 1,2 \tag{30}$$

Proof. The proof is given in Appendix B. □

From Theorem 2, it can be seen that the design of distributed observer parameters ensures stable convergence of state observation error and bounded disturbance observation error.

Based on Theorem 2, a parameter optimization model is proposed. Considering that the integral of time multiplied absolute error (ITAE) can evaluate the deviation between the controller and the expected dynamics, the minimization of ITAE of frequency is taken as the objective to suppress the ULFO. At the same time, in order to suppress LFOs, the objective is designed to minimize the absolute frequency deviation between region i and its neighbors. Thus, the final multi-objective optimization problem is established as shown in Equation (31).

$$\begin{aligned} &J = \min ITAE \\ &ITAE = \int_0^T t(\textstyle\sum_{i=1}^{n}|f_i - f_\infty| + \sum_{i=1}^{n}|\Delta f_i - \Delta f_j|)dt \end{aligned} \tag{31}$$

The range of control law parameter values is set to ensure that the values are reasonable, save computation and speed up the optimization search process:

$$\begin{aligned} &c_{i\min} \leq c_i \leq c_{i\max}(i = 1, \ldots, N) \\ &T_{fi\min} \leq T_{fi} \leq T_{fi\max}(i = 1, \ldots, N) \end{aligned} \tag{32}$$

Owing to the fact that fixed upper and lower bounds for (32) can slow the convergence speed, we propose an adaptive boundary determination method using Theorem 2. The upper and lower bounds of (32) are updated after each population update. For ℓ^{th} population with m current candidates, κ is estimated to the average value $\overline{\kappa}_\ell$, which is computed by using its definition in Theorem 2. Since a larger value of κ/P can accelerate the solution process, P is estimated to be $\overline{P}_\ell$ by the minimum value within the population by utilizing the following expression.

$$P_{\ell,a} = \frac{1}{2}\left(\sum_{i=1}^{N}\sum_{\beta=1}^{3}\frac{1}{2}e_{a,i,\beta}^2 + \sum_{i=1}^{N}\sum_{v=1}^{2}\frac{1}{2}\eta_{a,i,v}^2\right) \tag{33}$$

$$\overline{P}_\ell = \min[P_{\ell,1}, P_{\ell,2}, \ldots, P_{\ell,m}] \tag{34}$$

where $a = 1,2,\ldots,m$, $e_{a,i,\beta}$ is the minimum value of the time-domain curve for the βth error surface in region i corresponding to candidate solution a. $\eta_{a,i,v}$ is the minimum value of the time-domain curve for the v filtering error in region i corresponding to candidate solution a.

Based on the condition $\mu \geq \kappa/P$ in Theorem 2, μ is updated by the following equation.

$$\mu_\ell = \overline{\kappa}_\ell/\overline{P}_\ell \tag{35}$$

After computing μ_ℓ for the current population, constraint boundaries $c_{i\min}$ and $T_{fi\max}$ can be updated according to Theorem 2's definition $\mu_x = \min[2c_{i,k}, 2/T_{fi,v} - 2]$, which serve as constraints for the $(\ell+1)^{\text{th}}$ optimization:

$$\begin{cases} c_{i\min} = \frac{\mu_x}{2}, 2c_{i,k} < 2/T_{fi,v} - 2 \\ T_{fi\max} = \frac{\mu_x+2}{2}, 2c_{i,k} > 2/T_{fi,v} - 2 \end{cases} \tag{36}$$

To determine the constraint boundaries $c_{i\max}$ and $T_{fi\min}$ in the $(\ell+1)^{\text{th}}$ optimization, the following characteristics are considered. Since larger c_i values (within a certain range) improve system's dynamics, $c_{i\max}$ can be set moderately high. For T_{fi}, smaller values reduce filtering error. However, excessively small T_{fi} prolongs simulation time and reduces iteration efficiency. Thus, $T_{fi\min}$ should satisfy Theorem 2's requirement $1/T_{fi,v} - 1 > 0$ and avoid being too small.

As to solve the above optimization problem, we adopt the CGO algorithm [21], which employed the Sierpinski triangle algorithm to integrate chaos theory with game theory to improve search efficiency. The main concepts and equations of CGO are as follows.

Assume that in current population, there are m current candidate solutions with expression as follows.

$$W_i = \begin{bmatrix} w_i^1 & w_i^2 & \ldots & w_i^j & \ldots & w_i^d \end{bmatrix}, \text{ with } i = 1,2,\ldots,m \text{ and } j = 1,2,\ldots,d \tag{37}$$

where d is the number of parameters that need to be optimized, and W_i includes all parameters to be optimized (c_i, T_{fi}), which are randomly generated within the search space by Equation (38).

$$w_i^j(0) = w_{i,\min}^j + rand\left(w_{i,\max}^j - w_{i,\min}^j\right), \begin{cases} i = 1,2,\ldots,n \\ j = 1,2,\ldots,d \end{cases} \tag{38}$$

where $w_i^j(0)$ denotes the j-th decision variable of the i-th point in the search space, $w_{i,\max}^j$ and $w_{i,\min}^j$ represent the upper bound and lower bounds of decision variables, respectively, and $rand$ is a random number within the interval [0, 1].

The Sierpinski triangle algorithm is the main approach to update the candidate solutions W_i, which is formed by four seeds as follows.

$$S_i^1 = W_i + \rho_i \times (\omega_i \times GB - \xi_i \times MG_i), i = 1,2,\ldots,n \tag{39}$$

$$S_i^2 = GB + \rho_i \times (\omega_i \times W_i - \xi_i \times MG_i), i = 1,2,\ldots,n \tag{40}$$

$$S_i^3 = MG_i + \rho_i \times (\omega_i \times W_i - \zeta_i \times GB), i = 1,2,\ldots,n \tag{41}$$

$$S_i^4 = W_i\left(w_i^k = w_i^k + R\right), k = [1,2,\ldots,d] \tag{42}$$

where GB stands for the optimal candidate solution in the current state within the search space, MG_i is the average value of randomly selected initial qualified points, W_i represents the i-th candidate solution, ρ_i denotes a random number generated in a certain manner, ω_i and ξ_i are random integers that are either 0 or 1, and R is a random number within the range of [0, 1].

Each candidate solution is regarded as an independent player. These players select the optimal action and update their own states according to their current states and the states of other players. The pseudocode of the optimization process is provided as in Algorithm 2.

Algorithm 2. Optimization process of CGO

Input: $c_{i\min} \leq c_i \leq c_{i\max}(i = 1, \ldots, N)$, $T_{fi\min}$, $T_{fi\max}(i = 1, \ldots, N)$ and Equation (31)

Output: The optimal candidate solution GB (DFC parameters)

① First, generate random initial points $w_i^j(0)$ in the search space.

② Calculate the fitness (degree of proximity to the target) by substituting the initial solutions into the objective function.

③ Obtain the optimal candidate solution GB in the current state.

④ Enter the following loop:

```
While (t < maximum number of iterations)
  for i = 1: number of candidate solutions
    Generate MG_i by randomly selecting initial qualified points
    Generate a temporary triangle from W_i, MG_i, and GB
    Generate new seeds using the calculation formulas for the four seeds
    Calculate the fitness of the new seeds
      if the fitness value of the new seed is better than that of the worst candidate solution
        Replace the worst candidate solution with the new seed
      end
      If there is a better candidate solution, update GB
  end
    t = t + 1
end
```

⑤ Obtain the optimal solution GB

So far, the optimization of parameters using CGO has been completed.

5. Case Study and Analysis

In this subsection, the control effect of the proposed method on the dynamic frequency stabilization of the power system is verified by two multi-regional interconnected test systems.

5.1. Case Study and Analysis via a Two-Region Interconnected System

This subsection verifies the effectiveness and superiority of PV DRO and DFC, respectively, based on a two-region interconnected arithmetic system shown in Figure 10. Within each region, four thermal units and six PV units are set up, where each thermal unit has a capacity of 70 MW and each PV unit has a capacity of 20 MW [22,23].The equivalent model parameters are calculated based on [24], and listed in Tables 1 and 2, and the coefficient of tie-line between the two regions is $T_{12} = T_{21} = 0.2$ p.u.

Table 1. Parameters of six photovoltaic systems.

Number	T_V/s	T_I/s	Kp/(p.u.)
1	1.1	0.056	−13
2	1.3	0.054	−10
3	1.7	0.051	−20
4	1.2	0.055	−16.67
5	1.3	0.052	−20
6	1.4	0.053	−20

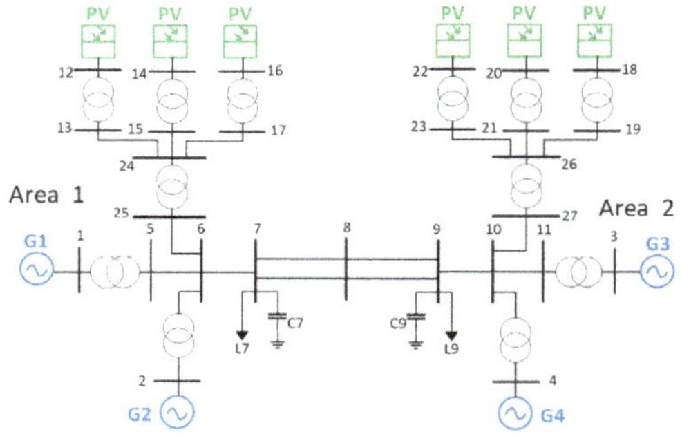

Figure 10. Two-region interconnected test system.

Table 2. Parameters of the aggregated photovoltaic system, and the region's inertia and damping.

T_V (s)	T_I (s)	K_p/(p.u.)	D/(p.u./Hz)	M (s)
1.363	0.053	−16.5	0.024	19.92

At t = 1 s, a step load of 0.06 p.u. is applied to Region 1. The frequency deviation curve of Region 1 without additional dynamic frequency controller is shown in Figure 11a. It can be observed that the system frequency exhibits two oscillatory modes, namely, a low-frequency oscillation at 0.4 Hz and an ultra-low-frequency oscillation at 0.059 Hz. Both oscillations are of weak damping.

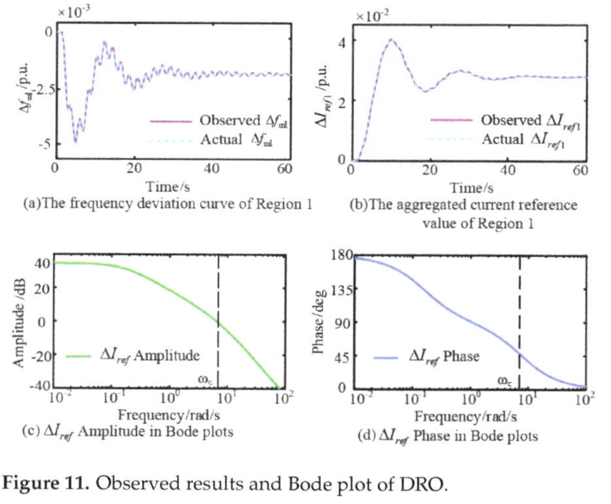

Figure 11. Observed results and Bode plot of DRO.

(1) Observer Performance Analysis

To test the performance of the proposed observer, Figure 11a,b illustrate the comparison between the observed and actual values of the two aggregated states Δf_{m1} and ΔI_{ref1} in Region 1. During the observation process, the maximum error of the state observer was 1.6×10^{-4} p.u., and the steady-state tracking error converged to 6×10^{-5} p.u. Figure 11c,d show a frequency domain analysis of the observation of signals ΔI_{ref1} using Bode plots. The DRO can effectively perform observations in the low-frequency band, and its amplitude gain attenuates at a rate of −40 dB/dec in the high-frequency band, ensuring that the observation results are not affected by high-frequency noise from the sensors.

Figure 12a,b show the comparison between the actual and observed values of disturbances Δp_{L1} and Δp_{L2} in two regions, respectively. As shown in Figure 12a, the observed values rapidly track the actual values in approximately 0.2 s after the disturbance occurs. Figure 12b further demonstrates that the steady-state error converges to 1.5×10^{-4} p.u through the global time-domain response. The accuracy and speed of the observation results demonstrate the good observation performance of the proposed photovoltaic DRC.

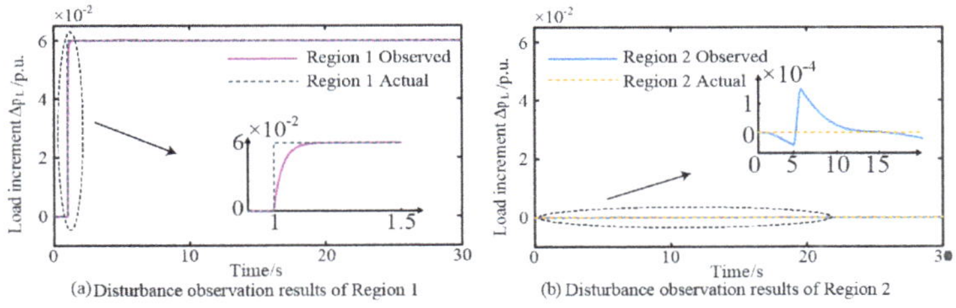

Figure 12. Disturbance observation results.

(2) Dynamic frequency stabilization control effect

To test the performance of the proposed DFC, Figure 13 shows the dynamic frequency response process of the proposed DFC under a 0.06 p.u. step disturbance of load in Region 1. Figure 13a,b show the control effects after optimizing the controller parameters Via CGO. The simulation results indicate that the proposed control method effectively limits the system's maximum frequency deviation to within 0.17 Hz. And, within 12 s after the disturbance, it effectively suppresses both low-frequency and ultra-low-frequency oscillations. The steady-state frequency deviation ultimately converges to 1×10^{-8} Hz.

Figure 13c,d present the control effects when the controller parameters are set based on experience. When the controller parameters are not optimized Via the CGO algorithm, neither the frequency nor the tie-line power variations have fully converged at t = 60 s. Through comparison, it is evident that the control effect of the controller without CGO is far inferior to that with CGO.

Figure 13e is the ITAE convergence curve during the CGO process, from which it can be seen that the ITAE value decreases steadily during optimization.

Furthermore, we also randomly selected 100 sets of parameters for Spearman correlation analysis, and the results are presented in Figure 13f. The analysis results indicate that $c_{i,1}$ and $T_{fi,2}$ are the controller parameters that have a relatively significant impact on the ITAE evaluation. Their correlation coefficients are 0.570 and 0.444, respectively, which are much higher than those of other controller parameters. This finding helps us accurately identify the key parameters and provides data support for the subsequent optimization of control performance.

(3) Comparative analysis with centralized control

Comparative experiments are conducted between the proposed method and the centralized frequency control [23]. A maximum communication time lag $\tau_{max} = 2.5$ s was set between the two regions. At t = 1 s, a 0.06 p.u. step load is applied to Region 1. Figure 14 shows the resulting frequency response curves of both regions, along with the dynamic variation in the total output power of the tie-line.

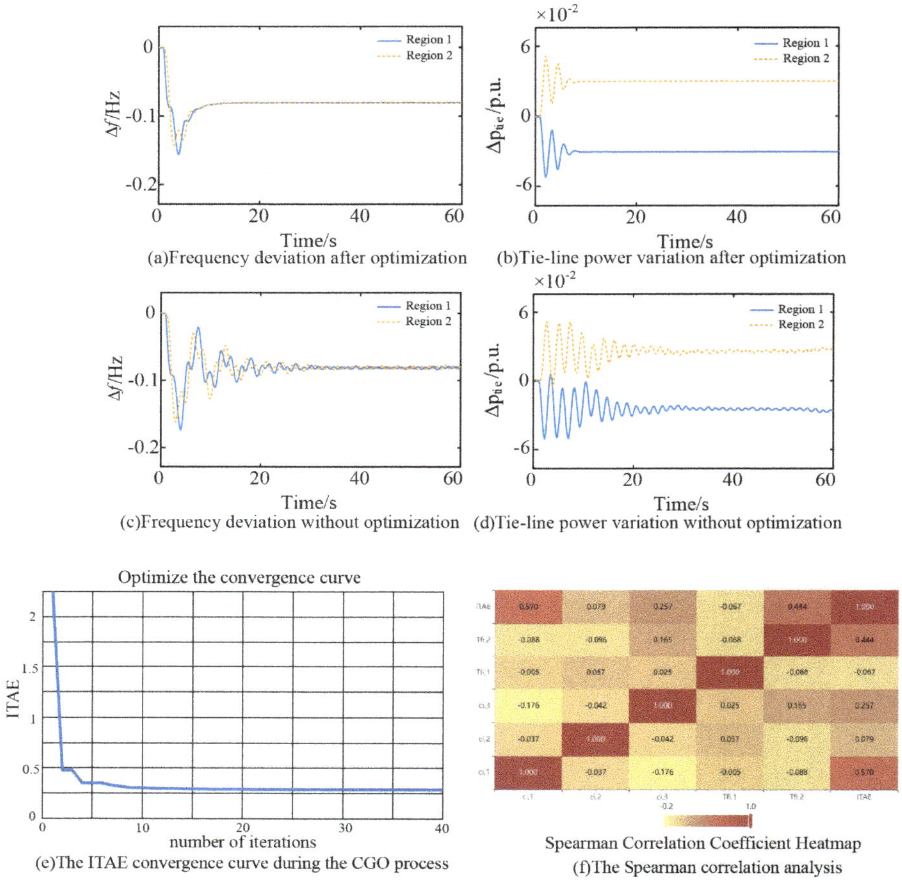

Figure 13. Controller and CGO Effect Verification.

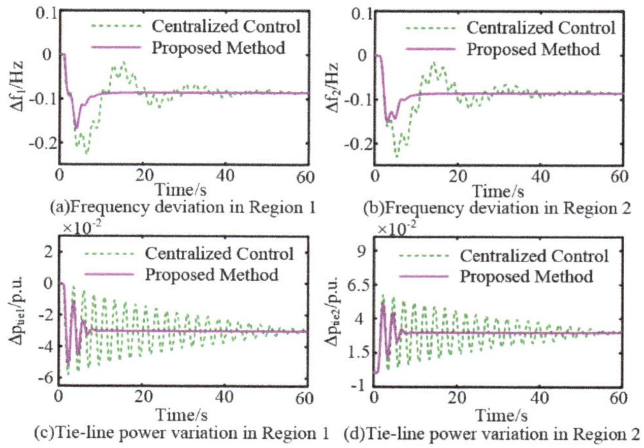

Figure 14. Dynamic response processes of the system under centralized and the proposed approaches.

As shown in Figure 14a,b, compared with the centralized control, the proposed method reduces the maximum frequency deviation from 0.23 Hz to 0.17 Hz, with a reduction of 26.1%. The settling time is shortened from 50.6 s to 12 s with a reduction of 76.3%.

To quantify the performance of regional coordinated control, the inter-area frequency deviation index is defined as $E_F = \int_0^T t\left(\sum_{i \neq j} |\Delta f_i - \Delta f_j|\right) dt$. The experimental data in Figure 14 show that, during the system's gradual frequency recovery, the centralized frequency regulation yields a large inter-area frequency cumulative deviation with $E_F = 0.222$. Since centralized control requires a double data communication task from the actor and the control center, which amplifies the time delay, which compromises the effect of centralized control.

Under the same communication condition, the proposed controller achieves a significantly smaller E_F of merely 0.014. This indicates the proposed DFC with reduced communication latency can effectively minimize inter-area frequency difference and gain good control performance.

5.2. Case Study of the Four-Region System

This subsection tests the synergy effect of the proposed method using a four-region interconnected system. It also analyzes the impact on the synergy effect under different communication conditions. The corresponding parameter settings are shown in Table 3. Among them, T_V is the time constant of the equivalent voltage outer loop of the photovoltaic aggregated PV inverter in the local area, T_I is the time constant of the equivalent current inner loop of the photovoltaic inverter in the local area, K_p is the equivalent droop coefficient of the photovoltaic cluster in the local area, M is the equivalent inertia of the local area, and D is the equivalent damping of the local area.

Table 3. Four region frequency system parameters.

Region	T_V/s	T_I/s	K_p/(p.u.)	D/(p.u./Hz)	M/s
1	1	0.056	−20	0.0166	16.7
2	1.6	0.054	−18.52	0.0178	22.2
3	2	0.054	−20	0.016	19.3
4	1	0.056	−20	0.0166	16.7

During the design phase of the observer, matrices F_i, T_i, K_i, and H_i are the to-be-designed parameters of the state observer, and K_{di} is the to-be-designed parameter of the disturbance observer

During the design phase of the controller, we have considered parameters as follows.

① The adjacency weight $a_{i,j}$, the reference signal weight $a_{i,0}$: these parameters are the design parameters to build error surface $e_{i,1}$. They are used in the calculation of Equation (10).

② Filter time parameter T_{fi}: they are design parameters to calculate error surfaces $e_{i,2}$ and $e_{i,3}$, which is used in Equation (13).

③ Parameters $c_{i,1}$, $c_{i,2}$ and $c_{i,3}$ (control law design parameters): these parameters enable the system states to evolve in accordance with various virtual control laws and ensure the stable convergence of the defined error surface. They are used in Equations (20), (24) and (28), respectively.

Figure 15 illustrates the communication topology among the interconnected system regions under both strong and weak communication conditions. The dashed line representing inter-area communication with a maximum communication time lag of 2.5 s.

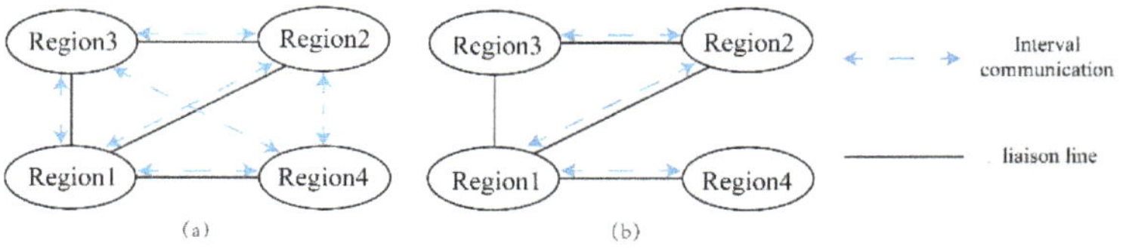

Figure 15. Communication topology of interconnected systems: (**a**) strong communication and (**b**) weak communication.

(1) Comparative Analysis with Decentralized Control

Using the strong interconnection communication topology shown in Figure 15a, the control effects of the traditional decentralized control [14] are compared with the proposed method under good communication environment. At t = 1 s, a step disturbance of load with an amplitude of 0.06 p.u. is synchronously applied to regions 1 and 4. The corresponding frequency response curves of each region are shown in Figure 16, and the dynamic curves of the total output power of the tie-line is illustrated in Figure 17.

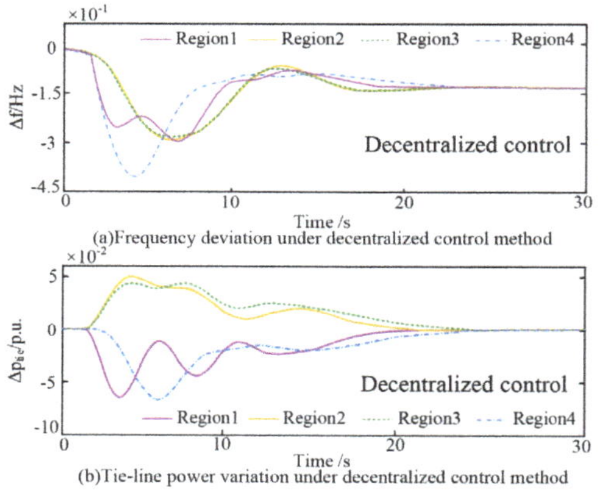

Figure 16. Dynamic response process of four-region interconnected system by using decentralized control.

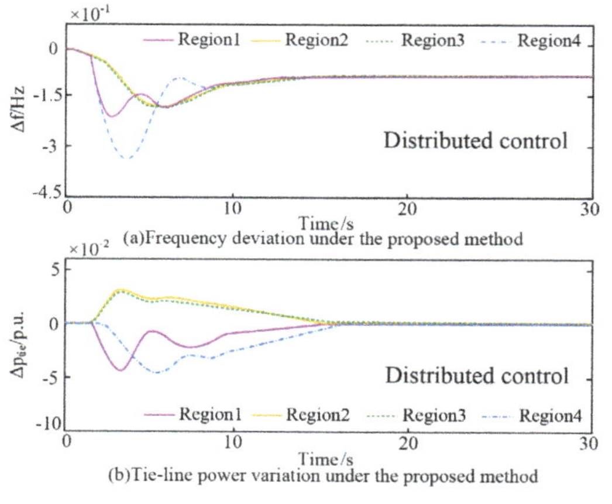

Figure 17. Dynamic response process of four-region interconnected system by using the proposed DFC.

As can be seen in Figure 16, after the load disturbance, each region presents independent response characteristics, resulting in significant frequency differences between regions 1 and 4. The maximum instantaneous frequency difference between the two reaches 0.19 Hz. Stability was regained after 23.46 s following the load disturbance. Given that region 4 is only interconnected with region 1 and exhibits the largest frequency deviation, it was selected for observation. Its maximum frequency offset ultimately reached 0.41 Hz. The amplitude of the maximum power fluctuation of the entire tie-line reaches 0.07 p.u.

In contrast, the method in this paper realizes inter-region dynamic synergy through the multi-agent consensus algorithm. The frequency deviation of regions 1–4 converges synchronously, the maximum frequency offset of region 4 is suppressed to 0.34 Hz, and is quickly suppressed within 13.4 s. Compared with decentralized control, this approach resulted in a 17.1% reduction in frequency deviation and a 42.9% shortening of the settling time.

To facilitate the comparison of the performance among centralized control, decentralized control, and the control method proposed in this paper, we have compiled Table 4 to compare the key parameters. This table contrasts the three methods in terms of settling time and maximum frequency deviation, allowing for a clear identification of the performance differences between the centralized control, decentralized control, and the method proposed in this paper.

Table 4. Comparison of key performance metrics.

Metrics	Two-Region Interconnected System		Four-Region Interconnected System	
	Centralized Control	**The Method in This Paper**	**Decentralized Control**	**The Method in This Paper**
Settling time/s	50.6	12	23.46	13.4
Maximum frequency offset/Hz	0.23	0.17	0.41	0.34

(2) Controller performance analysis under different communication conditions

This subsection evaluates the control performance of the proposed control method under both strong and weak communication topologies. Using the topologies shown in Figure 15a,b as examples, a 0.06 p.u. step load is applied to regions 1 and 4 at t = 1 s and t = 20 s, respectively. Under the conditions of strong communication and random communication delays, the frequency response curves of each region and the dynamic process of the total output power of the contact line are shown in Figure 18a,b. Under the proposed controller, the dynamic process of the system transitions to the steady state rapidly. And, with the inter-region frequency offsets stabilized, the tie-line power fluctuation also tends to stabilize.

Under strong communication conditions, with the communication delay set to fixed values of 1.5 s and 1 s, the frequency response curves of each region are shown in Figure 18c,d. It can be observed that all frequencies can converge quickly and stably. However, the convergence speed and convergence effect under the condition of fixed time delay are better than those under random time delay. Moreover, as the fixed time delay decreases, both the convergence speed and convergence effect improve.

To demonstrate the transient characteristics of the controller, under the condition of strong communication random time delay, we present the output u_i of the four-area controller within the time period of 0–8 s. Figure 18e shows that the controllers responded rapidly to the disturbance. The clipping module effectively suppressed the signal surge in the first two cycles, ensuring stable oscillation control. Subsequently, the control signals fell below the clipping amplitude by the third cycle. This effective response resulted in satisfactory performance by t = 8 s, as further evidenced by the frequency response in Figure 18a.

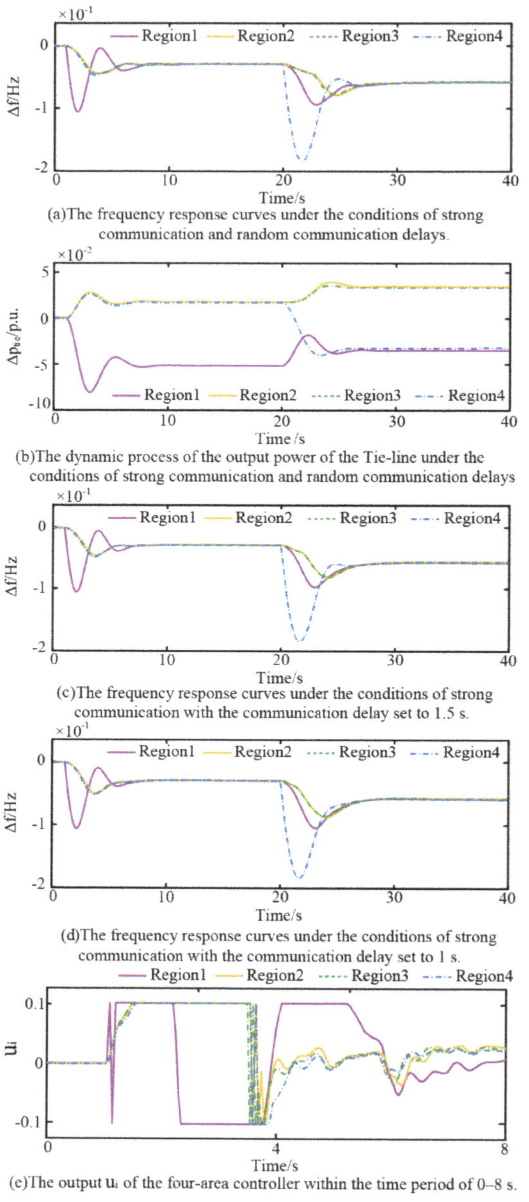

Figure 18. Dynamic of four-region interconnected test system under strong communication.

Figure 19a,b demonstrate the dynamic response curve of the system with the weak communication topology shown in Figure 15b using the proposed controller. It is shown that the frequency oscillation can also be narrowed down and eventually converge to a steady state.

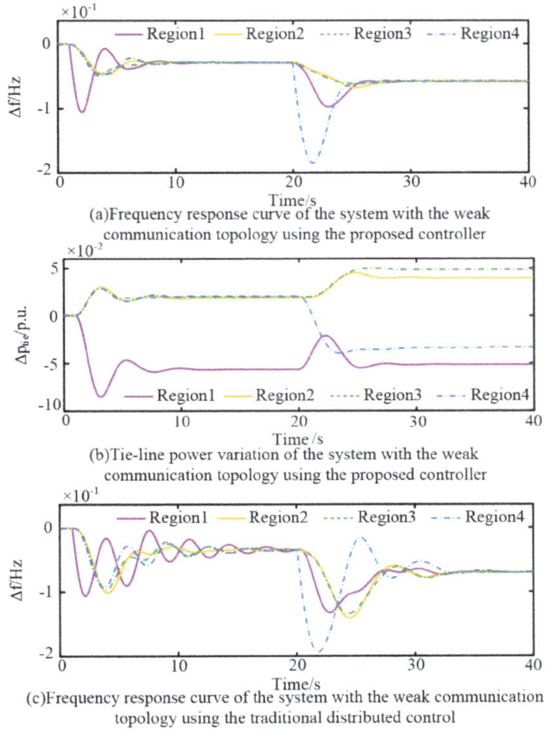

Figure 19. Dynamic of four-region interconnected test system under weak communication.

Figure 19c shows the frequency dynamic diagram obtained by using the traditional distributed control method under weak communication conditions. After the first disturbance at t = 1 s, the frequency under the traditional distributed control failed to fully converge even by t = 20 s. After the disturbance at t = 20 s, it took 18 s to reach a stable state. The convergence speed of the traditional distributed control is much slower than that of the control method proposed in this paper. It can be clearly seen that the control effect is far inferior to that of the control method proposed in this paper.

Table 5 shows the change in outflow power Via the tie-line after the perturbation in regions 1 and 4 under strong and weak communications, respectively. Analysis of the data in Table 5 reveals that under strong communication conditions (Figure 15a), all regions achieve full interconnection. As shown in Table 5, the tie-line power support received by Region 1 and Region 4 is remarkably similar under these conditions (The additional power flowing into Region 4 Via the tie-line is 95.3% of that into Region 1). This indicates that with favorable communication, the proposed coordinated control mechanism can achieve the allocation of cross regional power with the goal of minimizing ITAE in Equation (31), allowing the load disturbance to be shared more evenly across a broader area. Consequently, the overall frequency response performance is further optimized, evidenced by faster recovery and reduced oscillations as depicted in Figure 18a.

Table 5. Power support by tie-lines.

Test Environment	Region 1 Power Outflow Change/p.u.	Region 4 Power Outflow Change/p.u.
Weak communication	−0.0501	−0.0326
Strong communication	−0.0339	−0.0323

In contrast, under weak communication conditions, a significant disparity exists in the power support received by different regions (The additional power flowing into Region 4 via the tie-line is 65.1% of that into Region 1). However, as shown in Figure 19a, the proposed controller can also effectively suppress both low-frequency and ultra-low-frequency oscillations across all four regions. This phenomenon demonstrates that even in extreme situations characterized by non-ideal communication topologies, resulting in information asymmetry and uneven power allocation, and significant max delay conditions of up to 2.5 s, the controller can still maintain dynamic frequency stability, demonstrating the superiority of the distributed control algorithm based on state–disturbance observation and dynamic surface consensus control proposed in this paper.

6. Conclusions

This paper addresses the dynamic frequency stability challenges in wide-area interconnected power systems by proposing a distributed photovoltaic frequency controller based on a multi-agent consensus algorithm and dynamic surface control methodology. The principal conclusions are summarized as follows:

(1) The PV-based control strategy proposed in this paper can suppress both low-frequency oscillations and ultra-low-frequency oscillations simultaneously and also achieves favorable performance even under weak communication conditions.
(2) A three-layer distributed cloud control framework of terminal–edge–collaboration is constructed, which realized double high-precision observation of state variables and external perturbations through decentralized regional observer. The maximum error of the state observer is 1.6×10^{-4} p.u. and the steady state tracking error converges to 6×10^{-5} p.u.
(3) A distributed frequency controller (DFC) is proposed using the MAC and DSC method, which can effectively limit the maximum frequency deviation of the system to 0.17 Hz, and the steady state frequency deviation converges to 1×10^{-8} p.u.
(4) The parameter adjustment strategy based on CGO is proposed to achieve frequency deviation suppression. Simulation results show that the frequency deviation can be suppressed to ± 0.17 Hz under the condition of communication delay of 2.5 s.
(5) The test results show that compared with that of the traditional centralized control, the proposed dynamic frequency active support method can reduce the frequency deviation by 26.1% and the regulation time by 76.3%. And compared with that of the decentralized control, the proposed method can reduce the frequency deviation by 17.1% and the regulation time by 42.9%.
(6) The PV-based control method proposed in this study can effectively suppress low-frequency and ultra-low-frequency oscillations, significantly enhancing the power grid's capacity to accommodate high-proportion clean energy and its stability.

There remains room for further research regarding the finite-time convergence of the control method and we will conduct in-depth studies on this issue in the future. In addition, in our follow-up studies, we will also carry out in-depth research on issues including computational requirements, communication infrastructure needs, and potential integration challenges with existing control systems.

Author Contributions: Conceptualization, Y.Z. (Yichen Zhou), Y.G. and Y.L. (Yifei Liu); methodology, Y.Z. (Yichen Zhou), Y.L. (Yifei Liu) and Y.T.; software, Y.G. and Y.L. (Yifei Liu); validation, Y.Z. (Yichen Zhou), Y.G. and Y.L. (Yifei Liu); formal analysis, Y.G. and Y.T.; investigation, Y.Z. (Yichen Zhou), Y.G., Y.L. (Yifei Liu), L.T. and Y.L. (Yuyan Liu); resources, Y.T., L.T., Y.L. (Yuyan Liu) and J.Y.; data curation, Y.G., Y.Z. (Yifei Zhang) and R.C.; writing—original draft preparation, Y.L. (Yifei Liu); writing—review and editing, Y.Z. (Yichen Zhou), Y.G., Y.T., Y.L. (Yifei Liu), L.T., Y.Z. (Yifei Zhang), Y.L. (Yuyan Liu), X.Z., J.Y. and R.C.; visualization, Y.T., L.T., Y.L. (Yuyan Liu) and J.Y.; supervision, Y.Z. (Yichen Zhou) and Y.T.; project administration, Y.T.; funding acquisition, Y.Z. (Yichen Zhou). All authors have read and agreed to the published version of the manuscript.

Funding: This research was funded by China Southern Grid Digital Grid Funding, grant number DPGCSG-2024-KF-21. The APC was funded by China Southern Grid Digital Grid Funding.

Data Availability Statement: The original contributions presented in this study are included in the article. Further inquiries can be directed to the corresponding author.

Acknowledgments: The authors extend their sincere gratitude to North China Electric Power University for the support and resources provided during this study.

Conflicts of Interest: Author Yujia Tang was employed by the company National Energy Power Grid Technology R&D Centre, Guangzhou 510663, China. Author Tu Liang was employed by the company National Energy Power Grid Technology R&D Centre, Guangzhou 510663, China. Author Yuyan Liu was employed by the company National Energy Power Grid Technology R&D Centre, Guangzhou 510663, China. Author Jiawei Yu was employed by the company National Energy Power Grid Technology R&D Centre, Guangzhou 510663, China. The remaining authors declare that the research was conducted in the absence of any commercial or financial relationships that could be construed as a potential conflict of interest. The authors declare that this study received funding from China Southern Grid Digital Grid Funding, grant number DPGCSG-2024-KF-21. The funder had the following involvement with the study: study design, interpretation of data.

Appendix A

This appendix presents the proof of Theorem 1, and the key steps are as follows.

Theorem A1. *By reasonably designing the parameter matrices F_i, T_i, K_i, and H_i to satisfy Equation (9), the state observation error can be guaranteed to converge asymptotically. Additionally, by designing K_{di}, the disturbance observation error can be guaranteed to be a bounded quantity and arbitrarily small within the closed set $\Omega_{\tilde{d}}$. The larger the value of $\lambda_{min}(K_i^l)$, the smaller the disturbance observation error.*

$$\begin{aligned} & F_i = A_i - H_iC_iA_i - K_{i1}C_i \\ & K_{i2} = F_iH_i \\ & T_i = I_i - H_iC_i \\ & T_iE_i = 0 \end{aligned} \tag{A1}$$

Proof. Let $K_i = K_{i1} + K_{i2}$ be the state of the system. Combining the dynamic Equation (5) with Equations (6) and (7), we obtain the following relationship for the state observation error $\tilde{x}_i(t) = x_i(t) - \hat{x}_i(t)$ of the designed observer:

$$\begin{aligned} \dot{\tilde{x}}_i(t) \ & = (A_i - H_iC_iA_i - K_{i1}C_i)\tilde{x}_i(t) \\ & +(I_i - T_i - H_iC_i)B_iu_i(t) \\ & +(I_i - H_iC_i)E_id_i(t) \\ & +(A_i - H_iC_iA_i - K_{i1}C_i - F_i)z_i(t) \\ & +[(A_i - H_iC_iA_i - K_{i1}C_i)H_i - K_{i2}]y_i(t) \end{aligned} \tag{A2}$$

When the design matrix parameters satisfy the conditions of Equation (9), Equation (A2) can be simplified to $\dot{\tilde{X}}_i(t) = F_i\tilde{X}_i(t)$. At this point, the state observation can be decoupled from the input. As long as the design of K_{i1} ensures that F_i is a Hurwitz matrix, the state observation error will be progressively stable.

For the disturbance observation error, its dynamic equation is:

$$\begin{bmatrix} \dot{\tilde{x}}_i \\ E_i\dot{\tilde{d}}_i \end{bmatrix} = \begin{bmatrix} F_i & 0 \\ K_{\mathrm{d}i}(F_i - A_i) & -K_{\mathrm{d}i} \end{bmatrix} \begin{bmatrix} \tilde{x}_i \\ E_i\tilde{d}_i \end{bmatrix} - \begin{bmatrix} 0 \\ E_i\dot{d}_i \end{bmatrix} \tag{A3}$$

Define the Lyapunov function:

$$V_{do} = \frac{1}{2}\left(E_i\tilde{d}_i\right)^T \cdot E_i\tilde{d}_i \tag{A4}$$

For power systems, disturbances are bounded. Assume that $\|Ed\| \le D$ and the first-order derivative of the disturbance $\dot{d}_i$ is bounded. Differentiate Equation (A4) and combine it with Equation (A3) and Young's inequality to obtain:

$$\begin{aligned} \dot{V}_{do} &= \left(E_i\tilde{d}_i\right)^T\left[K_{\mathrm{d}i}(F_i - A_i)\tilde{x}_i - K_{\mathrm{d}i}E_i\tilde{d}_i - E_i\dot{d}_i\right] \\ &\le -\left(E_i\tilde{d}_i\right)^T K_{\mathrm{d}i}E_i\tilde{d}_i + \tfrac{1}{2}\left(E_i\tilde{d}_i\right)^T E_i\tilde{d}_i + \tfrac{1}{2}\iota^2 = -2\rho V_{do} + C_d \end{aligned} \tag{A5}$$

where $\left\|K_{\mathrm{d}i}(F_i - A_i)\tilde{x}_i - E_i\dot{d}_i\right\| \le \iota$, $C_d = 1/2\iota^2$ and $\rho = \lambda_{\min}(K_{\mathrm{d}i}) - 1/2$, $\lambda_{\min}(\cdot)$ denote the smallest eigenvalues of the matrix. Design to satisfy $\lambda_{\min}(K_{\mathrm{d}i}) > 1/2$, then $-2\rho V_{do} < 0$.

Solve the first-order linear differential equation shown in Equation (A5) to obtain:

$$0 \le V_{do}(t) \le \frac{C_d}{2\rho} + \left[V_{do}(0) - \frac{C_d}{2\rho}\right]e^{-2\rho t} \tag{A6}$$

Global consistency is ultimately bounded. Combining Equation (A4), we obtain:

$$\|E_i\tilde{d}_i\| \le \sqrt{\frac{C_d}{\rho} + \left[2V_{do}(0) - \frac{C_d}{\rho}\right]e^{-2\rho t}} \tag{A7}$$

Therefore, according to Equation (A7), there exists a positive constant $\zeta_{\tilde{d}} > \sqrt{C_d/\rho}$ such that the interference observation error converges to the closed set $\Omega_{E\tilde{d}} = \{E\tilde{d} \in R^4 \mid \|E\tilde{d}\| < \zeta_{\tilde{d}}, \zeta_{\tilde{d}} > \sqrt{C_d/\rho}\}$, indicating that the designed DRO interference observation error is bounded.

It can be seen from Equation (A7) that when designing the DRO, the value of C_d/ρ can be controlled through $K_{\mathrm{d}i}$, thereby ensuring that the interference observation error is arbitrarily small. Q.E.D. □

Appendix B

Theorem A2. *For each regional controlled system (1) under the action of DRO-based DFC, if the design parameters $c_{i,k}$ and $T_{fi,v}$ satisfy Equation (30), then for any $P > 0$ there exists a bounded compact set Ω_i such that when $\mu \ge \kappa/P$, all signals in the closed-loop system are semi-globally uniformly ultimately bounded, where $\mu = min\left[2c_{i,k}, 2/T_{fi,v} - 2\right]$, $\kappa = 1/2 \cdot \sum_{i=1}^{N}\left(\sum_{v=1}^{2} A_{i,v}^2 + D_{SF,i}^2 + D_{DCO,i}^2\right)$.*

$$c_{i,k} > 0, 1/T_{fi,v} - 1 > 0, i = 1,\ldots,N, v = 1,2 \tag{A8}$$

Proof. For the interconnected system, define the Lyapunov function as:

$$V=\sum_{i=1}^{N}V_{i,3}=\sum_{i=1}^{N}\sum_{\beta=1}^{3}\frac{1}{2}e_{i,\beta}^{2}+\sum_{i=1}^{N}\sum_{\nu=1}^{2}\frac{1}{2}\eta_{i,\nu}^{2} \tag{A9}$$

Differentiating Equation (A9) and incorporating the error surfaces (12), (14), the filter (13), and the constructed control laws (20), (24), (28) with Young's inequality yields:

$$\dot{V}\leq-\sum_{i=1}^{N}\sum_{\beta=1}^{3}c_{i,\beta}e_{i,\beta}^{2}-\sum_{i=1}^{N}\sum_{\nu=1}^{2}\left(1/T_{\mathrm{fi},\nu}-1\right)\eta_{i,\nu}^{2}+\frac{1}{2}\sum_{i=1}^{N}\sum_{j\in N_i}a_{ij}\tilde{d}_{j,1}^{2}(t_\tau)+\frac{1}{2}\sum_{i=1}^{N}\left(c_{i,3}+2\right)^{2}\tilde{x}_{i,4}^{2}+\frac{1}{2}\sum_{i=1}^{N}\left(\psi_{i,1}^{2}+\psi_{i,2}^{2}+\tilde{g}_{i,4}^{2}+\tilde{d}_{i,1}^{2}\right) \tag{A10}$$

Assuming stability, all observation error terms in Equation (A10) are quadratic. For analytical convenience, let the maximum observation error term $D_{DCO,i}^{2}$ represent the observation error for region i. For any $P>0$, there exists a set $\Omega_{i,m}=\left\{\sum_{\gamma=1}^{i}\frac{1}{2}\left(\sum_{\beta=1}^{m}e_{\gamma,\beta}^{2}+\sum_{\beta=1}^{m-1}\eta_{\gamma,\beta}^{2}\right)+\sum_{\beta\in N_i}e_{\beta,m+1}^{2}\leq 2P\right\}(m=1,2)$, where $|\dot{\alpha}_{i,k}|$ has an upper bound in the bounded compact set $\Omega_{i,m}$, denoted as $A_{i,m}$ [25]. This implies $|\psi_{i,m}|\leq A_{i,m}$, with $A_{i,m}$ being a positive constant. Within this set, the maximum error value for regional frequency deviation $d_{i,1}$ is $D_{SF,i}$. Defining $\Omega_i=\cap_{m=1}^{2}\Omega_{i,m}$, we can further bound Equation (A10) on the compact set Ω_i:

$$\dot{V}\leq-\sum_{i=1}^{N}\sum_{\beta=1}^{3}c_{i,\beta}e_{i,\beta}^{2}-\sum_{i=1}^{N}\sum_{\nu=1}^{2}\left(\frac{1}{T_{\mathrm{fi},\nu}}-1\right)\eta_{i,\nu}^{2}+\frac{1}{2}\sum_{i=1}^{N}\sum_{\nu=1}^{2}A_{i,\nu}^{2}+\frac{1}{2}\sum_{i=1}^{N}D_{SF,i}^{2}+\frac{1}{2}\sum_{i=1}^{N}D_{DCO,i}^{2} \tag{A11}$$

Define: $\kappa=1/2\cdot\sum_{i=1}^{N}\left(\sum_{\nu=1}^{2}A_{i,\nu}^{2}+D_{SF,i}^{2}+D_{DCO,i}^{2}\right)$. Let the design parameters $c_{i,\beta}$ and $\tau_{i,\nu}$ satisfy $c_{i,\beta}>0$ and $1/\tau_{i,\nu}-1>0$. Denoting $\mu=\min\left[2c_{i,\beta},2/T_{\mathrm{fi},\nu}-2\right](i=1,\ldots,N,\nu=1,2,\beta=1,\ldots,3)$, it follows that $\kappa>0$ and $\mu>0$.

Since: $\mu V=\mu\sum_{i=1}^{N}\sum_{\beta=1}^{3}1/2\cdot e_{i,\beta}^{2}+\mu\sum_{i=1}^{N}\sum_{\nu=1}^{2}1/2\cdot\eta_{i,\nu}^{2}$, Rearranging Equation (A11) yields:

$$\dot{V}\leq-\mu V+\kappa \tag{A12}$$

Setting $\mu>\kappa/P$, we have $-\mu V+\kappa<\kappa(1-V/P)$. When $V\geq P$, it follows that $\kappa(1-V/P)\leq 0$, implying $\dot{V}<0$. Thus, $V\leq P$ is an invariant set, and all signals in the closed-loop system are semi-globally uniformly ultimately bounded. Q.E.D. □

References

1. National Grid ESO. *Technical Report on the Low Frequency Demand Disconnection (LFDD) Event*; National Grid ESO: Warwick, UK, 2020.
2. Islam, M.R.; Azam, M.S.; Hossen, M.S.; Islam, M.S.; Worku, M.Y.; Shahriar, M.S.; Shafiullah, M. Power system stability enhancement through optimal PSS design. *e-Prime—Adv. Electr. Eng. Electron. Energy* **2024**, *9*, 100735. [CrossRef]
3. Xu, T.; Sun, S.; Zhou, Z.; Feng, H.; Deng, H.; Chen, Y.; Li, B. The principle and influence analysis of suppressing ultra-low frequency oscillation by using additional GPSS control. In Proceedings of the 2023 3rd International Conference on Intelligent Power and Systems (ICIPS), Shenzhen, China, 20–22 October 2023; pp. 250–254.
4. Sun, W.; Ma, Z. Mechanistic Analysis of Ultra-Low-Frequency Oscillation Accompanied by Low-Frequency Oscillation in Hydropower System. In Proceedings of the 2024 3rd International Conference on Energy and Electrical Power Systems (ICEEPS), Guangzhou, China, 14–16 July 2024; pp. 564–567.
5. Zhong, W.; Li, Z.; Tang, Z.; Du, Y.; Yang, X.; Feng, C.; Li, G. An ultra-low frequency oscillation suppression method for hydropower units based on GPSS parameter optimization. In Proceedings of the 2023 10th International Forum on Electrical Engineering and Automation (IFEEA), Nanjing, China, 3–5 November 2023; pp. 726–731.

6. Yang, L.; Huang, W.; Zhang, D.; He, P.; Meng, X.; Cao, P.; Yang, B. Comprehensive suppression strategies for low frequency and ultra-low frequency oscillations of the Yunnan power grid under asynchronous interconnection. *Power Syst. Prot. Control* **2021**, *49*, 133–140. [CrossRef]
7. Guang, X.; Fang, Y.; Li, Z.; Wu, X.; Liu, F.; Li, W. Phenomenon and Mechanism of Inter-area Tie-line Power Oscillation in Ultra-low Frequency Oscillation. *Autom. Electr. Power Syst.* **2020**, *44*, 69–77.
8. Cao, T.; Ye, Z.; Wu, Q.; Wan, X.; Wang, J.; Li, D. A Review of Adaptive Control Methods for Grid-Connected PV Inverters in Complex Distribution Systems. *Energies* **2025**, *18*, 473. [CrossRef]
9. Ali, J.S.; Qiblawey, Y.; Alassi, A.; Massoud, A.M.; Muyeen, S.M.; Abu-Rub, H. Power System Stability with High Penetration of Renewable Energy Sources: Challenges, Assessment, and Mitigation Strategies. *IEEE Access* **2025**, *13*, 39912–39934. [CrossRef]
10. National Energy Administration. Notice of the National Energy Administration on Improving the Grid-Related Security Capabilities of New Energy and New Grid-Connected Entities to Serve the High-quality Development of New Power Systems. 2024-09-30. 00019705/2024-00320. Available online: https://zfxxgk.nea.gov.cn/2024-09/30/c_1310787050.htm (accessed on 4 September 2025).
11. Pan, W.; Li, Y.; Cao, Y.; Xin, J. Frequency control of grid-connection system based on VSC-HVDC for large-scale centralized wind farm. *Electr. Power Autom. Equip.* **2015**, *35*, 94–99. [CrossRef]
12. Wu, X.; Wang, J.; Wang, S.; Xu, Y. Research on Multi-band Robust Supplementary Control of PV for Coordinated Suppression of Ultra-low Frequency/Low Frequency Oscillations. *Proc. CSEE* **2024**, 1–19. Available online: https://link.cnki.net/urlid/11.2107.tm.20241202.1138.002 (accessed on 4 September 2025).
13. Heins, T.; Joševski, M.; Gurumurthy, S.K.; Monti, A. Centralized Model Predictive Control for Transient Frequency Control in Islanded Inverter-Based Microgrids. *IEEE Trans. Power Syst.* **2023**, *38*, 2641–2652. [CrossRef]
14. Shangguan, X.C.; He, Y.; Zhang, C.K.; Jin, L.; Yao, W.; Jiang, L.; Wu, M. Control Performance Standards-Oriented Event-Triggered Load Frequency Control for Power Systems Under Limited Communication Bandwidth. *IEEE Trans. Control Syst. Technol.* **2022**, *30*, 860–868. [CrossRef]
15. Pijian, C.; Wen, T.; Yi, H. Decentralized load frequency control using DE optimized linear auto disturbance rejection controller for multi-area power systems. In Proceedings of the 2015 34th Chinese Control Conference (CCC), Hangzhou, China, 28–30 July 2015; pp. 217–222.
16. Le, J.; Zhao, L.; Wang, C.; Zhou, Q.; Wang, Y. Leader-follower Optimal Selection Method for Distributed Control System in Active Distribution Networks. *CSEE J. Power Energy Syst.* **2024**, *10*, 314–323. [CrossRef]
17. Li, Z.; Cheng, Z.; Si, J.; Zhang, S.; Dong, L.; Li, S.; Gao, Y. Adaptive Power Point Tracking Control of PV System for Primary Frequency Regulation of AC Microgrid with High PV Integration. *IEEE Trans. Power Syst.* **2021**, *36*, 3129–3141. [CrossRef]
18. Zhang, Z.; Du, E.; Teng, F.; Zhang, N.; Kang, C. Modeling Frequency Dynamics in Unit Commitment with a High Share of Renewable Energy. *IEEE Trans. Power Syst.* **2020**, *35*, 4383–4395. [CrossRef]
19. Wang, B.; Zhu, S.; Cai, G.; Yang, D.; Chen, Z.; Ma, J.; Sun, Z. Sparse Measurement-Based Modelling Low-Order Dynamics for Primary Frequency Regulation. *IEEE Trans. Power Syst.* **2024**, *39*, 681–692. [CrossRef]
20. Alhelou, H.H.; Golshan, M.E.H.; Hatziargyriou, N.D. Deterministic Dynamic State Estimation-Based Optimal LFC for Interconnected Power Systems Using Unknown Input Observer. *IEEE Trans. Smart Grid* **2020**, *11*, 1582–1592. [CrossRef]
21. Talatahari, S.; Azizi, M. Optimization of constrained mathematical and engineering design problems using chaos game optimization. *Comput. Ind. Eng.* **2020**, *145*, 106560. [CrossRef]
22. Wang, M.; Guo, J.; Ma, S.; Wang, T.; Zhang, X.; Luo, K.; Wang, G. Review of Transient Frequency Stability Analysis andFrequency Regulation Control Methods for Renewable Power Systems. *Proc. CSEE* **2023**, *43*, 1672–1694. [CrossRef]
23. Jietan, Z.; Linan, Q.; Pestana, R.; Fengkui, L.; Libin, Y. Dynamic frequency support by photovoltaic generation with "synthetic" inertia and frequency droop control. In Proceedings of the 2017 IEEE Conference on Energy Internet and Energy System Integration (EI2), Beijing, China, 26–28 November 2017; pp. 1–6.
24. Shi, Q.; Li, F.; Cui, H. Analytical Method to Aggregate Multi-Machine SFR Model with Applications in Power System Dynamic Studies. *IEEE Trans. Power Syst.* **2018**, *33*, 6355–6367. [CrossRef]
25. Yoo, S.J. Connectivity-Preserving Consensus Tracking of Uncertain Nonlinear Strict-Feedback Multiagent Systems: An Error Transformation Approach. *IEEE Trans. Neural Netw. Learn. Syst.* **2018**, *29*, 4542–4548. [CrossRef] [PubMed]

MDPI AG
Grosspeteranlage 5
4052 Basel
Switzerland
Tel.: +41 61 683 77 34

Symmetry Editorial Office
E-mail: symmetry@mdpi.com
www.mdpi.com/journal/symmetry

www.ingramcontent.com/pod-product-compliance
Lightning Source LLC
Chambersburg PA
CBHW061824210926
51850CB00006B/799
9783725864386